Water Treatment: Advanced Principles and Practices

Water Treatment: Advanced Principles and Practices

Editor: Vincent Emerson

R CALLISTO REFERENCE

www.callistoreference.com

Callisto Reference,
118-35 Queens Blvd., Suite 400,
Forest Hills, NY 11375, USA

Visit us on the World Wide Web at:
www.callistoreference.com

ISBN: 978-1-64116-128-2 (Hardback)

Cataloging-in-Publication Data

Water treatment : advanced principles and practices / edited by Vincent Emerson.
    p. cm.
Includes bibliographical references and index.
ISBN 978-1-64116-128-2
1. Water--Purification. 2. Sewage--Purification.
3. Water treatment plants. I. Emerson, Vincent.
TD430 .W38 2019
628.162--dc23

Table of Contents

Preface

Water treatment is a process that involves the treatment of water to render it acceptable for specific uses like drinking, irrigation, industrial water supply, etc. It involves either removal or reduction of the contaminants. Some of the contaminants of water include suspended solids, various microbes and minerals such as iron and magnesium. Different physical, chemical and biological processes such as filtration, disinfection, coagulation, etc. are used to treat water. Some of the key functional areas of water treatment include drinking water production, wastewater treatment, domestic water treatment, desalination and ultrapure water production. This book is a compilation of chapters that discuss the most vital concepts and emerging trends in the field of water treatment. The various advancements in treatment methods are glanced at and their applications as well as ramifications are looked at in detail. The extensive content herein provides the readers with a thorough understanding of the subject.

The information shared in this book is based on empirical researches made by veterans in this field of study. The elaborative information provided in this book will help the readers further their scope of knowledge leading to advancements in this field.

Finally, I would like to thank my fellow researchers who gave constructive feedback and my family members who supported me at every step of my research.

Editor

Onion membrane: an efficient adsorbent for decoloring of wastewater

Samaneh Saber-Samandari[*] and Jalil Heydaripour

Abstract

Background: Recently, researchers have tried to design synthetic materials by replicating natural materials as an adsorbent for removing various types of environmental pollutants, which have reached to the risky levels in nature for many countries in the world. In this research, the potential of onion membrane obtained from intermediate of onion shells for adsorption of methylene blue (MB) as a model cationic dye was exhibited.

Methods: Before and after adsorption, the membrane was characterized by Fourier transform infrared spectroscopy (FTIR) and optical and scanning electron microscopy in order to prove its dye adsorption capability. The various experimental conditions affecting dye adsorption were explored to achieve maximum adsorption capacity.

Results: The dye adsorption capacity of the membrane was found to be 1.055 $g.g^{-1}$ with 84.45% efficiency after one hour and 1.202 $g.g^{-1}$ with 96.20% efficiency after eight hours in contact with the dye solution (0.3 $g.L^{-1}$). Moreover, the kinetic, thermodynamic and adsorption isotherm models were employed to described the MB adsorption processes. The results show that the data for adsorption of MB onto the membrane fitted well with the Freundlich isotherm and pseudo-second-order kinetic models. In addition, the MB adsorption from room temperature to ~50°C is spontaneous and thermodynamically favorable.

Conclusions: Evidently, the high efficiency and fast removal of methylene blue using onion membrane suggest the synthesis of polymer-based membranes with similar physical and chemical properties of onion membrane as a valuable and promising wastewater decoloring agents in water treatment.

Keywords: Adsorption, Cationic dye, Methylene blue, Onion membrane, Water treatment

Background

Many industrial production activities (e.g. dye, cosmetic, plastics, food, textile, planting, and mining) result in water pollution since they produce pollutants such as water-coloring agents and toxic heavy metals that are extremely harmful to people and the environment even at low concentrations. The removal of these compounds from wastewaters before discharging them into the environment is of great importance, since many dyes and their degradation products are toxic and carcinogenic, posing a serious hazard to the environment. The conventional methods used to treat colored effluents include photocatalytic degradation, microbiological decomposition, electrochemical oxidation, membrane filtration, and adsorption techniques [1]. Among these, adsorption is the most widely used method because of its efficiency, low cost, easy operation, simple design, less energy intensiveness, and non-toxicity. Recently, numerous approaches to develop adsorbents that are more effective have been studied [2]. Among these, adsorbents containing natural and synthetic polymeric materials, industrial by-products, agricultural wastes and biomass were applied for removal of dyes from aqueous solution [3-6].

Dyes can be classified into cationic, anionic and nonionic dyes. Cationic dyes are basic dyes while the anionic dyes include direct, acid, and reactive dyes [7]. However, cationic dyes are widely used in acrylic, wool, nylon, and silk dyeing, they considered as toxic colorants and can cause harmful effects such as allergic dermatitis, skin irritation, mutations and cancer [8]. Cationic dyes carry a positive charge in their molecule, furthermore they are water soluble and yield

---

* Correspondence: samaneh.saber@gmail.com
Department of Chemistry, Eastern Mediterranean University, TRNC via Mersin 10, Gazimagusa, Turkey

colored cations in solution [9]. Methylene blue, rhodamine B, and brilliant green are representative examples of cationic dyes [10]. Methylene blue (MB) is an important basic dye and widely used in the textile industry. Acute exposure to MB may cause increased heart rate, shock, vomiting, cyanosis, jaundice, quadriplegia, heinz body formation, and tissue necrosis in humans [11].

The main aim of this study is to exhibit the ability of dried onion membrane for removal of MB from aqueous solution. However, many researchers have studied the dye and metal adsorption capacity of several biomasses such as rice, corn and coconut husks [12-14], papaya seeds [15], watermelon [3] and onion skin [4], but to the best of our knowledge, there is no study relating the adsorption properties of onion membrane obtained from intermediate of onion shells. In this study, the adsorption of MB was confirmed using FTIR and optical and electron microscopy. In addition, the effect of various factors such as contact time, initial dye concentration, pH, temperature and adsorbent dose on the adsorption rate was examined. Finally, the adsorption of MB was analyzed by employing the adsorption kinetic, isotherm models, and thermodynamics.

Experimental procedures
Material
The onions were purchased from local markets in Famagusta (North Cyprus). Hydrochloric acid 37% (Merck), sodium hydroxide (Mediko Kimya), potassium chloride (Mediko Kimya), potassium hydrogen phthalate (Merck), sodium hydrogen carbonate (Aldrich), and potassium hydrogen phosphate (Merck) were used to prepare the buffer solutions with different pH values. Finally, MB (Aldrich) with a molecular formula of $C_{16}H_{18}ClN_3S$ was used without further purification.

Adsorbent preparation
The onions were peeled, chopped and then the membranes were removed from the leaves. 1 gram of onion membrane was obtained from approximately 250 g onion. The membranes (~4 g/kg of onion) were rinsed and washed with distilled water to remove impurities. Then, it was dried at 60°C in an oven for one day. After that, the dried adsorbent was kept in the desiccator to avoid moisture adsorption.

Swelling measurements
The swelling property of the membranes were investigated by immersing 0.06 g of membrane (15 × 15 mm^2) in 250 mL of distilled water at room temperature (±20°C) in atmospheric conditions until swelling equilibrium was reached. Following the removal from

the water, they were blotted with filter paper and weighed. Then, the swelling capacity was calculated using the following equation [16,17]:

$$Swelling\ (\%) = \frac{W_2 - W_1}{W_1} \times 100 \qquad (1)$$

where W_1 (g) and W_2 (g) are the weights of the dried and swollen membranes, respectively.

Dye adsorption
The initial aqueous MB solutions were prepared by dissolving 0.075 g of MB in 250 mL of deionized water (0.3 g.L^{-1}). Then, the 0.06 g of dried onion membranes (15 × 15 mm^2) were immersed in a prepared MB solution and shaken at 300 rpm for 10 hours at room temperature. During this period, 2 mL of solution was taken for further analysis frequently. Finally, the collected solutions were filtered and the amount of non-adsorbed dye ions in the solutions was determined spectrophotometrically using a UV-visible spectrophotometer at a wavelength of 668 nm [6]. The MB adsorption amount, capacity, and efficiency were calculated by the following equations:

$$adsorption\ amount\ (g.L^{-1}) = C_i - C_e \qquad (2)$$

$$adsorption\ capacity\ (g.g^{-1}) = q_t = \frac{C_i - C_e}{W} \times V \qquad (3)$$

$$adsorption\ efficency\ (\%) = \frac{C_i - C_e}{C_i} \times 100 \qquad (4)$$

where W is the mass of adsorbent in g, V is the volume of MB solution in L, and C_i and C_e are the initial and equilibrium concentrations in g.L^{-1}, respectively.

In addition, the effect of variable conditions such as time, pH, adsorbent amount, and initial adsorbate concentration on the MB adsorption behavior of onion membranes was examined. For each case, one parameter was changed and analyzed and the other factors were kept constant. For instance, the influence of pH on adsorption was calculated by immersing 0.06 g of onion membrane in 250 mL of MB buffer solutions (0.3 g.L^{-1} concentration) with different pH values (3–11) and then shaken at room temperature (20°C) for eight hours (480 min).

Furthermore, the pH at point of zero charge (pHzpc), which shows the point that the acidic or basic functional groups do no contribute to the pH of the solution, was determined using the standard technique [18]. For this purpose, 50 mL of 0.01 mol.L^{-1} NaCl solution was placed in a 250 ml flask. The pH of the solutions in each flask was adjusted from 3–11 by adding either sodium hydroxide or hydrochloric acid solutions (0.1 mol.L^{-1}). Then, 0.06 g of onion membrane (15 × 15 mm^2) was immersed in each solution at 20°C and was allowed to equilibrate in an isothermal shaker. After a contact time of 24 hours, the suspensions were filtered through filter paper and the final pH values of supernatant were measured again using a pH meter. Lastly, the final pH values were plotted against the initial pH values. The pH at which the curve crosses the line final pH = initial pH was taken as the pHzpc of the onion membrane [19].

Methods of characterization

A UV/VIS spectrophotometer (Lambda 25 UV/VIS, Perkin-Elmer, Llantrisant, UK) was used to determine the MB adsorption amounts by the onion membranes. The pH values of the solutions, which were used to investigate the effect of pH on adsorption, were checked by a pH meter (WTW InoLab, accuracy ± 0.1). To prove the adsorption of MB by the onion membranes, the FTIR (Perkin-Elmer, Llantrisant, UK) spectra of membranes before and after adsorption were observed in the range of 500–4000 cm^{-1}. Images of the membranes were also taken with an optical microscope (MT9000 Polarizing Microscope, Meiji Techno Co. Ltd., Japan with Invenio 3S, Delta Pix Camera) and a scanning electron microscope (AIS2100, Seron Technology, Korea) operated at an acceleration voltage of 15 kV to study the change in surface morphology of the membrane after dye adsorption.

Results and discussion

Swelling properties of onion membrane

The dynamic swelling behavior of the onion membranes over 10 hours (600 min) of immersion in water is shown in Figure 1. The swelling percentage increased up to 1,106% and then plateaued, with no big differences in water uptake with further increases in time. Initially, the water molecules were in contact with the membrane, then, they attacked and penetrated into the onion membrane cells. Obviously, this swelling system cannot continue forever, and by the increasing membrane-water interaction, the osmotic pressure difference might be reduced. Finally, the osmotic force at the equilibrium state was balanced with an elasticity force. It should be noted that the elasticity force prevents the deformation of the onion membrane cells by the stretching balance of the cells.

Dye adsorption properties of onion membrane

The MB adsorption capacity of the onion membranes was examined. Figure 2 shows the preparation of the membranes for MB adsorption (Figure 2a) and color changes in the membrane and dye solutions before and after adsorption during the first and eight hours of contact time (Figure 2b-d). In addition, the scanning electron microscopy (SEM) images of the membranes before and after adsorption of MB reflect their surface morphology (Figure 2e and f). As can be seen, before adsorption the membrane

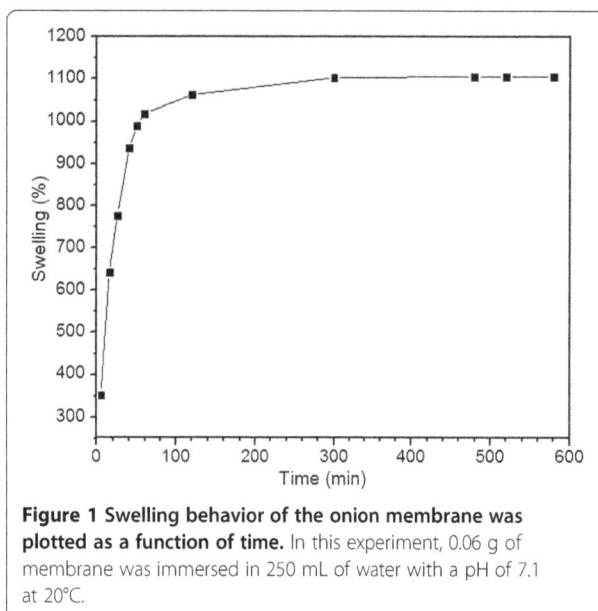

Figure 1 Swelling behavior of the onion membrane was plotted as a function of time. In this experiment, 0.06 g of membrane was immersed in 250 mL of water with a pH of 7.1 at 20°C.

has a smooth surface, whereas it exhibits a coarse surface due to the presence of MB molecules after adsorption. In addition to this distinctive change in the surface of the membrane, the optical microscope images of the membrane before and after MB adsorption (Figure 2g and h) also revealed the adsorption of MB by the onion membrane.

The FTIR spectrum of the onion membrane after adsorption of the dye was compared with the spectra of pure MB and dried onion membrane before adsorption in Figure 2i. The several peaks in the spectrum of dye-adsorbed membrane related to the MB and membrane are merged and slightly shifted. The peaks at 2,894 cm^{-1}, 2,900 cm^{-1}, and 1,744 cm^{-1} due to the C-H stretching in an aromatic methoxyl group and C = O stretching of carbonyl group of onion membrane, respectively, became stronger and were shifted to 2,916 cm^{-1}, 2,850 cm^{-1}, and 1,731 cm^{-1} in the dye adsorbed membrane [20]. The sharp and strong peak at 1,592 cm^{-1} of MB due to the presence of C = C vibration and N-H bending was merged with a peak at 1,599 cm^{-1} of membrane and showed a broader peak at 1,598 cm^{-1} in the spectrum of the dye-adsorbed membrane [21]. Like the two peaks at 1,443 cm^{-1} and 1,488 cm^{-1}, which indicates C = N stretching, the peaks at 1,220 cm^{-1} and 1,250 cm^{-1} of C = C stretching in aromatic rings of MB are merged and form broad peaks at 1,416 cm^{-1} and 1,232 cm^{-1}, respectively, in the spectrum of the dye-adsorbed membrane [18]. Finally, the peaks corresponding to the C-O stretching of the onion membrane and the C-S bending of the MB rings appeared at 1,008 cm^{-1} and 954 cm^{-1}, respectively [22]. The FTIR results accompanied with the supportive results of the SEM and the optical microscope confirmed the adsorption of MB by the onion membrane.

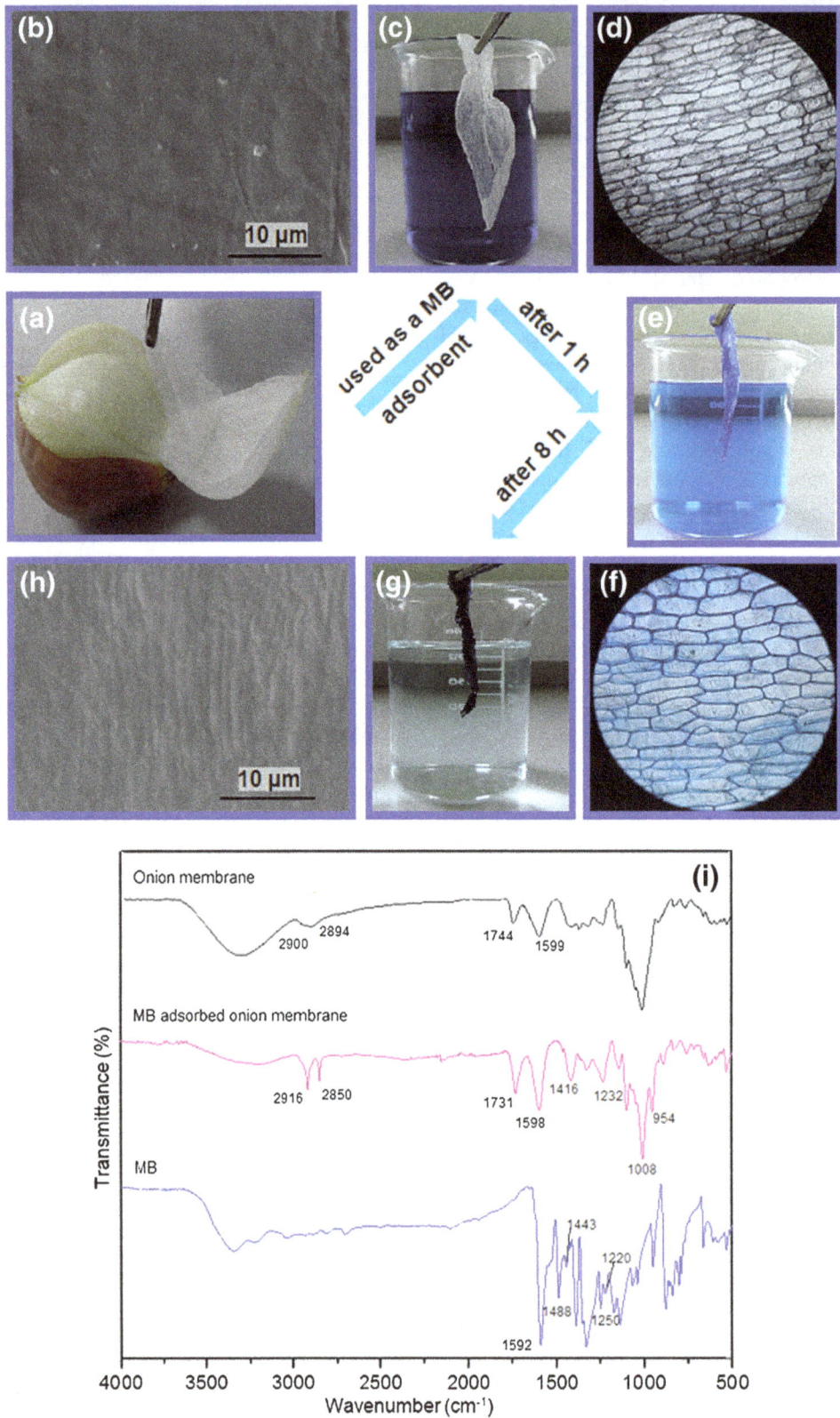

Figure 2 (a-d) Digital photographs, (e and f) SEM images, (g and h) optical microscope images, and (i) FTIR spectra of the onion membrane before and after dye adsorption are shown.

Onions contain protein (with –COOH and –NH$_2$ groups), sugars, carbohydrate, and vitamins A, B$_6$, and C (with –OH groups), minerals, and over 80% water [23]. As can be seen from Scheme 1, the onion membrane has several anionic groups such as –COOH and –OH. On the other hand, MB is a cationic dye consisting of = S– and –N (CH$_3$)$_2$ can become charged species and have ionic and dipole–dipole interactions with anionic groups in the surface of the onion membrane [24]. In addition, the = N– and –N (CH$_3$)$_2$ groups in the structure of MB can have hydrogen bonds with hydrogen atom of –COOH and –OH groups of the onion membrane. Therefore, the adsorbent can uptake MB very fast with high efficiency through the strong electrostatic attraction between the surface groups on the membrane and the cationic MB.

Effect of time on adsorption

The adsorption performance of the onion membrane was evaluated by a batch equilibration technique as a function of time. As shown in Figure 3, the adsorption capacity of MB ions on the membrane increased rapidly during the first hour of contact and then became slower until equilibrium was reached after eight hours (480 min). The maximum adsorption was 1.055 g.g^{-1} with 84.45% efficiency after the first hour and 1.202 g.g^{-1} with 96.20% efficiency after eight hours. This behavior can be attributed to the larger surface area of the onion membrane at the initial stage of the adsorption process. Subsequently, as the surface sites became saturated, adsorption did not increase significantly with further contact time.

Adsorption kinetics

In order to investigate the possibility of using an adsorbent for a particular separation task and to determine the adsorption efficiency as well as the adsorption rate, the kinetic mechanism of the adsorption process was considered. The adsorption kinetics of the MB ions with the onion membranes was investigated using three kinetic models: pseudo-first-order, pseudo-second-order, and intra-particle diffusion, which are given in the following equations, respectively.

$$\log(q_e - q_t) = \log q_e - \frac{k_1}{2.303}t \tag{5}$$

$$\frac{t}{q_t} = \frac{1}{k_2 q_e^2} + \frac{t}{q_e} \tag{6}$$

$$q_t = k_3(t)^{\frac{1}{2}} + C_i \tag{7}$$

where t is the time (min) and q_e, q_t, (g.g^{-1}) and q_e^2 are the amounts of MB adsorbed by the onion membrane at

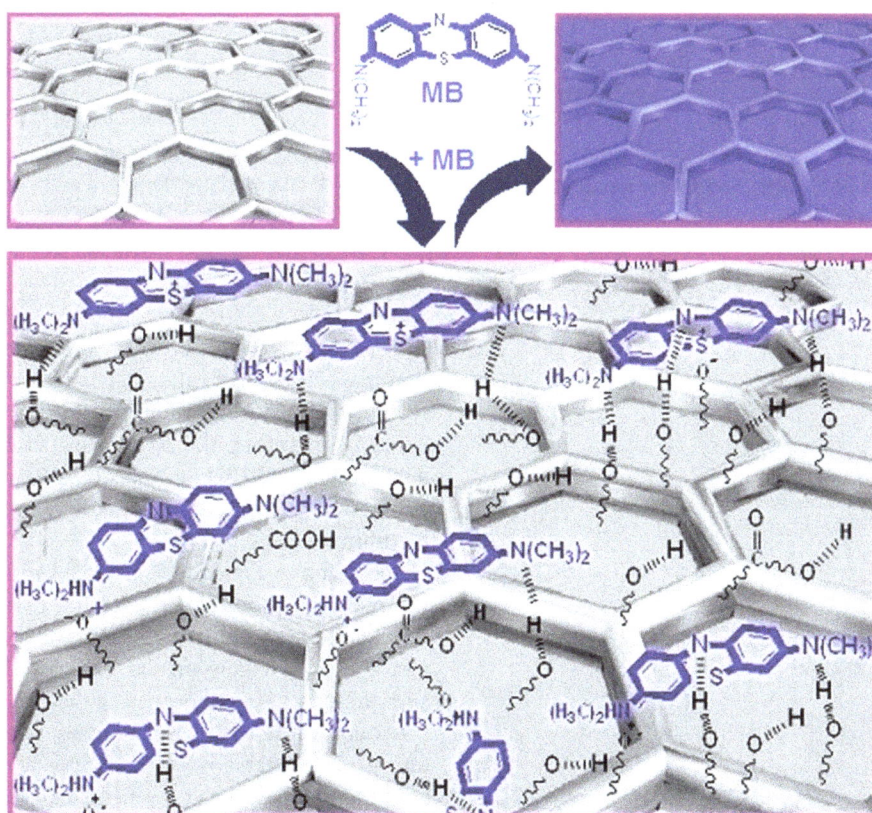

Scheme 1 Schematic illustration for the adsorption mechanism of MB by onion membrane.

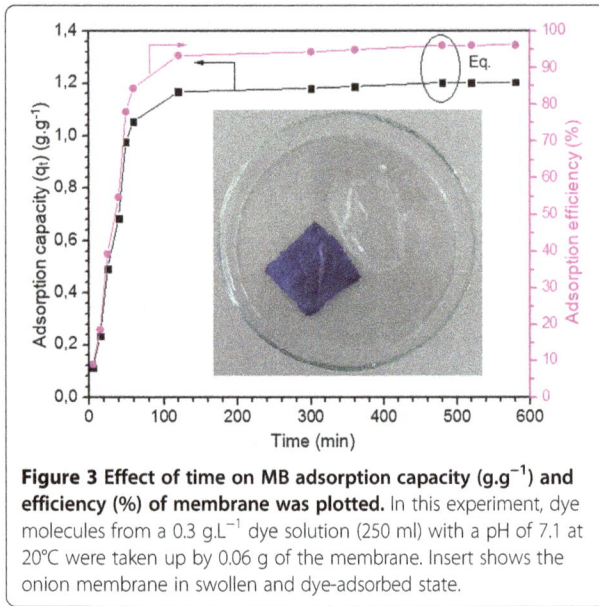

Figure 3 Effect of time on MB adsorption capacity (g.g⁻¹) and efficiency (%) of membrane was plotted. In this experiment, dye molecules from a 0.3 g.L⁻¹ dye solution (250 ml) with a pH of 7.1 at 20°C were taken up by 0.06 g of the membrane. Insert shows the onion membrane in swollen and dye-adsorbed state.

equilibrium, at time t, and at maximum adsorption capacity, respectively. k_1 (min⁻¹), k_2 (g.g⁻¹.min⁻¹), and k_3 (g.g⁻¹.min⁻⁰·⁵) are the adsorption rate constants of the pseudo-first-order, pseudo-second order and the intra-particle diffusion models, respectively. In addition, C_i (g.g⁻¹) is the intra-particle diffusion constant, which is directly proportional to the boundary layer thickness. As shown in Table 1, the theoretical equilibrium adsorption capacity (1.3166 g.g⁻¹) using the pseudo-second-order model compared well with the experimental data (1.2025 g.g⁻¹), with a better R^2 value. Moreover, Figure 4a

Table 1 Comparison of kinetic models for adsorption of MB using onion membrane

Kinetic models and parameters	MB
q_e exp. (g.g⁻¹)	1.2025
Pseudo-first order	
K_1 (min⁻¹)	0.0278
q_e cal. (g.g⁻¹)	0.6831
R^2	0.8255
Pseudo-second order	
$K_2 \times 10^{-4}$ (g.g⁻¹.min⁻¹)	0.0205
q_e cal. (g.g⁻¹)	1.3166
R^2	0.9893
Intraparticle diffusion	
$K_{3(1)}$ (g.g⁻¹.min⁻¹ᐟ²)	0.1747
C_1 (g.g⁻¹)	0.3567
R^2	0.9414
$K_{3(2)}$ (g.g⁻¹.min⁻¹ᐟ²)	0.0086
C_2 (g.g⁻¹)	1.0266
R^2	0.7264

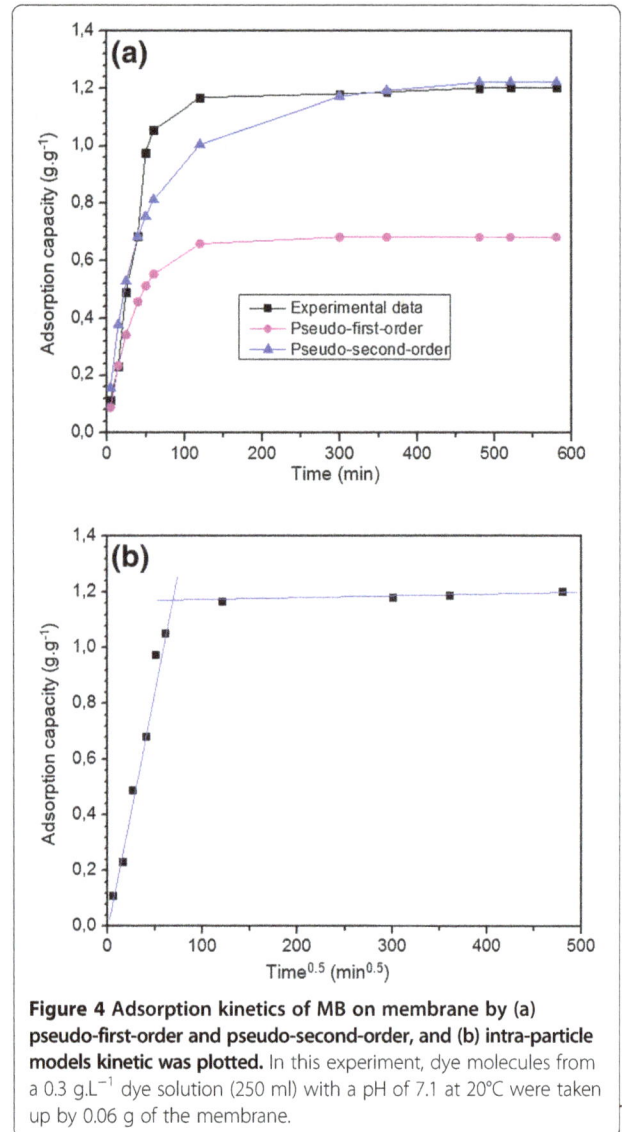

Figure 4 Adsorption kinetics of MB on membrane by (a) pseudo-first-order and pseudo-second-order, and (b) intra-particle models kinetic was plotted. In this experiment, dye molecules from a 0.3 g.L⁻¹ dye solution (250 ml) with a pH of 7.1 at 20°C were taken up by 0.06 g of the membrane.

shows the agreement between the experimental adsorption capacities with the calculated values of pseudo-second-order, which were obtained using the data in Table 1. Therefore, the agreement between experimental data and pseudo-second-order can prove the physical adsorption of MB on a highly heterogeneous onion membrane.

According to Equation 7, if the intra-particle diffusion is the main rate-controlling step, the plot of q_t versus $t^{0.5}$ should be linear and pass through the origin. However, the plot shown in Figure 4b did not pass through the origin and presented multilinearity, indicating the presence of two steps in the adsorption process. The intra-particle diffusion parameters for these steps are summarized in Table 1. The first linear segments can be attributed to the dye transfer from the solution onto the external surface or boundary layer of the onion membrane. The second step could be attributed to the final apparent equilibrium process, which reflects the intra-particle diffusion slowing down due to lowering the dye concentration in the solution [5]. As seen from the data in Table 1, the intra-particle diffusion constants of the two linear segments are not similar and the first step comprises the bigger $k_{3(1)}$ value (0.1747 $g.g^{-1}.min^{-0.5}$) and the higher correlation coefficient 0.9414. This observation indicates that the adsorption of dye onto the onion membrane at the first section occurs more rapidly due to the availability of adsorption centers, then this is followed by the slow diffusion, which takes up to eight hours.

Effect of pH on adsorption

The pH value of the solution plays a significant role in the adsorption capacity of the adsorbate onto the adsorbent. As can be seen in Figure 5, with an increase in the initial pH of the MB solution from 3 to 8, adsorption capacity and efficiency increases rapidly and then increases slowly with a further increase in the pH. The maximum adsorption capacity was obtained at 1.245 $g.g^{-1}$, with 99.64% efficiency at pH 11. This result can be explained by the electrostatic interaction between the cationic MB species and the surface of the adsorbent, which should be a negatively charged species. The lower adsorption at acidic pH levels was probably due to the presence of an excess of H^+ ions competing with the dye cations for adsorption sites [19]. In order to confirm these results, the pHzpc of the onion membrane was determined. In this study, pHzpc value was 5.9, which at this pH the adsorbent surface has net electrical neutrality. At a pH below the pHzpc, the surface of the adsorbent is positive, and at a pH above the pHzpc, the surface of the adsorbent becomes more negatively charged by losing protons. Therefore, the adsorption

of the MB reached its maximum value in the higher pH because of strong electrostatic attractions between the negatively charged surface of the onion membrane and the cationic MB.

Effect of adsorbent amount on adsorption

In order to find the influence of various onion membrane amounts on their dye uptake capacity, an adsorption of 250 mL MB solutions (0.3 $g.L^{-1}$) was examined using five different membrane doses ranging from 0.015 to 0.12 g for eight hours in the atmospheric conditions. As shown in Figure 6, with an increase in the amount of membrane to 0.12 g, the adsorption amount and consequently the adsorption efficiency increases to 98.83%. This is most likely due to an increase in the numbers of adsorption sites at the adsorbent surface area, and as a result, increases the removal efficiency of MB. However, it is reasonable to observe a decrease in adsorption capacity to 0.617 $g.g^{-1}$ by increasing the adsorbent dose, which is a denominator of the fraction in Equation 3.

Effect of the initial concentration of dye solution on adsorption

Figure 7 shows the maximum capacity of an onion membrane in the adsorption of MB using adsorbent-adsorbate solutions with different initial dye concentrations (0.1-0.9 $g.L^{-1}$). It is clear that the MB adsorption capacity of an onion membrane increases with an increase in the initial concentration of the solution. This is most likely due to a greater availability of adsorbate ions in the vicinity of the onion membrane and a high driving force for mass transfer before the adsorption-desorption equilibrium was reached [25]. However, in this study, unlike the adsorption amount and capacity, the efficiency of the dye removal decreased,

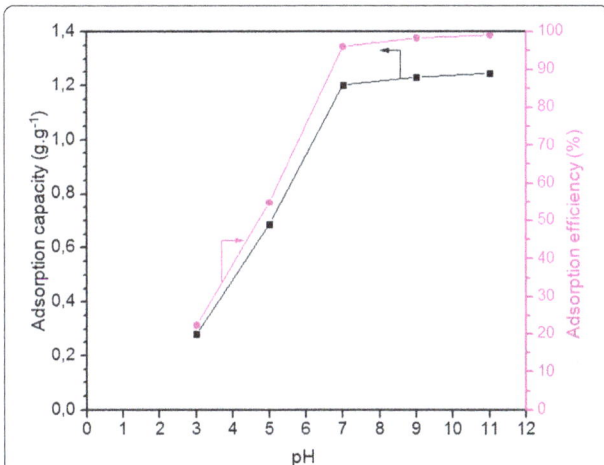

Figure 5 Effect of pH on the MB adsorption capacity ($g.g^{-1}$) and efficiency (%) of membrane was plotted. In these experiments, 0.06 g of membrane adsorbed dye molecules from a 0.3 $g.L^{-1}$ dye solution (250 ml) at 20°C for 8 hours.

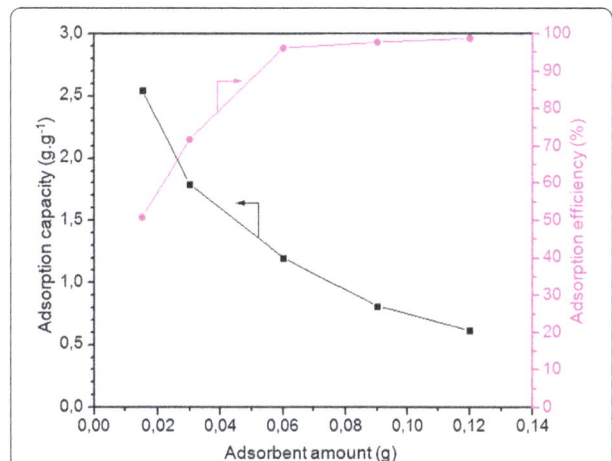

Figure 6 Effect of adsorbent amount on the MB adsorption capacity ($g.g^{-1}$) and efficiency (%) of membrane was plotted. In these experiments, the membrane adsorbed dye molecules from a 0.3 $g.L^{-1}$ dye solution (250 ml) with a pH of 7.1 at 20°C for 8 hours.

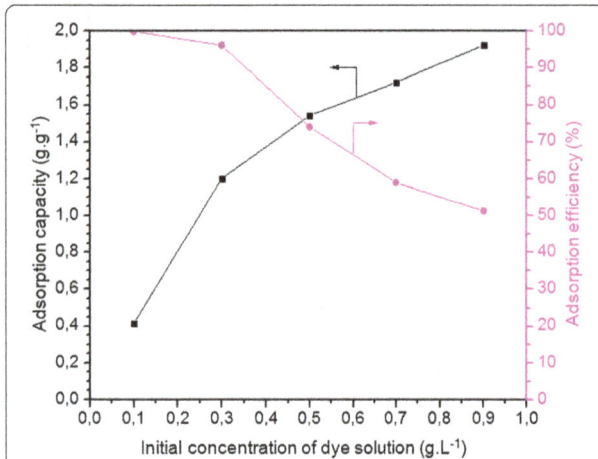

Figure 7 Effect of initial dyes concentration on the MB adsorption capacity (g.g^{-1}) and efficiency (%) of membrane was plotted. In these experiments, 0.06 g of the membrane adsorbed dye molecules from a dye solution (250 ml) with a pH of 7.1 at 20°C for 8 hours.

process. The Langmuir model, the most popular, has been widely used to describe single-solute systems. The Langmuir model is based on the assumption that adsorbates produce monolayer coverage on the outer surface with uniform energies of adsorption, which is structurally homogeneous [30]. In the Freundlich model, the adsorption of an adsorbate occurs on a heterogeneous surface via multilayer adsorption with non-uniform distribution of heat of adsorption. The theoretical Langmuir and Freundlich isotherm models are represented by the following equations:

$$\frac{C_e}{q_e} = \frac{C_e}{q_m} + \frac{1}{q_m k_L} \tag{8}$$

$$\log q_e = \log k_F + \frac{1}{n} \log C_e \tag{9}$$

where q_e (g.g^{-1}) is the amount of dye adsorbed at equilibrium time, q_m (g.g^{-1}) is the maximum adsorption capacity, and C_e (g.L^{-1}) is the equilibrium dye concentration. k_L and k_F (L.g^{-1}) are the Langmuir and Freundlich adsorption equilibrium constant. 1/n is the empirical Freundlich constant. As it is clear from Figure 9, the calculated values of q_e belong to the Freundlich model is in agreement with the experimental value, which revealed that the Freundlich model is more suitable than the Langmuir model for describing the adsorption. This is confirmed by the correlation coefficient (R^2) of the Freundlich isotherm model (0.9922), which is greater than R^2 (0.9878) of the Langmuir model (Table 2). The empirical Freundlich constant (1/n) which can be obtained from the linear plot of $\log q_e$ versus $\log C_e$, is an indicator of the favorability and surface affinity for the solute. When the 1/n values are in the range 0.1-1,

which may be attributed to the saturation of adsorption sites on the adsorbent surface [9]. The maximum MB adsorption capacity of an onion membrane is 1.9230 g.g^{-1} with the lowest efficiency at 51.28%, which is significantly higher than the other adsorbents [26-29] listed in Figure 8.

Isotherm of adsorption

The adsorption isotherm generally illustrates the interaction of an adsorbate with the adsorbent and also it can indicates the adsorption capacity of the adsorbent. Therefore, two isotherm models, Langmuir and Freundlich were investigated to find a more suitable model for the design

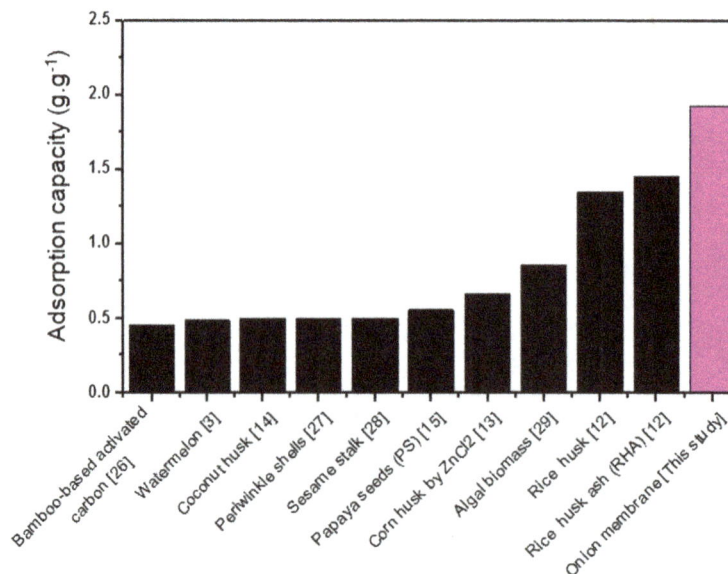

Figure 8 The maximum MB adsorption capacities of onion membrane were compared with several biomass adsorbents.

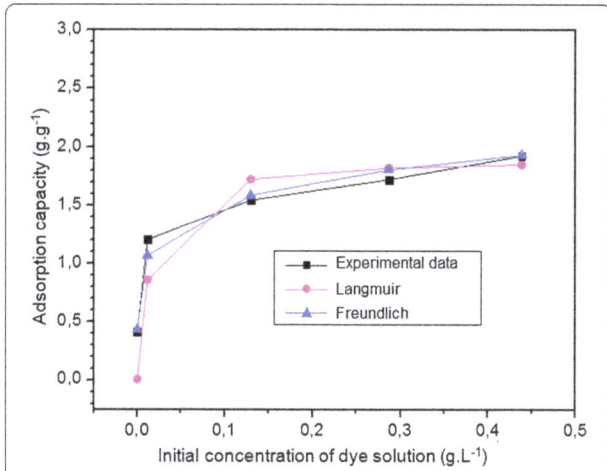

Figure 9 Adsorption isotherm of MB on onion membrane by Langmuir and Freundlich were plotted. In these experiments, 0.06 g of the membrane adsorbed dye molecules from a dye solution (250 ml) with a pH of 7.1 at 20°C for 8 hours.

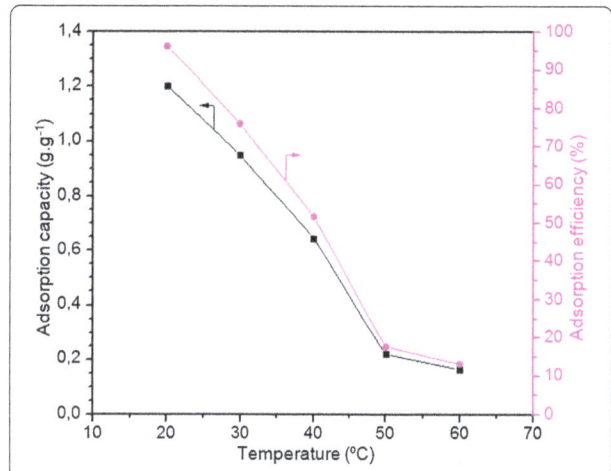

Figure 10 Effect of temperature on the MB adsorption capacity (g.g^{-1}) and efficiency (%) of membrane was plotted. In these experiments, 0.06 g of membrane adsorbed dye molecules from a 0.3 g.L^{-1} dye solution (250 ml) with a pH of 7.1 for 8 hours.

the adsorption process is favorable. In addition, if the n is below one, then the adsorption is a chemical process; otherwise, the adsorption is a physical process. In this study, the value of 1/n is 0.24, lying between 0.1 and 1, indicating that adsorption of MB ions by an onion membrane is favorable with physisorption.

Effect of temperature on adsorption

The effect of temperature on the adsorption capacity of onion membrane (0.06 g) was studied using 250 mL of MB solution in different temperatures (20°C-60°C) for eight hours (480 min) in the atmospheric conditions. As shown in Figure 10, increasing the temperature leads to a decrease in the MB adsorption capacity of the membrane (0.164 g.g^{-1}, with 13.15% efficiency) after eight hours of contact time. This can be attributed to a weakening of the adsorptive forces between the active sites on the sorbent and the dye

molecules due to the degradation of the onion membrane in the high temperature. The results suggest that a high temperature is not suitable for adsorbing MB when the onion membrane is adsorbent. Therefore, it is better to let the temperature of industrial outcome solutions decrease to 20°C to achieve maximum adsorption capacity.

Adsorption thermodynamics

The effect of temperature can be further investigated by calculating the thermodynamic parameters, including standard enthalpy change ($\Delta H°$), the Gibbs free energy change ($\Delta G°$), and standard entropy change ($\Delta S°$). They have been determined using the following equations:

$$\Delta G^0 = -RT \ln K_d \tag{10}$$

The distribution ratio (K_d), can be calculated using the below equation:

$$K_d = \frac{q_e}{C_e} \tag{11}$$

Then, the relation between $\Delta G°$, $\Delta H°$, and $\Delta S°$ can be expressed by the following equations:

$$\Delta G° = \Delta H° - T\Delta S° \tag{12}$$

Table 2 Comparison of isotherm models for adsorption of MB using onion membrane

Isotherm models and parameters	MB
Langmuir	
q_{max} (g.g^{-1})	1.9054
q_e cal. (g.g^{-1})	1.8471
K_L (L.g^{-1})	72.387
R^2	0.9878
Fruendlich	
q_e cal. (g.g^{-1})	1.9312
K_F (L.g^{-1})	2.2074
n	6.1690
R^2	0.9922

Table 3 Thermodynamic parameters for adsorption of MB using onion membrane

Thermodynamic constant	Temperature (K)				
	293	303	313	323	333
$\Delta G°$ (kJ.mol^{-1})	−10.79	−7.533	−4.273	−1.013	2.247
$\Delta H°$ (kJ.mol^{-1})	−106.3				
$\Delta S°$ (kJ.mol^{-1}.K^{-1})	−0.326				

Standard enthalpy ($\Delta H°$) and entropy ($\Delta S°$) were determined from the slope and intercept of the plot of $\ln K_d$ vs $1/T$, which came from the Van's Hoff equation:

$$\ln K_d = \frac{\Delta S°}{R} - \frac{\Delta H°}{RT} \qquad (13)$$

Table 3 presents the thermodynamic parameters of MB adsorption by the onion membrane in different contact temperatures. It is clear that the values of free energy ($\Delta G°$) at all temperatures except 60°C were negative. This confirms that the adsorption of dye at these temperatures is spontaneous and thermodynamically favorable. The negative value of $\Delta H°$ confirms the exothermic nature of adsorption, which is supported by the decrease in the adsorption capacity of the onion membrane. Besides, the negative value of $\Delta S°$ resulted from the decreased randomness at the solid–liquid interface during the adsorption of MB ions onto the onion membrane.

Conclusion

The present study confirms the high potential of onion membranes with special physical and chemical characteristics for quick and efficient removal of MB from aqueous solutions. The amount of dye adsorbed varied with time, temperature, pH, adsorbent dosage, and initial dye concentration. The adsorption experiments indicated that onion membrane have a high MB adsorption capacity (1.9230 g.g^{-1}) when 0.06 g of adsorbent was immersed in 250 ml of dye solution (0.9 g.L^{-1}) with a pH of 7.1 at 20°C. The adsorption of MB by the onion membrane agreed with the pseudo-second-order model. Moreover, analysis of the equilibrium isotherms using the Langmuir and Freundlich isotherms showed that the Freundlich model fitted well with the experimental data. The thermodynamic studies suggested that the adsorption reaction was an exothermic and spontaneous process. Finally, the results suggest the synthesis of polymer-based membranes with similar physical and chemical properties of onion membrane as valuable and highly efficient adsorbents, which can be applied for dye removal in water treatment processes.

Competing interests
The authors declare that they have no competing interests.

Authors' contributions
SS-S as a first and corresponding author has contributed to performing the experiments, acquisition of data, analysis and interpretation of data, writing the manuscript. JH as a second author has contributed to performing the experiments, acquisition of data and preparing the tables and figures. All authors read and approved the final manuscript.

Acknowledgements
The authors thank Dr. Saeed Saber-Samandari of the Amirkabir University of Technology for his help with the SEM. The authors would like to acknowledge Mr. Ehsan Bahramzadeh (PhD student in Eastern Mediterranean University) for his help in preparing the manuscript.

References
1. Bayramoglu G, Adiguzel N, Ersoy G, Yilmaz M, Arica MY. Removal of textile dyes from aqueous solution using amine-modified plant biomass of a. caricum: equilibrium and kinetic studies. Water Air Soil Pollut. 2013;224:1.
2. Crini G. Recent developments in polysaccharide-based materials used as adsorbents in waste water treatment. Prog Polym Sci. 2005;30:38.
3. Lakshmipathy R, Sarada NC. Adsorptive removal of basic cationic dyes from aqueous solution by chemically protonated watermelon (Citrullus Lanatus) rind biomass. Desalin Water Treat. 2013;52:6175.
4. Santhi T, Manonmani S. Adsorption of methylene blue from aqueous solution onto a waste aquacultural shell powders (Prawn Waste). Sustain Environ Res. 2012;22:45.
5. Constantin M, Asmarandei I, Harabagiu V, Ghimici L, Ascenzi P, Fundueanu G. Removal of anionic dyes from aqueous solutions by an ion-exchanger based on pullulan microspheres. Carbohydr Polym. 2013;91:74.
6. Saber-Samandari S, Saber-Samandari S, Nezafati N, Yahya K. Efficient removal of lead (II) ions and methylene blue from aqueous solution using Chitosan/Fe-Hydroxyapatite nanocomposite beads. J Environ Manage. 2014;146:481.
7. Mishra G, Tripathy M. A critical review of the treatment for decolorization of textile effluent. Colourage. 1993;40:35.
8. Eren E. Investigation of a basic dye removal from aqueous solution onto chemically modified Unye Bentonite. J Hazard Mater. 2009;166:88.
9. Salleh MAM, Mahmoud DK, Karim WAWA, Idris A. Cationic and anionic dye adsorption by agricultural solid wastes: a comprehensive review. Desalination. 2011;280:1.
10. Saber-Samandari S, Gulcan HO, Saber-Samandari S, Gazi M. Efficient removal of anionic and cationic dyes from an aqueous solution using pullulan-graft-polyacrylamide porous hydrogel. Water Air Soil Pollut. 2014;225:2177.
11. Vadivelan V, Kumar KV. Equilibrium, kinetics, mechanism, and process design for the sorption of methylene blue onto rice husk. J Colloid Interface Sci. 2005;286:90.
12. Sharma P, Kaur R, Baskar C, Chung WJ. Removal of methylene blue from aqueous waste using rice husk and rice husk ash. Desalination. 2010;59:249.
13. Khodaie M, Ghasemi N, Moradi B, Rahimi M. Removal of Methylene Blue from Wastewater by Adsorption onto ZnCl$_2$ Activated Corn Husk Carbon Equilibrium Studies. J Chemistr. 2013;383985.
14. Khodaie M, Ghasemi N, Moradi B, Rahimi M. Removal of Methylene Blue from Wastewater by Adsorption onto ZnCl$_2$ Activated Corn Husk Carbon Equilibrium Studies. J Chemistr. 2013; Article ID 383985, 6 pages, http://dx.doi.org/10.1155/2013/383985.
15. Hameed BH. Evaluation of papaya seeds as a novel nonconventional low-cost adsorbent for removal of methylene blue. J Hazard Mater. 2009;162:939.
16. Saber-Samandari S, Gazi M, Yilmaz O. Synthesis and characterization of chitosan-graft-poly (N-Allyl Maleamic Acid) hydrogel membrane. Water Air Soil Pollut. 2013;224:1624.
17. Sharma K, Kaith BS, Kumar V, Kalia S, Kumar V, Swart HC. Water retention and dye adsorption behavior of Gg-Cl-Poly(Acrylic Acid-Aniline) based conductive hydrogels. Geoderma. 2014;232–234:45.
18. Sartape AS, Patil SA, Patil SK, Salunkhe ST, Kolekar SS. Mahogany fruit shell: a new low-cost adsorbent for removal of methylene blue dye from aqueous solutions. Desalin Water Treat. 2013;53:98.
19. Saka C, Sahin O, Celik MS. The removal of methylene blue from aqueous solutions by using microwave heating and pre-boiling treated onion skins as a new adsorbent. Energ Source Part A. 2012;34:1577.
20. Saka C, Sahin O. Removal of methylene blue from aqueous solutions by using cold plasma- and formaldehyde-treated onion skins. Color Technol. 2011;127:246.
21. Chowdhury A, Bhowal A, Datta S. Equilibrium, thermodynamic and kinetic studies for removal of copper (II) from aqueous solution by onion and garlic skin. Water. 2012;4:37.
22. Pathani D, Sharma S, Singh P. Removal of Methylene Blue by Adsorption onto Activated Carbon Developed from Ficus Carica Bast. Arab J Chem. 2013; Received 9 May 2012; accepted 17 April 2013. doi:10.1016/j.arabjc.2013.04.021.
23. Suleria HAR, Butt MS, Anjum FM, Saeed F, Khalid N. Onion: nature protection against physiological threats. Crit Rev Food Sci. 2015;55:50.

24. Al-Ghouti MA, Khraisheh MAM, Ahmad MNM, Allen S. Adsorption behaviour of methylene blue onto Jordanian diatomite: a kinetic study. J Hazard Mater. 2009;165:589.
25. Saber-Samandari S, Saber-Samandari S, Gazi M. Cellulose-Graft-Polyacrylamide/Hydroxyapatite composite hydrogel with possible application in removal of Cu(II) ions. React Funct Polym. 2013;73:1523.
26. Hameed BH, Din ATM, Ahmad AL. Adsorption of methylene blue onto bamboo-based activated carbon: kinetics and equilibrium studies. J Hazard Mater. 2007;141:819.
27. Bello OS, Adeogun IA, Ajaelu JC, Fehintola EO. Adsorption of methylene blue onto activated carbon derived from periwinkle shells: kinetics and equilibrium studies. Chem Ecol. 2008;24:285.
28. Maiti S, Purakayastha S, Ghosh B. Production of low-cost carbon adsorbents from agricultural wastes and their impact on dye adsorption. Chem Eng Comm. 2008;195:386.
29. Rubin E, Rodriguez P, Herrero R, Vicente MES. Adsorption of methylene blue on chemically modified algal biomass: equilibrium, dynamic and surface data. J Chem Eng Data. 2010;55:5707.
30. Mahmoodi NM, Arami M, Bahrami H, Khorramfar S. The effect of pH on the removal of anionic dyes from colored textile wastewater using a biosorbent. J Appl Polym Sci. 2011;120:2996.

2

Performance evaluation and modeling of a submerged membrane bioreactor treating combined municipal and industrial wastewater using radial basis function artificial neural networks

Seyed Ahmad Mirbagheri[1], Majid Bagheri[1*], Siamak Boudaghpour[1], Majid Ehteshami[1] and Zahra Bagheri[2]

Abstract

Treatment process models are efficient tools to assure proper operation and better control of wastewater treatment systems. The current research was an effort to evaluate performance of a submerged membrane bioreactor (SMBR) treating combined municipal and industrial wastewater and to simulate effluent quality parameters of the SMBR using a radial basis function artificial neural network (RBFANN). The results showed that the treatment efficiencies increase and hydraulic retention time (HRT) decreases for combined wastewater compared with municipal and industrial wastewaters. The BOD, COD, NH_4^+-N and total phosphorous (TP) removal efficiencies for combined wastewater at HRT of 7 hours were 96.9%, 96%, 96.7% and 92%, respectively. As desirable criteria for treating wastewater, the TBOD/TP ratio increased, the BOD and COD concentrations decreased to 700 and 1000 mg/L, respectively and the BOD/COD ratio was about 0.5 for combined wastewater. The training procedures of the RBFANN models were successful for all predicted components. The train and test models showed an almost perfect match between the experimental and predicted values of effluent BOD, COD, NH_4^+-N and TP. The coefficient of determination (R^2) values were higher than 0.98 and root mean squared error (RMSE) values did not exceed 7% for train and test models.

Keywords: Combined wastewater, Submerged membrane bioreactor, Treatment efficiency, Artificial neural network, Radial basis function

Introduction

The membrane bioreactor (MBR), especially the submerged membrane bioreactor (SMBR), has been extensively investigated and applied for municipal and industrial wastewater treatment. There are more than 2200 MBR installations in operations or under construction worldwide and most of them are for municipal wastewater treatment [1,2]. Earlier studies have already shown that MBRs can be operated at much higher efficiency than of what is needed for municipal wastewater [3,4]. Treatment performances were generally good, and

deterioration of the performance was not observed [5]. Rosenberger et al. [3] studied aerobic treatment of municipal wastewater in an MBR for 535 day. The pilot plant comprised an anoxic zone to enable denitrification. The hydraulic retention time (HRT) varied between 10.4 and 15.6 hours. Treatment performance was very stable and on a high level. The chemical oxygen demand (COD) was reduced by 95%. Nitrification was complete and up to 82% of the total nitrogen could be denitrified. The excellent capability of SMBRs in the treatment of municipal wastewater has decreased the HRT to the minimum possible amount compared with conventional activated sludge processes. In other words, it seems that SMBRs are over designed for the treatment of municipal wastewater.

* Correspondence: bagherimajead@yahoo.com
[1]Department of Civil Engineering, K.N. Toosi University of Technology, Vanak square, Tehran, Iran
Full list of author information is available at the end of the article

The interest of using MBR instead of classical activated sludge system for the treatment of industrial wastewater was demonstrated [6,7]. Zhao et al. [8] used a laboratory-scale anaerobic/anoxic/oxic membrane bioreactor system to treat heavily loaded and toxic coke plant wastewater and operated for more than 500 days. Treatment performance, acute toxicity assessment, and dissolved organic characteristics of the system were investigated. When the HRT of the system was 40 hours, the removal efficiencies of COD, phenol, $NH_3 - N$, total nitrogen (TN) and acute toxicity were 89.8%, 99.9%, 99.5%, 71.5% and 98.3%, respectively. A desirable treated wastewater is water that is not only low in organic or mineral components, and free from biological entities such as bacteria, pathogens, and viruses but also cost efficient and reliable [9,10]. HRT plays an important role in the removal of pollutants in activated sludge processes coupled with membranes [11]. The amount of HRT for most of industrial wastewaters is higher than 2 days. As a result, the treatment of industrial wastewater is more expensive than treatment of municipal wastewater by considering the important role of HRT in efficiency and the cost of wastewater treatment.

The components in the industrial wastewater are in huge amount; for instance, high amount of COD and biochemical oxygen demand (BOD) [12], ammonia, suspended solid or heavy metal [13] and sometimes shock loading will happen. High strength of industrial wastewater results in the low biodegradability characteristic of wastewater [14-16]. A solution used to increase the biodegradation of the slowly biodegradable compounds is the adsorbent addition in the bioreactor [17-19]. Generally, BOD/COD equal to 0.5 is considered as readily biodegradable or easily treatable [20-24]. The BOD/COD ratio for the industrial wastewaters is not equal to 0.5 and varies from 0.117 to 0.773. The idea of combined municipal and industrial wastewater is an approach to set this ratio to 0.5 and improve the other criteria for a desirable wastewater treatment such as the total BOD to total phosphorous (TBOD/TP) ratio and the influent BOD and COD concentrations.

Treatment process models are essential tools to assure proper operation and better control of wastewater treatment plants. Considerable effort has been devoted to the modeling of activated sludge processes (ASPs) since early 1970s [25]. Some deterministic models have been developed basing on the fundamental biokinetics such as activated sludge model number one (ASM1) [26,27]. Following ASM1, ASM2, ASM2d and ASM3 models were developed. The ASM2 [28] models extended the capabilities of ASM1 to involve the biological phosphorus and nitrogen removals. Whereas, ASM3 [29,30] introduced an alternative concept to the previous ASM biokinetics and aimed at simplifying the model application. Despite the availability of ASM models, the diagnosis of the process interactions and modeling of ASP in an SMBR is still difficult [30,31]. Parameter estimation and calibration of ASM models require expertise and significant effort. Moreover, calibration has to be performed for each specific treatment system. Therefore, application of ASM models to real systems can be cumbersome and problematic [25,32].

In recent years, artificial neural networks (ANNs) have been used for monitoring, controlling, classification and simulation of ASPs. ANN is a non-parametric model which utilizes interconnected mathematical nodes or neurons to form a network that can model complex functional relationships [33]. So far, different types of neural network architectures and their performances have been studied for the purpose of neuroidentification [34-37]. It includes radial basis functions (RBFs), multi-layer perceptrons (MLPs), recurrent neural networks (RNNs), and echo-state networks (ESNs). In the literature to date, a limited number of applications of ANNs have been made to SMBRs for modeling of a plant operation [31,38]. Geissler et al. [31] used an ANN model to predict the filtration performance in a submerged capillary hollow fiber membrane treating municipal wastewater. The training procedure for the ANN was conducted based upon pilot-studies with an MBR system using a novel submerged capillary module. Good correlations were found between the predicted and measured permeability using ANN. Cinar et al. [38] have also proposed an ANN model for a SMBR treating cheese whey and evaluated its performance at different sludge residence time. The results of the training procedure for effluent total phosphate, COD, ammonia, nitrate were successful. However, the results of the testing procedure for effluent total phosphate were not as good as for the effluent ammonia and nitrate, although they were better than the results of effluent COD. Up till now, there have not been any investigations on treating combined municipal and industrial wastewater by SMBRs for the purpose of optimizing HRT or performance improvement. Furthermore, no attempt has been made on the modeling of the combined municipal and industrial treatment systems.

In order to achieve the objective of this study, it was decided to employ a type of RBF, which is most commonly used in classification problems [39]. RBFs have been successfully applied for solving dynamic system problems, because they can predict the behavior directly from input/output data [40-42]. The radial basis function artificial neural network (RBFANN) was applied to model the effluent quality parameters of an SMBR treating combined municipal and industrial wastewater. The influent concentration of parameters, HRT, mixed liquor volatile suspended solids (MLVSS), total dissolved solids (TDS) and pH were inputs of the RBFANN models.

Sensitivity analyses were performed to determine the effect and importance order of each input parameter on the changes of effluent concentrations.

Material and methods
Pilot plant configuration
An SMBR was used in order to treat combined municipal and industrial wastewater in this study. Figure 1 shows the schematic diagram of the hollow fiber SMBR. The SMBR consisted of a storage tank, an anaerobic reactor, an anoxic reactor and an oxic reactor as simultaneous aeration/filtration reactor. The storage tank was made of plastic measuring 0.7 by 0.7 meter and total volume of 0.49 m^3. The influent pump established a continuous influent wastewater flow from feeding tank to the anaerobic reactor. The anaerobic, anoxic and oxic reactors were made of Plexiglas with total volume of 0.06 m^3, 0.1 m^3 and 0.24 m^3, respectively. The anaerobic reactor measuring 0.4 by 0.3 meter was located 1.2 meter above the ground level to establish a continuous flow to anoxic reactor. The anoxic reactor measuring 0.5 by 0.4 meter was located 0.9 meter above the ground level to establish a continuous flow to oxic reactor. The oxic reactor measuring 0.8 by 0.5 meter performed a simultaneous aeration/filtration treating role in the SMBR system. A small portion of the sludge in the oxic reactor was recirculated back into the anoxic reactor using a recirculation pump, where it was mixed with the effluent of anaerobic reactor. This recirculation is a key feature of the activated sludge process. The temperature

was kept about 25°C and the HRT of the SMBR system varied during the experiments. The SMBR consisted of a polypropylene hollow fiber membrane with a nominal pore size of 0.04 μm. The overall membrane surface area was 8 m^2 per module. The maximum permitted pressure for the hollow fiber membrane was about 30 kPa. A pressure gauge was installed on the suction path to turn off the suction pump and open the backwash path when the trans-membrane pressure exceeded permitted limit and membrane foiling occurred. Table 1 shows the detailed specification of hollow fiber membrane.

Municipal wastewater characteristics
The pilot plant was located in Ekbatan wastewater treatment plant in Tehran, Iran, which has been operating since 1988. Influent wastewater analysis for the wastewater treatment plant was carried out for a four month period. According to the results obtained from raw wastewater analysis, the maximum values were selected as critical design parameters. Table 2 shows the critical values of the influent wastewater characteristics to Ekbatan wastewater treatment plant.

Industrial wastewater characteristics
The industrial wastewaters are defined by high strength wastewaters because of the high concentration of their components. Table 3 shows the characteristics of high strength wastewater for different industries. The COD, BOD and total suspended solids (TSS) are three most high concentration components of industrial wastewaters

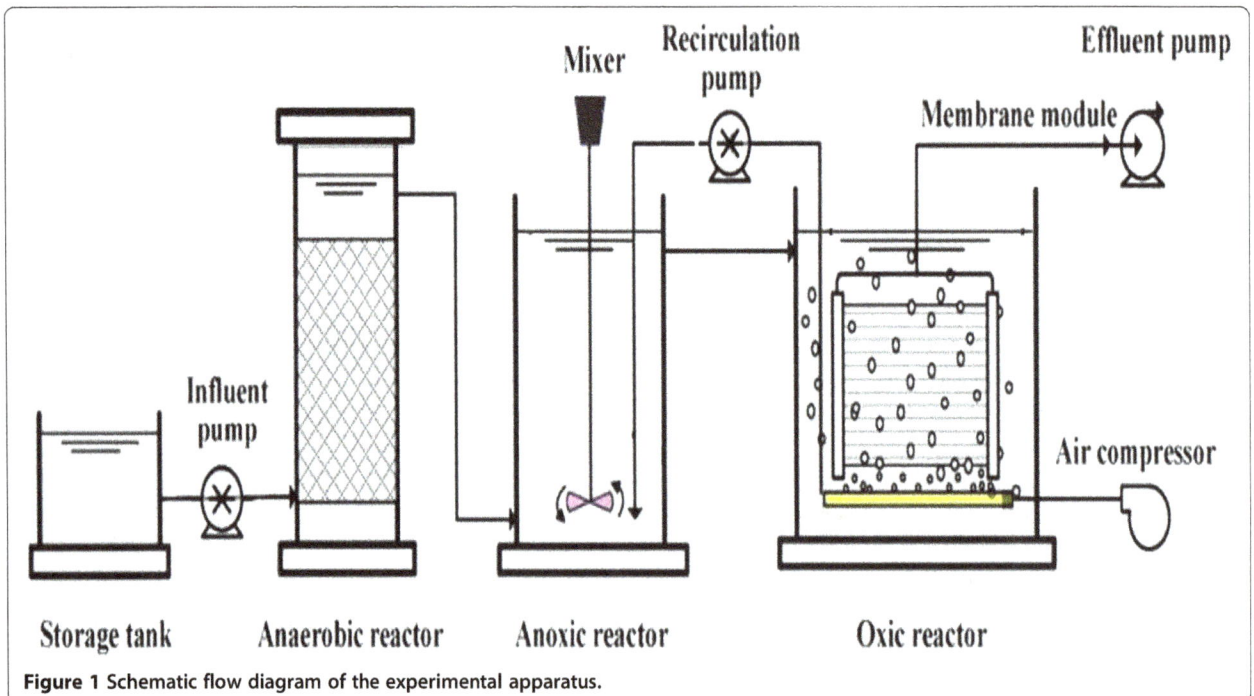

Figure 1 Schematic flow diagram of the experimental apparatus.

Table 1 Specifications of the hollow fiber membrane used in this study

Description	Value
Material	Polypropylene
Capillary Thickness	40 ~ 50 μm
Capillary Outer Diameter	450 μm
Capillary Pore Diameter	0.01 ~ 0.2 μm
Gas permeation	$7.0 * 10^{-2}$ cm^3/cm^2 · S · cm Hg
Porosity	40 ~ 50%
Lengthways strength	120,000 kPa
Designed flux	6 ~ 9 L/M^2/H
Area of membrane module	8 m^2/module
Operating Pressure	−10 ~ −30 kPa
Flow rate	1.0 ~ 1.2 m^3/ day

compared with municipal wastewater. In the current research, the high strength wastewater was simulated by increasing the influent COD, BOD, and TSS concentration to 2000, 1300 and 5000 mg/L, respectively. The simulated wastewater had the characteristics close to Beverage wastewater [12], and its TSS concentration was increased greatly. The TDS concentration was increased to 4500 mg/L and pH was different for each experiment.

Combined municipal and industrial wastewater characteristics

Much work has been performed to study the performance of SMBRs in the treatment of municipal and industrial wastewater, including influences of biological processes and membrane fouling. A factor which influences membrane performance in an optimum treatment of wastewater is decreasing HRT while keeping effluent components lower than discharge limits. Moreover, wastewater with BOD/COD ratio equal to 0.5 is considered as readily biodegradable. It has been shown biodegradability greater than 0.5 for spent caustic wastewater after treatment by using wet air oxidation [43]. If the ratio value is less than 0.5, the wastewater needs to have physical or chemical treatment before a biological treatment takes place [43,44]. Table 3 shows that for the most of industries this ratio is not equal to 0.5 and HRT of industrial wastewater

Table 2 Municipal wastewater characteristics in the critical conditions

Parameter	Value	Parameter	Value
Temperature (°C)	25.8	Org-N (mg/L)	16.8
DO (mg/L)	0	TKN (mg/L)	39.9
BOD$_5$ (mg/L)	180	TS (mg/L)	810
COD (mg/L)	380	TDS (mg/L)	630
NO$_3^-$–N (mg/L)	0.96	TSS (mg/L)	180
NH$_4^+$–N (mg/L)	23.1	TP (mg/L)	16.54

is noticeably higher than HRT of municipal wastewater. Therefore, the municipal and industrial wastewater were combined in proportions that the BOD/COD ratio approached 0.5 for the produced combined wastewater.

Analytical methods

Temperature, pH, dissolved oxygen (DO), BOD, COD, TSS, TDS, mixed liquor suspended solids (MLSS) and MLVSS concentration, TP, NH$_4^+$–N and NO$_3^-$–N were measured in this study. The pH and temperature were measured using a digital pH meter. A dissolved oxygen meter (YSI 5000) was utilized to determine DO. Biodegradability was measured by 5-day biochemical oxygen demand test according to the standard methods [45]. The seed for BOD test was obtained from the Ekbatan wastewater treatment plant [46]. The COD was determined according to the standard methods [45]. Weekly analyses included mixed liquor volatile and total suspended solids (MLVSS, MLSS) in the oxic reactor [47]. MLSS and MLVSS were determined at the Ekbatan wastewater treatment plant laboratory at the temperature of 550°C [20]. The TP and NH$_4^+$–N were measured with aid of a spectrophotometer (The Hach DR 5000 UV–vis Laboratory Spectrophotometer) at the wastewater treatment plant laboratory.

RBFANN; background and methodology

ANNs can be used for monitoring, controlling, classification and simulation of experimental data. ANNs are mathematical models simulating important parameters based on past observations in complex systems. There are many types of ANNs for modeling and function approximation of the engineering problems [35]. The two well-known ANN models, namely RBFANN and multilayer perceptron artificial neural network (MLPANN) have been used in engineering applications to model or approximate properties [48]. An RBFANN is the most commonly used neural network for pattern recognition problems, and it is also widely used for fault diagnosis. The RBFANN has the advantages of a fast learning process, a learning stage without any iteration of updating weights, robust ability, and adaptation capability compared with other ANNs such as MLP, RNN and ESN [41,49]. RBFANNs have a very strong mathematical foundation rooted in regularization theory for solving ill-conditioned problems [48]. Tomenko et al. [50] reported that RBFANNs produced better results than multiple regression in a wetland treatment system. Madaeni et al. [51] modeled O$_2$ separation from air in a hollow fiber membrane module with glassy membrane using ANN. They found RBFANN as the best network with minimum training error and checked generalization capability of their network with some unseen data (test data).

Table 3 Characteristics of high strength wastewater for different industries

Industry	HRT (day)	COD (mg/L)	BOD (mg/L)	BOD/COD	NH_4 -N (mg/L)	TSS (mg/L)	So_4^{2-} (mg/L)	po_4^3 (mg/L)	Oil (mg/L)	Phenol (mg/L)
Tannery [14]		2000	-	-	-	-	400	-	-	-
Tannery [12]		16000	5000	0.313	450	-	-	-	-	-
Textile [15]	2	6000	700	0.117	20	-	-	120	-	-
Textile [16]	0.7-4	4000	500	0.125	4.8	-	200	2	-	-
Dyeing [21]		1300	250	0.192	100	200	-	-	40	-
Textile [22]	0.58	1500	500	0.333	50	140	-	7	-	-
Wheat starch [10]		35000	16000	0.457	-	13300	-	-	-	-
Dairy [12]		3500	2200	0.629	120	-	-	-	-	-
Beverage [12]		1800	1000	0.556	-	-	-	-	-	-
Palm oil [23]	0.8	67000	34000	0.507	50	24000	-	-	100000	-
Pet food [24]	2.9	21000	10000	0.476	110	54000	-	200	-	-
Dairy product [18]		880	680	0.773	-	2480	-	-	-	-
Phenolic [19]	0.42	797	-	-	131	-	-	-	-	37.3
Pharmaceutical [11]	1	6300	3225	0.51	-	-	-	-	-	-

Shahsavand and Pourafshari Chenar [52] found RBFANN a better choice for the prediction of gas separation performance in a membrane system because of its powerful noise filtering capability. The RBFANNs could offer more successful solutions operating on limited data (less time consuming, supplying additional information on existing relationships), thus indicating that the use of hyperspheres to divide up the pattern space into various classes is more advantageous when dealing with data described by a small number of variables [39].

The structure of the basic RBFANN consisted of one input layer, one output layer, and one hidden layer. A single-output RBFANN with N hidden layer nodes can be described by Eq. (1) and (2).

$$Y = \sum_{n=1}^{N} w_n \theta_n(X) \qquad (1)$$

where X and Y are the input and output of the network, $X = (x1, x2,..., xm)T$, wn is the connecting weights between nth hidden node and the output layer, θn is the output value of the nth hidden node, and

$$\theta_n(X) = e^{\left(-x-\mu_n/\sigma_n^2\right)} \qquad (2)$$

where μn is the center vector of nth hidden node, $x - \mu_n$ is the Euclidean distance between x and μn, and σn is the radius of the nth hidden node.

As its name implies, radially symmetric basis function is used as activation functions of hidden nodes [39,48]. The transformation from the input nodes to the hidden nodes is a non-linear one, and training of this portion of

the network is generally accomplished by an unsupervised fashion. The training of the network parameters (weight) between the hidden and output layers occurs in a supervised fashion based on target outputs [48]. For a RBFANN, the capabilities of the final network are determined by the parameter optimization algorithms and the structure size. However, the number of hidden nodes in these RBFANNs is often assumed to be constant [53]. The transfer function of the RBFANN hidden in layer unit is symmetrical RBF (such as Gaussian function). RBF network has approximation to nonlinear continuous function, and it can learn at high speed and undertake a wide range of data fusion and high-speed processing data in parallel [54]. In this study, the RBFANN applies different network functions such as newrbe and newrb to the input data. The newrb function designs a radial basis neural network and the newrbe function designs an exact radial basis ANN.

Model development

Simulation model of operational parameters was established based on the theory of feed forward artificial neural networks, namely RBFANN using the mathematical software program MATLAB. The operating parameters, including influent BOD, COD, NH_4^+–N or TP as well as HRT, MLVSS, TDS and pH were the input variables of the RBFANN models. These variables were used to train RBFANN in order to simulate effluent BOD, COD, NH_4^+–N and TP. Experimental data over 90 days (30 total data points) were used in RBFANN modeling process. Table 4 shows the statistical characteristics of the measured

Table 4 Characteristics of measured variables used for modeling by RBFANN

Input variable no.	Input variable	Value	Output variable	Value
1	Influent conc.		Effluent conc.	
	BOD (mg/L)	500–600	BOD (mg/L)	5.5–172.3
	COD (mg/L)	1000–12000	COD (mg/L)	11–396.5
	NH_4^+–N (mg/L)	21–27	NH_4^+–N (mg/L)	0.2–3.1
	TP (mg/L)	15–16.4	TP (mg/L)	1.4–6.4
2	HRT (h)	3–11		
3	MLVSS (mg/L)	4120–5990		
4	TDS (mg/L)	500–4900		
5	pH	6.2–7.6		

process variables used in this study to model effluent BOD, COD, NH_4^+–N and TP by RBFANN.

In order to obtain convergence within a reasonable number of cycles, the input and output data should be normalized and scaled to the range of 0–1 by Eq. (3) [55,56]:

$$x_{ni} = \frac{x_i - x_{min}}{x_{max} - x_{min}} \tag{3}$$

where xi is the initial value, xmax and xmin are the maximum and minimum of the initial values, and xni is the scaled value. After the training and testing of the ANN, the output data were scaled to the real-world values through the Eq. (4).

$$x_i = x_{ni}(x_{max} - x_{min}) + x_{min} \tag{4}$$

In the current research, the developed networks consisted of three layers including one input layer that comprised five neurons (including influent BOD, COD, NH_4^+–N or TP, HRT, MLVSS, TDS and pH), one hidden layer consisting of five neurons (which were constant due to the application of newrbe function) and the output layer that had one output neuron (which was effluent BOD or COD, NH_4^+–N or TP). The RBFANN applied network function of newrbe (design exact radial basis network) to the input data. The newrbe function creates a two-layer network with biases for the both layers (Figure 2). The first layer has radial basis transfer function (radbas). Consequently, it calculates its weighted inputs with Euclidean distance weight function (dist) and its net input with product net input function (netprod). The second layer has linear transfer function (purelin). Consequently, it calculates its weighted inputs with Dot product weight function (dotprod) and its net input with sum net input function (netsum). For the newrbe function, the center is selected by using the fixed centers selected at random method [57]. This function can produce

a network with zero error on training vectors. The stopping criterion for the newrbe function is when the outputs are exactly the matrix of target class vectors when the inputs are the matrix of input vectors [58]. The RBFANN was designed in an innovative loop that can apply newrbe function to the input data for user defined number of times in order to minimize error.

The network was trained until the network average root mean squared error (RMSE) was minimum and coefficient of determination (R^2) approached 1. Other parameters for network were chosen as the default values of the software. The performances of the ANN models were measured by R^2 and RMSE between the predicted values of the network and the experimental values, which were calculated by Eq. (5) and (6), respectively [59].

$$R^2 = 1 - \frac{\sum_{i=1}^{n}\left(y_i^* - y_p^{(i)}\right)^2}{\sum_{i=1}^{n}(y_i^* - \bar{y})^2} \tag{5}$$

$$RMSE = \sqrt{\frac{1}{n}\sum_{i=1}^{n}\left(y_p^{(i)} - y_i^*\right)^2} \tag{6}$$

where y is the average of y over the n data, and y_i^* and $y_p^{(i)}$ are the ith target and predicted responses, respectively.

Results and discussion
BOD removal efficiency and modeling outcomes
The experiments were executed from HRT of 1 to 7 h in order to optimize HRT for the municipal wastewater. The influent BOD and MLVSS concentration varied from 140 to 180 mg/L and 3380 to 5470 mg/L, respectively. The removal performance of the SMBR increased from the HRT of 1 to 5 h so the effluent decreased to 5 mg/L with removal efficiency of 97% at HRT of 5 h. The effluent BOD concentration increased and removal performance decreased after 5 h due to self-degradation of microorganisms [60]. Consequently, the kinetic constants including the half saturation coefficient (K_s), the maximum substrate utilization rate (k) and endogenous decay coefficient (K_d) were calculated as 113 mg L^{-1}, 2.05 d^{-1} and 0.036 d^{-1} different HRTs. The influent BOD concentration decreased from 175 mg/L to 25 mg/L for the effluent concentration at HRT of 1 h. It means that the effluent BOD concentration at HRT of 1 h meets discharge limits of municipal wastewater (BOD < 30 mg/L).

The HRT varied from 5 to 20 h in order to optimize HRT for the industrial wastewater. The influent BOD concentration varied from 1265 to 1360 mg/L and MLVSS concentration changed from 4350 to 7890 mg/L. The BOD removal efficiency increased from the HRT of 5 to 17 h in a trend like municipal wastewater. The effluent BOD decreased to 8.7 mg/L with removal

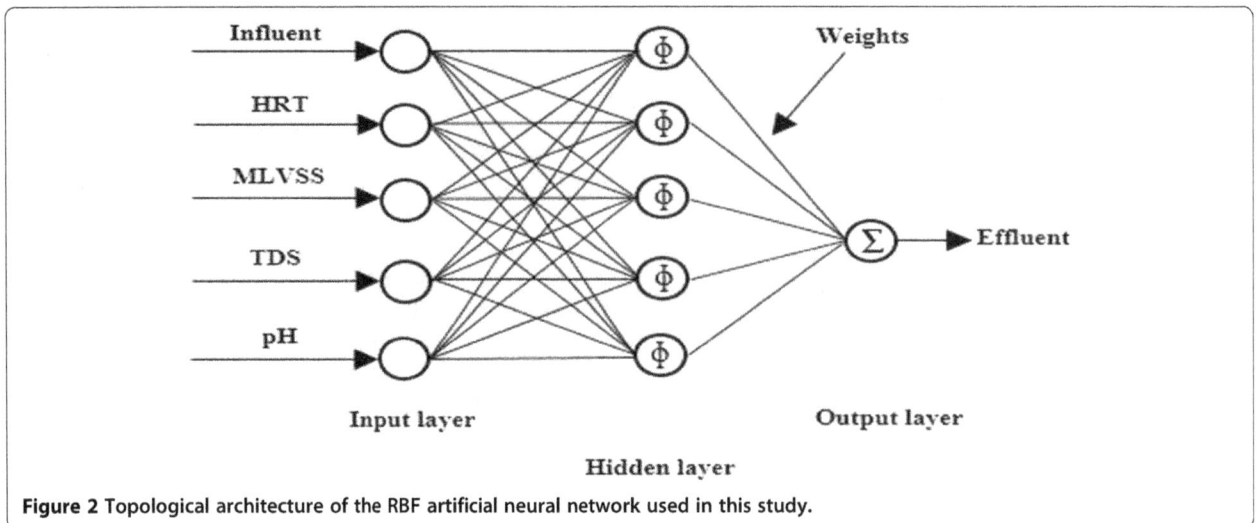

Figure 2 Topological architecture of the RBF artificial neural network used in this study.

efficiency of 99.34% at HRT of 17 h. Then efficiency decreased because of self-degradation [61]. Consequently, the kinetic constants were calculated as K_s equal to 163.55 mg L^{-1} and k equal to 3.56 d^{-1}, K_d equal to 0.013 d^{-1}. At HRT of 13 h the effluent BOD concentration decreased to 42.1 mg/L with removal efficiency of 96.9%. It was concluded that the HRT of 13 h cannot meet the discharge limits and the HRT of 17 h was the optimal result for the treatment of industrial wastewater. The difference between the required HRT for the treatment of municipal and industrial wastewater denotes the noticeable difference in the cost of municipal and industrial wastewater treatment. Therefore, the idea of combined municipal and industrial wastewater was followed as a key to improve the efficiency and decrease the cost of wastewater treatment.

The concentration of components of combined municipal and industrial wastewater was between municipal and industrial wastewaters. The BOD/COD ratio for the combined wastewater was about 0.5 compared with municipal and industrial wastewaters, which changed from 0.38 to 0.5 and from 0.6 to 0.7, respectively [9]. The HRT varied from 3 to 11 h in order to decrease effluent BOD concentration to discharge limits. The influent BOD and MLVSS concentration for the combined wastewater varied from 500 to 600 and from 4120 to 5990 mg/L, respectively. The influent BOD concentration decreased from 557 mg/L to 5.5 mg/L for the effluent of BOD with removal efficiency of 99% at HRT of 9 h. The removal efficiency decreased by increasing HRT from 9 to 11 h because of self-degradation [60,61]. Consequently, the kinetic constants were calculated as K_s equal to 177.84 mg L^{-1} and k equal to 5.29 d^{-1}, K_d equal to 0.011 d^{-1}. At HRT of 7 h for the combined wastewater, influent BOD concentration decreased from 600 mg/L to 19 mg/L for the effluent of BOD with

removal efficiency of 96.9%. Therefore, the effluent BOD concentration met discharge limits at HRT of 7 h. It was observed that the performance of the SMBR increases when the influent BOD concentration is lower than 600 to 700 mg/L, which is in a good agreement with the findings of previous studies [3,24]. We concluded that by combining municipal and industrial wastewaters, the treatability of wastewater could be improved by setting BOD/COD ratio to 0.5 and reduction of wastewater strength by decreasing influent BOD concentration to lower than 700 mg/L.

In order to model the effluent BOD concentration by RBFANN, the influent BOD, HRT, MLVSS, TDS and pH were input variables of the neural network. The RBFANN applied network function of newrbe to the input data and the spread of RBF was considered equal to its default value, 1. A large spread results in a smooth function approximation, but, by contrast, a large spread can cause numerical problems [48]. The network function of newrbe selected 70% of normalized data to train and 30% to test RBFANN models [38]. The RBFANN was designed in an innovative loop that applied newrbe to the data for more than 30 times in order to minimize error. Optimal network was chosen on the basis of the minimum average error. Figure 3 shows the denormalized results of the effluent BOD modeling using the RBFANN according to training and testing data sets. The results of the training procedure by RBFANN were successful for the effluent BOD concentration. Figure 3 shows that the train and test models by RBFANN indicated an almost perfect match between the experimental and the predicted effluent BOD values compared with the results of models introduced by Cinar et al. [38]. The RMSE values for train and test (verification process) models were 8.67 and 4.56 mg/L, respectively, and the R^2 values were greater than 0.99 for both models. The

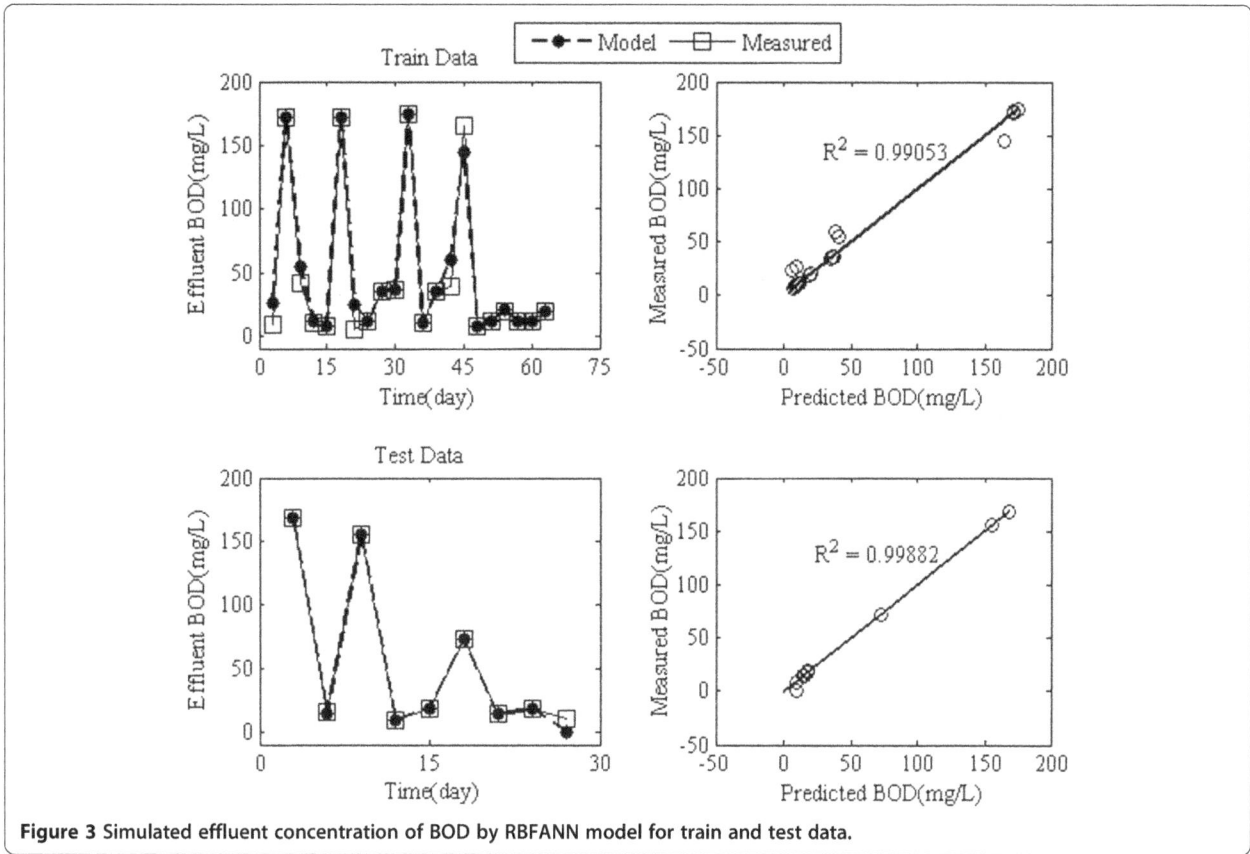

Figure 3 Simulated effluent concentration of BOD by RBFANN model for train and test data.

RBFANN predicted effluent BOD so accurate that the mean average error for train and test models did not exceed 5% and 3%, respectively.

The effluent BOD was modeled individually by considering different single and joint variables as inputs of neural network to examine the effect of each variable on the changes of effluent BOD concentration. The joint inputs to train the neural network were groups of two, three and four variables. Table 5 shows HRT among single input variables, and HRT and MLVSS among groups of two variables had the most considerable effect on the effluent BOD concentration. Furthermore, HRT, MLVSS and TDS among groups of three variables, and HRT, MLVSS, TDS and influent BOD between groups of four variables were determined to have the greatest effects on the effluent BOD concentration.

The sensitivity [62] of effluent BOD concentration to changes of an input variable such as HRT determines the influence and importance of HRT on the effluent BOD models. The effect of each variable on the RBFANN models to simulate effluent of BOD compared with other variables was determined by its importance order. Table 5 shows the importance order of each input variable and the joint variables for effluent BOD. The variable with higher rank of importance denoted not only a favorable match between experimental and simulated values by RBFANN

Table 5 Effect of different single and joint variables on the effluent BOD models

Input variable no.	R^2		RMSE (mg/L)		Importance order
	Train	Test	Train	Test	
1	0.804	0.863	35.62	32.80	5
2	0.991	0.999	8.15	2.79	1
3	0.907	0.995	24.76	6.43	2
4	0.705	0.674	48.71	42.27	4
5	0.804	0.863	35.62	32.8	3
2-1	0.973	0.961	17.39	16.28	4
2-3	0.998	0.999	2.98	2.15	1
2-4	0.996	0.978	6.82	2.46	3
2-5	0.995	0.999	6.65	1.76	2
2-3-1	0.992	0.999	6.17	2.41	2
2-3-4	0.998	1	4.08	0.12	1
2-3-5	0.996	0.998	5.62	4.23	3
2-3-4-1	0.998	0.998	3.69	3.07	1
2-3-4-5	0.997	0999	4.29	3.25	2
2-3-4-1-5	0.990	0.998	8.67	4.56	1

The numbers 1 to 5 refers to input variables identified in Table 4.

models but also low RMSE and high R^2 values. The variation of effluent BOD concentration was influenced by HRT, MLVSS, TDS, influent BOD concentration and pH, respectively. The results of this study show that the HRT and MLVSS have the greatest influence on the effluent BOD, which are in a good agreement with earlier experimental studies [63,64].

COD removal efficiency and modeling outcomes

To perform COD experiments for the municipal wastewater the influent COD concentration varied from 310 to 360 mg/L. By increasing HRT from 1 to 5 h the removal efficiencies increased and the best results obtained at HRT of 5 h. At this point, the effluent COD concentration reached to 8 mg/L with removal efficiency of 97.9%. The effluent COD concentration increased and removal efficiency decreased after 5 h [60]. Consequently, the kinetic constants were calculated as K_s equal to 96 mg L^{-1}, k equal to 2.31 d^{-1} and K_d equal to 0.043 d^{-1}. The influent COD concentration decreased from 354 mg/L to 53 mg/L for the effluent COD at HRT of 1 h. It means that the effluent COD concentration at HRT of 1 h meets discharge limits of the municipal wastewater (COD < 60 mg/L).

For the industrial wastewater the influent COD concentration varied from 2050 to 2120 mg/L. From the HRT of 5 to 17 h the removal efficiency increased and the best results obtained at HRT of 17 h. At this point, the influent COD concentration decreased from 2100 mg/L to 14.8 mg/L for the effluent COD. Consequently, the kinetic constants were calculated as K_s equal to 308 mg L^{-1}, k equal to 2.81 d^{-1}, K_d equal to 0.019 d^{-1}. The influent COD concentration decreased from 2055 mg/L to 71.2 mg/L for the effluent with removal efficiency of 96.5% at HRT of 13 h. Therefore, it was concluded that the HRT of 13 h cannot meet the discharge limits and HRT of 17 h was considered as the optimal result.

The COD experiments were performed for the combined municipal/industrial wastewater with the HRT varying from 3 to 11 h. The influent COD concentration for the combined wastewater varied from 1000 to 1200 mg/L. The influent COD concentration at HRT of 9 h decreased from 1130 mg/L to 10.3 mg/L for the effluent of COD with removal efficiency of 99.1%. By increasing HRT from 9 to 11 h the removal efficiency decreased [60,61]. Consequently, the kinetic constants were calculated as K_s equal to 113.2 mg L^{-1} k equal to 2.72 d^{-1}, K_d equal to 0.022 d^{-1}. The results indicated that at HRT of 7 h the influent COD concentration decreases from 1080 mg/L to 50 mg/L for the effluent with removal efficiency of 96%. Therefore, the effluent COD concentration at HRT of 7 h met discharge limits. It was observed that the performance of the SMBR increases

due to reduction of wastewater strength [9,24] when the influent COD concentration is lower than 1000 to 1100 mg/L.

In order to model the effluent COD by RBFANN, the influent COD, HRT, MLVSS, TDS and pH were the input variables of the neural network. The results of the effluent COD modeling using the RBFANN for train and test data were denormalized to compare the observed values of effluent COD concentration with simulated values. The results of the training procedure by RBFANN were successful for the effluent COD concentration. Figure 4 shows the train and test models by RBFANN indicated an almost perfect match between the experimental and the simulated effluent COD values compared with the results of models introduced by Cinar et al. [38]. The RMSE values for train and test (verification process) models were 25.62 and 9.12 mg/L, respectively. The values of R^2 for train and test models were 0.99 and 0.98, respectively. The RBF models predicted effluent COD concentration with a mean average error for train and test models, which did not exceed 7% and 3%, respectively.

Table 6 shows HRT among single input variables, and HRT and MLVSS among groups of two variables had the most considerable effects on the effluent COD. Furthermore, HRT, MLVSS and TDS among groups of three variables, and HRT, MLVSS, TDS and influent COD between groups of four variables were determined to have the greatest effect on the effluent COD. Table 6 also shows the importance order of each input and joint variable for effluent COD according to sensitivity analysis procedure. The variation of effluent COD concentration was influenced by HRT, MLVSS, TDS, influent COD concentration and pH, respectively. The results showed that the HRT and MLVSS have the greatest influence on the effluent BOD, which are in a good agreement with earlier experimental studies [63,64]. The results of sensitivity analyses were highly collaborated for both BOD and COD models performed by RBFANN. The variation of both effluent BOD and COD concentration to all input variables were the same in their simulation processes, which can be justified by similarities of their natures. As a result, to control the changes of effluent BOD and COD concentration the most effective variables are HRT and MLVSS.

NH_4^+–N removal efficiency and modeling outcomes

The influent NH_4^+–N concentration for the municipal wastewater varied from 18 to 24 mg/L. The results showed that the removal efficiency of NH_4^+–N is improved with increase of the HRT. The influent NH_4^+–N concentration decreased from 24 mg/L to 0.4 mg/L for the effluent at HRT of 7 h. The influent NH_4^+–N

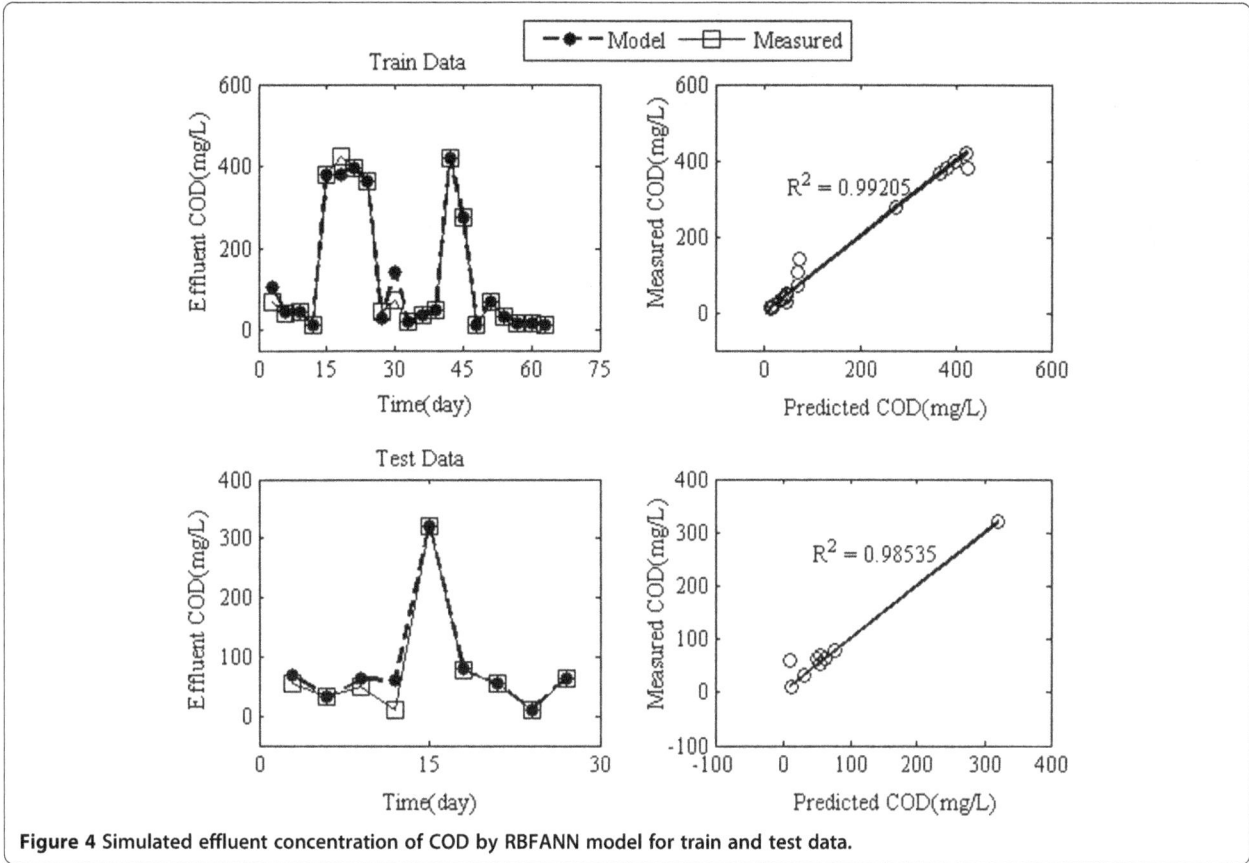

Figure 4 Simulated effluent concentration of COD by RBFANN model for train and test data.

concentration decreased from 23 mg/L to 0.8 mg/L at HRT of 5 h. As the optimal result, the effluent NH_4^+–N concentration with removal efficiency of 96.5% at HRT of 5 h met the discharge limit (NH_4^+–N <1 mg/L). Increasing the cell retention time increases Azotobacters and Orgasnotrophic bacteria and then followed by rapid removal of dissolved carbonaceous biochemical oxygen demand (CBOD) [20]. Rapid removal of dissolved CBOD increases the aeration time for the nitrification process [5,20]. Increases of the Azotobacter populations in activated sludge process occurs at relatively high hydraulic retention time [5,64]. Hence, the required HRT of municipal wastewater treatment was increased to 5 h.

For the industrial wastewater the influent NH_4^+–N concentration varied from 24 to 31 mg/L. The NH_4^+–N concentration decreased with increase of HRT for the industrial wastewater. And at HRT of 17 h the effluent N H_4^+–N concentration was equal to 0.2 mg/L with the removal efficiency of 99.35%. The effluent NH_4^+–N concentration at HRT of 13 h decreased from 30 mg/L to 0.7 mg/L for the effluent with the removal efficiency of 97.7%.

For the combined municipal and industrial wastewater the influent NH_4^+–N concentration varied from 21 to 27 mg/L. The effluent concentration at HRT of 7 h met

Table 6 Effect of different single and joint variables on the effluent COD models

Input variable no.	R^2		RMSE (mg/L)		Importance order
	Train	Test	Train	Test	
1	0.179	0.599	153.49	91.66	5
2	0.947	0.997	44.36	14.1	1
3	0.941	0.994	48.23	35.67	2
4	0.6	0.8	123	63.41	3
5	0.862	0.715	70	109.03	4
2-1	0.951	0.999	47	4.3	4
2-3	0.995	0.999	15.43	6.19	1
2-4	0.998	0.994	11.23	13.58	3
2-5	0.987	0.998	22.45	7.66	2
2-3-1	0.984	0.996	26.66	13.34	3
2-3-4	0.994	0.996	16.48	8.48	1
2-3-5	0.974	0.997	33.13	11.56	2
2-3-4-1	0.992	0.999	19.58	2.61	1
2-3-4-5	0.987	0.991	22.08	14.93	2
2-3-4-1-5	0.992	0.985	25.62	9.12	1

The numbers 1 to 5 refers to input variables identified in Table 4.

the discharge limits. The influent concentration decreased from 24 mg/L to 0.8 mg/L for the effluent with removal efficiency of 96.7%. The results of NH_4^+-N experiments for three systems showed that the required HRT for NH_4^+-N removal increases the required HRT for the treatment of municipal wastewater and does not have any effect on the required HRT for industrial and combined municipal and industrial wastewaters. The idea of combined wastewater is more useful because the treatment system is able to receive sewage with higher influent BOD and COD concentration without any increase in the HRT.

In order to model the effluent NH_4^+-N by RBFANN, the influent NH_4^+-N concentration, HRT, MLVSS, TDS and pH were the input variables of network. The results of the effluent NH_4^+-N modeling using the RBFANN for train and test data sets were denormalized to compare the observed values of effluent NH_4^+-N concentration with simulated values. The results of the training procedure by RBFANN were successful for the effluent NH_4^+-N. Figure 5 shows that the train and test models by RBFANN indicated an almost perfect match between the experimental and the simulated values of effluent NH_4^+-N concentration compared with the results of models introduced by Cinar et al. [38]. The RMSE values for train

and test (verification process) models were 0.06 and 0.14 mg/L, respectively. The R^2 values for train and test models were 0.98 and 0.99, respectively. The RBF models simulated effluent NH_4^+-N with a mean average error for train and test models, which did not exceed 2% and 5%, respectively.

Table 7 shows HRT among single input variables, and HRT and MLVSS among groups of two variables had the most considerable effect on the COD effluent. Furthermore, HRT, MLVSS and pH among groups of three variables, and HRT, MLVSS, pH and influent NH_4^+-N between groups of four variables were determined to have the greatest effect on the COD effluent. Table 7 also shows the importance order of each input variable and the joint variables for effluent NH_4^+-N according to sensitivity analysis procedure. The variation of effluent NH_4^+-N concentration was influenced by HRT, MLVSS, pH, influent NH_4^+-N and TDS, respectively. The current research shows that the HRT and MLVSS have the greatest influence on the changes of effluent NH_4^+-N, which are in a good agreement with earlier experimental studies [63,64]. The results of sensitivity analysis for the effluent NH_4^+-N models indicated that to control the changes of effluent NH_4^+-N concentration the most effective variables are HRT and MLVSS.

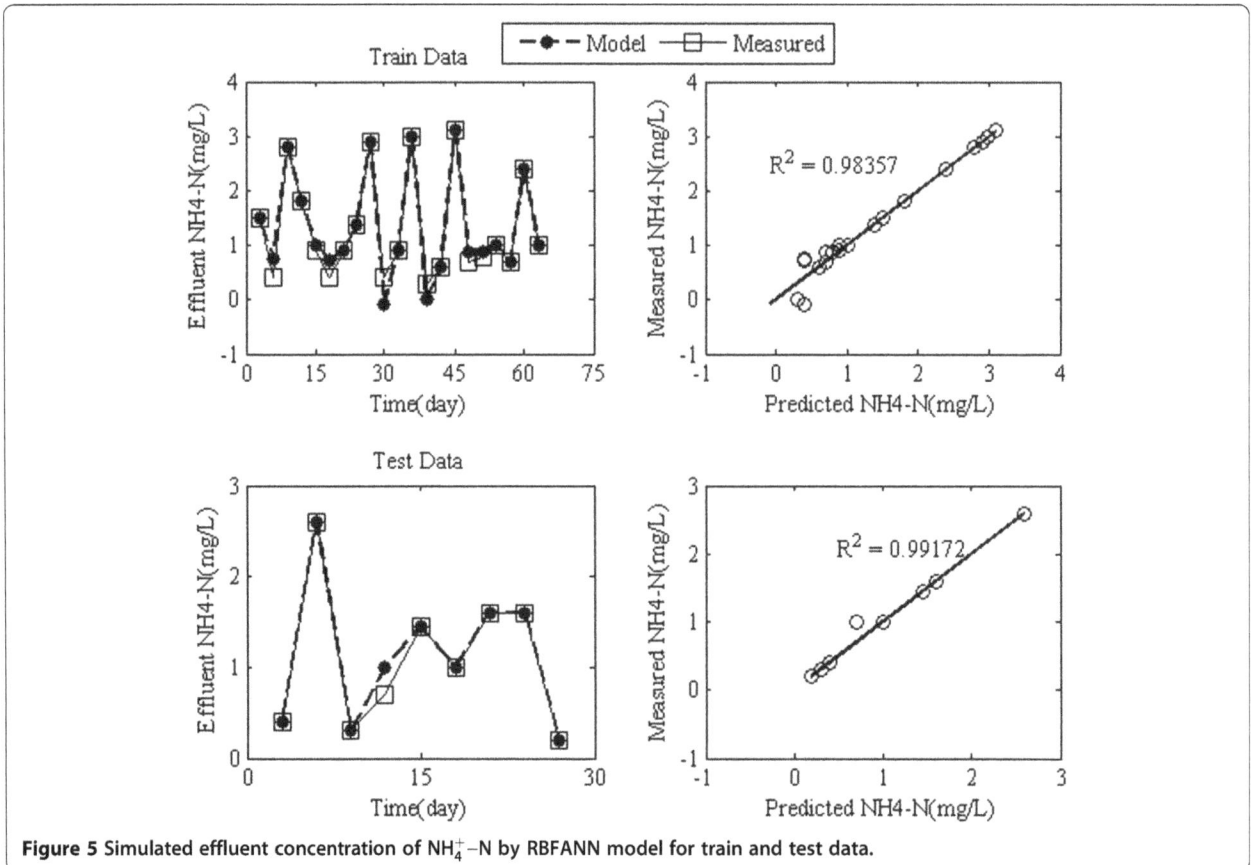

Figure 5 Simulated effluent concentration of NH_4^+-N by RBFANN model for train and test data.

Table 7 Effect of different single and joint variables on the effluent NH_4^+-N models

Input variable no.	R^2		RMSE (mg/L)		Importance order
	Train	Test	Train	Test	
1	0.59	0.44	0.78	0.62	5
2	0.983	0.998	0.16	0.12	1
3	0.923	0.985	0.35	0.17	2
4	0.54	0.6	0.81	0.52	4
5	0.78	0.94	0.56	0.29	3
2-1	0.99	0.993	0.13	0.13	3
2-3	0.99	0.994	0.13	0.09	1
2-4	0.99	0.964	0.12	0.23	4
2-5	0.99	0.992	0.14	0.11	2
2-3-1	0.987	0.995	0.15	0.09	3
2-3-4	0.99	0.996	0.13	0.05	2
2-3-5	0.988	1	0.14	0	1
2-3-5-1	0.99	0.998	0.14	0.06	1
2-3-5-4	0.97	0.996	0.2	0.1	2
2-3-5-1-4	0.98	0.991	0.06	0.14	1

The numbers 1 to 5 refers to input variables identified in Table 4.

Results of TP removal efficiency and modeling

The influent TP concentration for the municipal wastewater varied from 13.2 to 15.1 mg/L. The results showed that the removal efficiency of TP is improved with increase of the HRT from 1 to 5 h. Then, the removal efficiency decreased for the HRT of 5 to 7 h. The influent TP concentration decreased from 14.8 to 2 mg/L for the effluent at HRT of 5 h with removal efficiency of 86.5%. The influent TP concentration decreased from 14.9 to 5.8 mg/L for the effluent at HRT of 1 h and met discharge limit (effluent TP <6 mg/L). The effluent TP concentration depends on the TBOD/TP ratio so with a ratio less than 20, it is not possible to achieve the effluent TP concentration lower than 2 mg/L [65]. We observed that the TBOD/TP ratio for the influent of Ekbatan wastewater treatment plant was less than 20. Subsequently, the combination of municipal and industrial wastewater was an effective method to correct this problem.

The influent TP concentration for the combined wastewater varied from 15 to 16.4 mg/L. The influent TP concentration decreased from 16.2 mg/L to 5.7 mg/L for the effluent TP at HRT of 1 h. With increasing the HRT from 7 to 9 h the effluent TP concentration reached to lower than 1 mg/L with removal efficiency of 92%. The influent TP varied from 17.24 to 20.3 mg/L for the industrial wastewater. The high ratios of the TBOD/TP for the industrial wastewater allowed the effluent TP to be lower than 1 mg/L. The effluent TP concentration reached to 0.7 mg/L with the removal efficiency of 96.55% at HRT of 17 h.

In order to model the effluent TP by RBFANN, the influent TP concentration, HRT, MLVSS, TDS and pH were the input variables of the neural network. The results of the effluent TP modeling using the RBFANN for train and test data sets were denormalized to compare the observed values of effluent TP with simulated values. The results of the training procedure by RBFANN were successful for the effluent TP. Figure 6 shows that the train and test models by RBFANN indicated an almost perfect match between the experimental and the simulated effluent TP concentration compared with the results of models introduced by Cinar et al. [38]. The RMSE values for train and test (verification process) models were 0.32 and 0.17 mg/L respectively, and the value of R^2 was 0.99 for both models. The RBFANN models simulated effluent TP so accurate that the mean average error for train and test models did not exceed 5% and 3%, respectively.

Table 8 shows HRT among single input variables, and HRT and pH among groups of two variables had the most considerable effect on the effluent TP. Furthermore, HRT, pH and MLVSS among groups of three variables, and HRT, pH, MLVSS and influent TP concentration between groups of four variables were determined to have the greatest effect on the effluent TP. Table 8 also shows the importance order of each input variable and the joint variables for effluent TP according to sensitivity analysis procedure. The variation of effluent TP concentration was influenced by HRT, pH, MLVSS, influent TP concentration and TDS, respectively, which is in a good agreement with earlier experimental studies [63,64].

Conclusions

The current research was an effort to evaluate performance of an SMBR treating combined municipal and industrial wastewater compared with treating municipal and industrial wastewaters. The combined municipal and industrial wastewater showed more satisfactory treating features for wastewater treatment by an SMBR compared with the municipal and industrial wastewaters. Although the concentration of components in combined wastewater were almost half of the industrial wastewater, required HRT for combined wastewater was 7 h in comparison to 17 h for industrial wastewater. It was observed that treatment performance of the SMBR improves and HRT decreases noticeably by decreasing BOD and COD concentration to lower than 700 mg/L and 1000 mg/L, respectively. The results indicated that the combined wastewater improves treatment performance by increasing TBOD/TP ratio and setting BOD/COD ratio to 0.5. Therefore, effluent TP concentration was lower than 2 mg/L by increasing TBOD/TP ratio to more than 20. This study showed that it is possible to

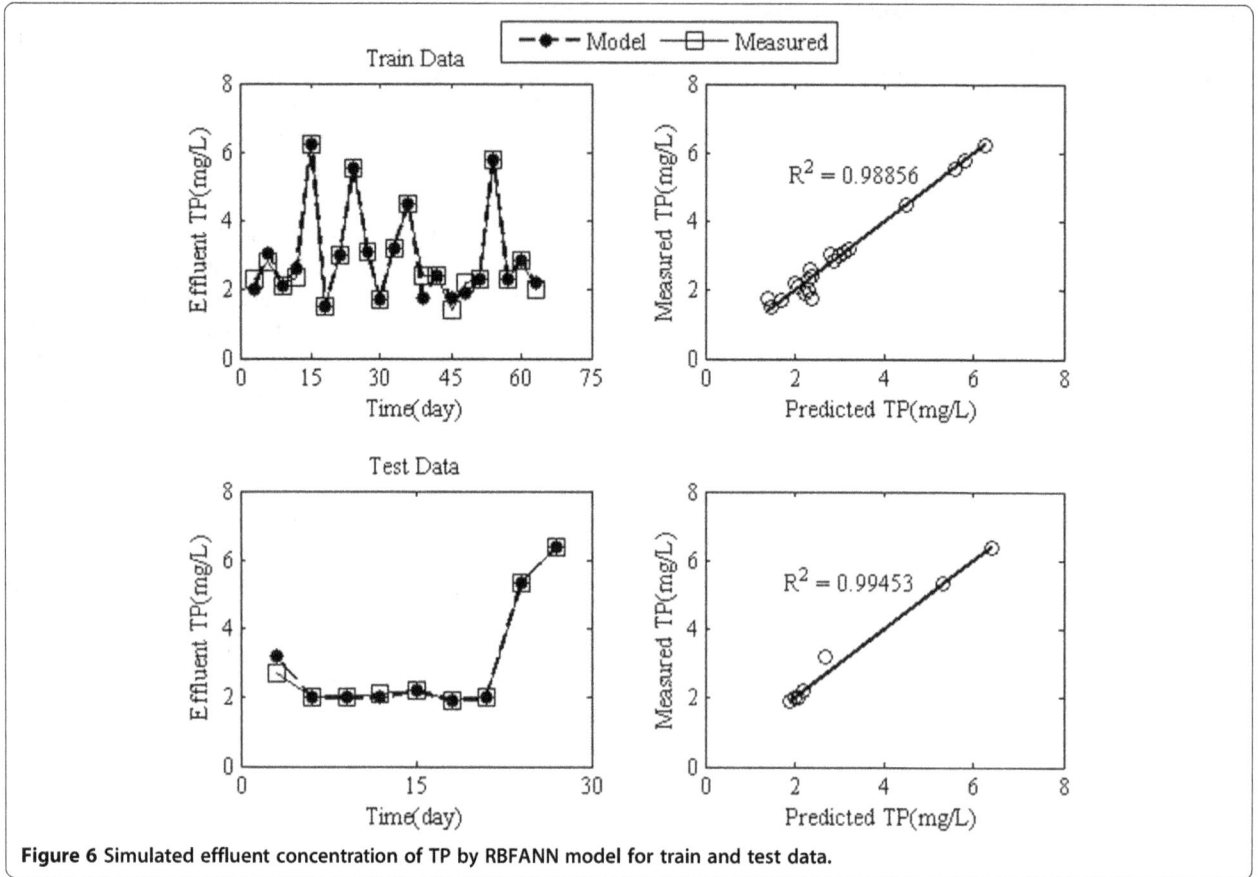

Figure 6 Simulated effluent concentration of TP by RBFANN model for train and test data.

Table 8 Effect of different single and joint variables on the effluent TP models

Input variable no.	R^2		RMSE (mg/L)		Importance order
	Train	Test	Train	Test	
1	0.34	0.3	1.15	1.88	5
2	0.972	0.992	0.35	0.14	1
3	0.878	0.978	0.7	0.29	2
4	0.51	0.8	1.39	0.82	4
5	0.77	0.96	0.99	0.4	3
2-1	0.975	0.97	0.37	0.37	4
2-3	0.99	0.99	0.22	0.16	2
2-4	0.92	0.99	0.53	0.09	3
2-5	0.995	0.997	0.11	0.11	1
2-5-1	0.99	0.998	0.2	0.13	2
2-5-3	0.991	1	0.14	0	1
2-5-4	0.96	0.999	0.35	0.09	3
2-5-3-1	0.99	0.998	0.2	0.1	1
2-5-3-4	0.99	0.993	0.23	0.11	2
2-5-3-1-4	0.988	0.994	0.32	0.17	1

The numbers 1 to 5 refers to input variables identified in Table 4.

achieve a cost efficient wastewater treatment by SMBR for the combined wastewater because required HRT was at the same range for BOD, COD and NH_4^+-N compared with the municipal and industrial wastewaters.

Treatment process models are efficient tools to assure proper operation and better control of wastewater treatment systems. ANNs have been successfully used for monitoring, controlling, classification and simulation of experimental data. In this study, an RBFANN was utilized in order to simulate effluent quality parameters of the SMBR treating combined municipal and industrial wastewater. An RBFANN is the most commonly used neural network for pattern recognition problems, it is also widely used for fault diagnosis and solving ill-conditioned problems. The RBFANN has the advantages of a fast learning process, a learning stage without any iteration of updating weights, robust ability, and adaptation capability compared with other ANNs such as MLP, RNN and ESN. The results showed that the training and testing procedure by RBFANN for effluent BOD, COD, NH_4^+-N and TP were successful. The train and test models showed an almost perfect match between the experimental and predicted values of effluent BOD, COD, NH_4^+-N and TP.

Abbreviations

MBR: Membrane bioreactor; SMBR: Submerged membrane bioreactor; HRT: Hydraulic retention time; COD: Chemical oxygen demand; TN: Total nitrogen; BOD: Biochemical oxygen demand; TBOD/TP: Total BOD to total phosphorous; ASP: Activated sludge process; ASM: Activated sludge model; ANN: Artificial neural network; RBF: Radial basis function; MLP: Multi-layer perceptron; RNN: Recurrent neural network; ESN: Echo-state network; RBFANN: Radial basis function artificial neural network; MLVSS: Mixed liquor volatile suspended solids; TDS: Total dissolved solids; TSS: Total suspended solids; DO: Dissolved oxygen; MLSS: Mixed liquor suspended solids; MLPANN: Multi-layer perceptron artificial neural network; RMSE: Root mean squared error; R^2: Coefficient of determination; K_s: Half saturation coefficient; k: Maximum substrate utilization rate.K_d, Endogenous decay coefficient; CBOD: Carbonaceous biochemical oxygen demand..

Competing interests

The authors declare that they have no competing interests.

Authors' contributions

SAM participated in design of the pilot plant and contributed in the modeling process. MB participated in the design of pilot, carried out the experiments and modeling and drafted the manuscript. SB participated in modeling process and helped to draft manuscript. ME participated in the experiments and helped to draft manuscript. ZB carried out the biological interpretation, participated in modeling process and helped to draft manuscript. All authors read and approved the final manuscript.

Acknowledgments

The authors are grateful to Ekbatan wastewater treatment plant for their technical and logistical assistance during this work which was supported by authors. And also, we wish to thank Ali Reza Jafari and Ali Morad Kamarkhani for technical help with modeling by artificial neural network.

Author details

[1]Department of Civil Engineering, K.N. Toosi University of Technology, Vanak square, Tehran, Iran. [2]Department and Faculty of Basic Sciences, PUK University, Kermanshah, Iran.

References

1. Yang W, Cicek N, Ilg J. State-of-the-art of membrane bioreactors: Worldwide research and commercial applications in North America. J Membr Sci. 2006;270:201–11.
2. Katsou E, Malamis S, Loizidou M. Performance of a membrane bioreactor used for the treatment of wastewater contaminated with heavy metals. Bioresour Technol. 2011;102:4325–32.
3. Rosenberger S, Krüger U, Witzig R, Manz W, Szewzyk U, Kraume M. Performance of a bioreactor with submerged membranes for aerobic treatment of municipal waste water. Water Res. 2002;36:413–20.
4. Ferrai M, Guglielmi G, Andreottola G. Modelling respirometric tests for the assessment of kinetic and stoichiometric parameters on MBBR biofilm for municipal wastewater treatment. Environ Model Softw. 2010;25:626–32.
5. Muller E, Stouthamer A, Van Verseveld HW, Eikelboom D. Aerobic domestic waste water treatment in a pilot plant with complete sludge retention by cross-flow filtration. Water Res. 1995;29:1179–89.
6. Zaloum R, Lessard S, Mourato D, Carriere J. Membrane bioreactor treatment of oily wastes from a metal transformation mill. Water Sci Technol. 1994;30:21–7.
7. Scholz W, Fuchs W. Treatment of oil contaminated wastewater in a membrane bioreactor. Water Res. 2000;34:3621–9.
8. W-t Z, Huang X, Lee D-j. Enhanced treatment of coke plant wastewater using an anaerobic–anoxic–oxic membrane bioreactor system. Sep Purif Technol. 2009;66:279–86.
9. Mutamim NSA, Noor ZZ, Hassan MAA, Olsson G. Application of membrane bioreactor technology in treating high strength industrial wastewater: a performance review. Desalination. 2012;305:1–11.
10. Sutton PM. Membrane bioreactors for industrial wastewater treatment: Applicability and selection of optimal system configuration. Proceedings Water Environ Federation. 2006;2006:3233–48.
11. Chang C-Y, Chang J-S, Vigneswaran S, Kandasamy J. Pharmaceutical wastewater treatment by membrane bioreactor process–a case study in southern Taiwan. Desalination. 2008;234:393–401.
12. Marrot B, Barrios-Martinez A, Moulin P, Roche N. Industrial wastewater treatment in a membrane bioreactor: a review. Environ Prog. 2004;23:59–68.
13. Barakat MA. New trends in removing heavy metals from industrial wastewater. Arab J Chem. 2011;4:361–77.
14. Artiga P, Ficara E, Malpei F, Garrido J, Mendez R. Treatment of two industrial wastewaters in a submerged membrane bioreactor. Desalination. 2005;179:161–9.
15. Badani Z, Ait-Amar H, Si-Salah A, Brik M, Fuchs W. Treatment of textile waste water by membrane bioreactor and reuse. Desalination. 2005;185:411–7.
16. Brik M, Schoeberl P, Chamam B, Braun R, Fuchs W. Advanced treatment of textile wastewater towards reuse using a membrane bioreactor. Process Biochem. 2006;41:1751–7.
17. Lesage N, Sperandio M, Cabassud C. Study of a hybrid process: Adsorption on activated carbon/membrane bioreactor for the treatment of an industrial wastewater. Chemi Eng Process: Process Intensif. 2008;47:303–7.
18. Katayon S, Megat Mohd Noor M, Ahmad J, Abdul Ghani L, Nagaoka H, Aya H. Effects of mixed liquor suspended solid concentrations on membrane bioreactor efficiency for treatment of food industry wastewater. Desalination. 2004;167:153–8.
19. Viero AF, De Melo TM, Torres APR, Ferreira NR. The effects of long-term feeding of high organic loading in a submerged membrane bioreactor treating oil refinery wastewater. J Membr Sci. 2008;319:223–30.
20. Tchobanoglous G, Burton FL, Stensel HD, Metcalf & Eddy. Wastewater Engineering; Treatment and Reuse. New York, NY: McGraw-Hill Education; 2003.
21. Feng F, Xu Z, Li X, You W, Zhen Y. Advanced treatment of dyeing wastewater towards reuse by the combined Fenton oxidation and membrane bioreactor process. J Environ Sci. 2010;22:1657–65.
22. Yigit N, Uzal N, Koseoglu H, Harman I, Yukseler H, Yetis U, et al. Treatment of a denim producing textile industry wastewater using pilot-scale membrane bioreactor. Desalination. 2009;240:143–50.
23. Yuniarto A. Ujang Z. Performance of bio-fouling reducer in submerged membrane bioreactor for palm oil mill effluent treatment. In International Conference & Exposition on Environmental Management and Technologies, PWTC, Kuala Lumpur: Noor ZZ; 2008.
24. Acharya C, Nakhla G, Bassi A. Operational optimization and mass balances in a two-stage MBR treating high strength pet food wastewater. J Environ Eng. 2006;132:810–7.
25. Moral H, Aksoy A, Gokcay CF. Modeling of the activated sludge process by using artificial neural networks with automated architecture screening. Comput Chem Eng. 2008;32:2471–8.
26. Gernaey KV, Van Loosdrecht M, Henze M, Lind M, Jørgensen SB. Activated sludge wastewater treatment plant modelling and simulation: state of the art. Environ Model Softw. 2004;19:763–83.
27. Henze M, Grady C, Gujer W, Marais G, Matsuo T. A general model for single-sludge wastewater treatment systems. Water Res. 1987;21:505–15.
28. Henze M, Gujer W, Mino T, Matsuo T, Wentzel M, Marais G. Wastewater and biomass characterization for the activated sludge model no. 2: biological phosphorus removal. Water Sci Technol. 1995;31:13–23.
29. Gujer W, Henze M, Mino T, Loosdrecht M. Activated sludge model no. 3. Water Sci Technol. 1999;39:183–93.
30. Chen L, Tian Y, Cao C, Zhang S, Zhang S. Sensitivity and uncertainty analyses of an extended ASM3-SMP model describing membrane bioreactor operation. J Membr Sci. 2012;389:99–109.
31. Geissler S, Wintgens T, Melin T, Vossenkaul K, Kullmann C. Modelling approaches for filtration processes with novel submerged capillary modules in membrane bioreactors for wastewater treatment. Desalination. 2005;178:125–34.
32. Mannina G, Cosenza A, Viviani G. Uncertainty assessment of a model for biological nitrogen and phosphorus removal: Application to a large wastewater treatment plant. Phys Chem Earth, Parts A/B/C. 2012;42:61–9.
33. Sha W, Edwards K. The use of artificial neural networks in materials science based research. Mater Des. 2007;28:1747–52.
34. Azmy AM, Erlich I, Sowa P. Artificial neural network-based dynamic equivalents for distribution systems containing active sources. IEE Proceedings-Generation, Trans Distrib. 2004;151:681–8.
35. Park J-W, Venayagamoorthy GK, Harley RG. MLP/RBF neural-networks-based online global model identification of synchronous generator. Ind Electron, IEEE Trans. 2005;52:1685–95.

36. Singh S, Venayagamoorthy G. Online identification of turbogenerators in a multimachine power system using RBF neural networks. St. Louis, Missouri, USA: Artificial Neural Networks in Engineering Conference (ANNIE) 2000; 2002. p. 485–90.

37. Venayagamoorthy GK. Online design of an echo state network based wide area monitor for a multimachine power system. Neural Netw. 2007;20:404–13.

38. Çinar Ö, Hasar H, Kinaci C. Modeling of submerged membrane bioreactor treating cheese whey wastewater by artificial neural network. J Biotechnol. 2006;123:204–9.

39. Suchacz B, Wesołowski M. The recognition of similarities in trace elements content in medicinal plants using MLP and RBF neural networks. Talanta. 2006;69:37–42.

40. Ferrari S, Bellocchio F, Piuri V, Borghese NA. A hierarchical RBF online learning algorithm for real-time 3-D scanner. Neural Netw, IEEE Trans. 2010;21:275–85.

41. Lee C-M, Ko C-N. Time series prediction using RBF neural networks with a nonlinear time-varying evolution PSO algorithm. Neurocomput. 2009;73:449–60.

42. Wang S, Yu D. Adaptive RBF network for parameter estimation and stable air–fuel ratio control. Neural Netw. 2008;21:102–12.

43. Kumfer B, Felch C, Maugans C. Wet air oxidation treatment of spent caustic in petroleum refineries. In: N.P.R.s. Association, editor. National Petroleum Refiner's Association Conference, Phoenix, AZ. 2010. p. 21–3.

44. Samudro G, Mangkoedihardjo S. Review on BOD, COD and BOD/COD ratio: a triangle zone for toxic, biodegradable and stable levels. Int J Acad Res. 2010;2:235–9.

45. Andrew D. Standard methods for the examination of water and wastewater. APHA-AWWA-WEF: Washington, D.C; 2005.

46. Elmolla ES, Chaudhuri M. Combined photo-Fenton–SBR process for antibiotic wastewater treatment. J Hazard Mater. 2011;192:1418–26.

47. L-y F, Wen X-h, Yi Qian Q-IL. Treatment of dyeing wastewater in two SBR systems. Process Biochem. 2001;36:1111–8.

48. Kashaninejad M, Dehghani A, Kashiri M. Modeling of wheat soaking using two artificial neural networks (MLP and RBF). J Food Eng. 2009;91:602–7.

49. Abyaneh HZ. Evaluation of multivariate linear regression and artificial neural networks in prediction of water quality parameters. J Environ Health Sci Eng. 2014;12:1–8.

50. Tomenko V, Ahmed S, Popov V. Modelling constructed wetland treatment system performance. Ecol Model. 2007;205:355–64.

51. Madaeni S, Zahedi G, Aminnejad M. Artificial neural network modeling of O2 separation from air in a hollow fiber membrane module. Asia-Pacific J Chem Eng. 2008;3:357–63.

52. Shahsavand A, Chenar MP. Neural networks modeling of hollow fiber membrane processes. J Membr Sci. 2007;297:59–73.

53. Eslamimanesh A, Gharagheizi F, Mohammadi AH, Richon D. Artificial neural network modeling of solubility of supercritical carbon dioxide in 24 commonly used ionic liquids. Chem Eng Sci. 2011;66:3039–44.

54. Liu S, Zhang Y, Ma P, Lu B, Su H. A Novel Spatial Interpolation Method Based on the Integrated RBF Neural Network. Procedia Environ Sci. 2011;10:568–75.

55. Sahoo GB, Ray C. Predicting flux decline in crossflow membranes using artificial neural networks and genetic algorithms. J Membr Sci. 2006;283:147–57.

56. Xi X, Cui Y, Wang Z, Qian J, Wang J, Yang L, et al. Study of dead-end microfiltration features in sequencing batch reactor (SBR) by optimized neural networks. Desalination. 2011;272:27–35.

57. Lowe D, Broomhead D. Multivariable functional interpolation and adaptive networks. Complex syst. 1988;2:321–55.

58. Demuth H, Beale M. Neural network toolbox user's guide. 2000.

59. Pendashteh AR, Fakhru'l-Razi A, Chaibakhsh N, Abdullah LC, Madaeni SS, Abidin ZZ. Modeling of membrane bioreactor treating hypersaline oily wastewater by artificial neural network. J Hazard Mater. 2011;192:568–75.

60. De la Cruz N, Gimenez J, Esplugas S, Grandjean D, De Alencastro L, Pulgarin C. Degradation of 32 emergent contaminants by UV and neutral photo-fenton in domestic wastewater effluent previously treated by activated sludge. Water Res. 2012;46:1947–57.

61. Qin L, Zhang G, Meng Q, Xu L, Lv B. Enhanced MBR by internal micro-electrolysis for degradation of anthraquinone dye wastewater. Chem Eng J. 2012;210:575–84.

62. Saltelli A, Ratto M, Andres T, Campolongo F, Cariboni J, Gatelli D, et al. Global sensitivity analysis: the primer. England: John Wiley & Sons Ltd, The Atrium, Southern Gate, Chichester, West Sussex PO19 8SQ; 2008.

63. Campos JC, Borges RMH, Oliveira Filho AM, Nobrega R, Sant'Anna G. Oilfield wastewater treatment by combined microfiltration and biological processes. Water Res. 2002;36:95–104.

64. Tay J-H, Zeng JL, Sun DD. Effects of hydraulic retention time on system performance of a submerged membrane bioreactor. Sep Sci Technol. 2003;38:851–68.

65. Kim SH, Baumann ER. Investigation of Chemical Phosphate Removal from an Oxidation Ditch by Field Evaluation. Environ Eng Res. 1997;2:207–16.

3

The estimation of per capita loadings of domestic wastewater in Tehran

Alireza Mesdaghinia[1], Simin Nasseri[1], Amir Hossein Mahvi[2,3], Hamid Reza Tashauoei[4] and Mahdi Hadi[1*]

Abstract

The amount of wastewater characteristics loading is one of the main parameters in the design of wastewater collection and treatment systems. The generation per capita per day (GPCD) of wastewater characteristics was estimated by analyzing the monthly data of nine wastewater treatment plants in Tehran, capital city of Iran. GPCD values were calculated from measured collected wastewater flow, the population and concentration data. The results indicated the values of 32.96 ± 1.91, 49.25 ± 2.49, 37.31 ± 2.44, 6.77 ± 0.53, 1.96 ± 0.11, 92.23 ± 5.68, 2.07 ± 0.39 and 128.96 ± 6.69 g/d.cap of GPCD for BOD_5, COD, TSS, TKN, P, TDS, ON and TS, respectively, for Tehran's wastewater. The per capita estimated for the wastewater production and treatment were determined to be 186.06 ± 7.85 and 136.72 ± 5.43 L/d.cap, respectively. It is estimated that about 504 m^3/d and 346 m^3/d of sludge, will be produced and waste as excrement raw sludge, respectively, in Tehran. Simple regression models were presented the relationships such as the change of collected and treated wastewater with population and changes of GPCD parameters with each other. It was revealed that the Tehran's wastewater may be classified as highly degradable, but during recent decades its Biodegradability Index (BI) has been reduced up to 15%. The new suggested revised per capita parameters can be used for design purposes in Tehran, and possibly, in areas with similar characteristics, substituting the classical values obtained from foreign textbooks. These values could help in designing more accurate treatment systems and may lower the required capacity for the treatment of wastewater up to 40% in Tehran.

Keywords: Domestic wastewater, Per capita loading, Wastewater treatment plant

Introduction

The amount of wastewater characteristics loading is one of the main parameters in the design of wastewater collection and treatment systems. The per capita loading of wastewater characteristics such as chemical oxygen demand (COD), Biological oxygen demand (BOD_5), nitrogen, phosphorus and solids have been considered as useful main functions in the design of wastewater collection systems and in the control of water resources pollution [1].

The pollutants per capita values can be used to estimate the present and the future pollution loading of wastewater produced from a population. These also are useful to estimate the equivalent population of an urban or industrial wastewater flow [2]. By expressing the wastewater pollution in terms of per capita values, the concept of pollution would be more understandable for the citizens and policy makers. However, the changes in mode and living standards [3] and the development of wastewater collection and treatment technologies, would suggest that a review and a re-examination of the pollution per capita loadings be made [1]. On the other hand, the discharge of these pollutants can cause considerable problems in environment [4]. Increased pollution of water-receiving bodies, and the imposition of restrictive limits by local administrations, led a need for new treatment technologies [5-9]. Thus it is very important that the properties of the discharged wastewater be assessed in order to be able to improve current technologies and provide adequate wastewater treatment [10].

Tehran, the capital city of Iran, has a population of over 7.5 million. There are nine public and 18 private wastewater treatment plant (WWTP) in Tehran those can treat more than 100 MCM of wastewater per year [11]. By construction of a new treatment plant system in the south of Tehran which can treat about 750 million cubic meters per year after project completion, the other

* Correspondence: hadi_rfm@yahoo.com
[1]Center for Water Quality Research (CWQR), Institute for Environmental Research (IER), Tehran University of Medical Sciences, Tehran, Iran
Full list of author information is available at the end of the article

treatment systems will be switched off gradually. Based on the current existing information of in-use domestic wastewater treatment plants in Tehran it may be possible to make an estimate of wastewater per capita parameters that can be used for design purposes and development programs of wastewater treatment systems in Tehran.

One of aims of the study reported in this paper is to re-establish the main wastewater pollution measures (including BOD_5, COD, TSS, TKN, P, TDS, ON(Organic Nitrogen) and TS) generation per capita per day (GPCD) according to the recorded data of nine public wastewater treatment systems in Tehran. The per capita loading of wastewater is an important parameter in the design process of a treatment plant's units. Iran's water and sewage utility (ABFA company) states that per capita water consumption in Tehran is currently about 378 liters per day [11]. Our study also sought to determine the amount of wastewater produced per capita per day or the conversion factor of water to wastewater. Pollutant discharge per capita (PDC; g/d.cap) with the wastewater treatment system in Tehran is another parameter that is important in the case of wastewater discharge in to receiving bodies. This parameter and the per capita of the producing sludge for domestic wastewater in Tehran would also be estimated in this study from wastewater treatment plants data. The later parameter may be useful in sludge treatment and management programs. The findings of this study can be used as basic data for the design of wastewater treatment systems in Tehran and possibly in areas with similar characteristics.

Materials and methods

The data were used in this study obtained from monthly reports (from 2007 to 2013) of nine wastewater treatment plants in Tehran including Sahebgharaniyeh, Mahallti, Zargandeh, Qeitariyeh, Qods, Shush, Ekbatan, Dowlatabadi and Jonoob. The data includes the quantitative information of wastewater production and characteristics data over 80 months from April 2007 to November 2013 for all domestic WWTPs in Tehran. The analysis of wastewater samples were done by the laboratory stuffs of WWTPs according to Standard Methods for the Examination of Water and Wastewater [12]. The parameters monitored were temperature, pH, Biological Oxygen Demand (BOD_5), Total Dissolved Solid (TDS), Total Solids (TS), Total Suspended Solid (TSS), Chemical Oxygen Demand (COD), Organic Nitrogen (ON), Total Phosphorous (TP) and Total Kjeldahl Nitrogen (TKN). Data analysis involved the data pre-processing and conducting some descriptive and analytical studies using Microsoft Excel 2007 and R software packages.

The study included data pre-processing and preparing them to make the estimates of desired parameters. Initial

data processing, although is a time consuming step, but is very important part in the success of statistical analysis [13]. At this step, the raw data of nine in-use WWTPs collected during the recent years was assessed. The raw data was included 720 instances, although there were some outlier values. The data that appear to be very distant from the normal data distribution may be classified as being outliers. In certain instances however, this outlying value may be correct and is a natural product of the variables distribution [14]. All examples with missing values were represented with NA (not available). In this study, we took a normal distribution with a cutoff of three times of standard deviations around the mean to detect the outliers. Thus, the data that was more than $\mu \pm 3SD$ was considered as outliers.

The descriptive statistics of raw (720 instances) and pre-processed (499 instances) data are summarized in Tables 1 and 2, respectively.

The estimation of pollutant generation per capita per day (GCPD) values was conducted according to the following equations:

$$GCPD_{BOD5} = ((Collect/30) \times BODin)/CurentPop$$
$$GCPD_{COD} = ((Collect/30) \times CODin)/CurentPop$$
$$GCPD_{TSS} = ((Collect/30) \times TSSin)/CurentPop$$
$$GCPD_{TKN} = ((Collect/30) \times TKNin)/CurentPop$$
$$GCPD_P = ((Collect/30) \times Pin)/CurentPop$$
$$GCPD_{TDS} = ((Collect/30) \times TDSin)/CurentPop$$
$$GCPD_{OrgN} = ((Collect/30) \times OrNin)/CurentPop$$
$$GCPD_{TS} = ((Collect/30) \times TSin)/CurentPop$$
$$GCPD_{Slu_P} = ((ProSlu/30)/CurentPop) \times 1000$$
$$GCPD_{Slu_E} = ((ExcSlu/30)/CurentPop) \times 1000$$
$$Coll_{PerP} = ((Collect/CurentPop)/30) \times 1000$$
$$Tret_{PerP} = ((Treat/CurentPop)/30) \times 1000$$

The pollutant discharge per capita (PDC; g/d.cap) with the wastewater treatment system was defined and determined in terms of pollutant generation per capita (GCPD; g/d.cap) and pollutant removal efficiency (PRE; %) of the wastewater treatment systems as follow [15]:

$$PDC = GCPD \times [(100-PRE)/100]$$

Results and discussion

The total yearly averaged population covered by Tehran's nine municipal wastewater treatment plants from 2007 to 2013 was determined to be 4,502,065 persons per year. In average, out of 22,778,632 m^3/year estimated collected wastewater from 2007 to 2013, only 19,749,770 m^3/year of it were treated. In other words, about three MCM per year (13% of estimated collected wastewater from 2007 to 2013) were discharged into the environment without adequate treatment. The mathematical relationships between

Table 1 The statistics of raw data of wastewater characteristic for Tehran WWTPs

Parameter	Mean	SD	SE	Max.	Min.	UB	LB
Nomin	300670	281473	24756	900000	10500	325427	275914
Collect	458191	807439	85130	7445021	39000	543321	373060
Treat	397265	796648	83993	7445021	11395	481258	313272
ProSlu	88.16	77.63	6.83	225.00	1.00	94.99	81.34
ExcSlu	64.15	57.37	5.05	188.00	1.00	69.20	59.11
CurentPop	90558	172233	18159	1459808	7000	108717	72399
Tin	21.83	3.81	0.42	29.50	12.90	22.24	21.41
pHin	7.96	0.34	0.04	9.00	6.91	7.99	7.92
BODin	171.85	54.42	5.93	352.00	15.00	177.78	165.92
CODin	259.60	79.69	8.74	507.00	28.80	268.34	250.87
TSin	697.97	171.35	19.03	1280.00	246.00	716.99	678.94
TSSin	198.65	69.09	7.61	400.00	30.00	206.26	191.04
TDSin	494.88	145.66	17.72	985.00	185.00	512.60	477.16
Pin	11.24	4.68	0.53	29.70	3.92	11.77	10.72
OrNin	11.30	6.05	1.99	28.80	0.97	13.29	9.31
TKNin	35.16	12.33	1.55	79.38	12.00	36.71	33.61
Tout	20.94	4.13	0.45	28.20	10.00	21.39	20.49
pHout	7.39	0.35	0.04	9.00	6.50	7.43	7.35
BODout	11.90	8.96	0.98	72.28	2.40	12.88	10.93
CODout	23.95	12.83	1.41	98.50	6.40	25.36	22.54
TSout	482.38	135.12	15.03	934.50	3.00	497.41	467.35
TSSout	14.25	14.04	1.55	114.00	1.00	15.80	12.70
TDSout	465.12	128.18	15.59	902.00	214.00	480.71	449.52
Pout	4.24	1.60	0.18	12.97	0.05	4.42	4.06
OrNout	0.93	0.75	0.09	5.00	0.00	1.02	0.83
TKNout	4.28	4.33	0.57	23.85	0.04	4.85	3.71

SD, Standard deviation; SE, 1.96 × Standard error; Max., Maximum of observation; Min., Minimum of observation; UB, Upper bound of 95% confidence interval; LB, Lower bound of 95% confidence interval.

wastewater flows and population in Tehran were shown in Figure 1.

According to the latest report of the basic operational items of Tehran's wastewater company [16], a capacity of 675,000 m^3/d had been allocated for the treatment of wastewater produced by 3,150,000 persons up to June 2014. This means that the per capita loading of wastewater considered by designers for the design of a wastewater treatment plant in Tehran, is about 215 L/d. By a roughly estimation according to Figure 1(b), it is expected that for a population of 3,150,000 persons, the capacity of 455,822 m^3/d is a sufficient capacity that needs to be allocated. This capacity is about 40% lower than current allocated value. Therefore, the overestimated per capita loading value of 215 L/d in Tehran, leads the system to be designed for a capacity more than that is required.

The pattern changes of Tehran wastewater quality contents from 1984 to 2013 is another issue were assessed in Table 3. The most of the wastewater quality parameters have not shown considerable changes during this period. However, averaged value of TSS parameter in the influent wastewater of Tehran's WWTPs was considerably decreased from 353.33 in 1984 to 198.65 mg/L in 2013. It is clear that TSS content of Tehran domestic wastewater has been decreased by 56% of its value in 1984. This result indicates that the per capita generation of TSS parameter could also be reduced similarly.

The estimated per capita loadings for domestic wastewater of Tehran, as summarized from the results of this investigation, are presented in Table 4. The GPCD values of 32.96 ± 1.91, 49.25 ± 2.49, 37.31 ± 2.44, 6.77 ± 0.53, 1.96 ± 0.11, 92.23 ± 5.68, 2.07 ± 0.39 and 128.96 ± 6.69 g/d.cap were estimated for BOD_5, COD, TSS, TKN, P, TDS, ON and TS, respectively. In a study conducted by Azimi and Ameri [20] on the estimation of per capita loadings for domestic wastewater of Saheb-Gharanieh treatment plant in Tehran (based on the data collected

Table 2 The statistics of pre-processed data of wastewater characteristic for Tehran WWTPs

	Mean	SD	SE	Max.	Min.	UB	LB
Nomin	300670	281473	24757	900000	10500	325427	275914
Collect	458190	807439	85130	7445021	39000	543321	373060
Treat	397265	796648	83993	7445021	11395	481258	313273
ProSlu	88.16	77.63	6.83	225.00	1.00	94.99	81.34
ExcSlu	64.15	57.37	5.05	188.00	1.00	69.20	59.11
CurentPop	90558	172233	18159	1459808	7000	108717	72399
Tin	21.83	3.81	0.42	29.50	12.90	22.24	21.41
pHin	7.96	0.34	0.04	9.00	6.91	7.99	7.92
BODin	171.85	54.42	5.93	352.00	15.00	177.78	165.92
CODin	259.60	79.69	8.74	507.00	28.80	268.34	250.87
TSin	697.97	171.35	19.03	1280.00	246.00	716.99	678.94
TSSin	198.65	69.09	7.61	400.00	30.00	206.26	191.04
TDSin	494.88	145.66	17.72	985.00	185.00	512.60	477.16
Pin	11.24	4.68	0.53	29.70	3.92	11.77	10.72
OrNin	11.30	6.05	1.99	28.80	0.97	13.29	9.31
TKNin	35.16	12.33	1.55	79.38	12.00	36.71	33.61
Tout	20.94	4.13	0.45	28.20	10.00	21.39	20.49
pHout	7.39	0.35	0.04	9.00	6.50	7.43	7.35
BODout	11.90	8.96	0.98	72.28	2.40	12.88	10.93
CODout	23.95	12.83	1.41	98.50	6.40	25.36	22.54
TSout	482.38	135.12	15.03	934.50	3.00	497.41	467.35
TSSout	14.25	14.04	1.55	114.00	1.00	15.80	12.70
TDSout	465.12	128.18	15.59	902.00	214.00	480.71	449.52
Pout	4.24	1.60	0.18	12.97	0.05	4.42	4.06
OrNout	0.93	0.75	0.09	5.00	0.00	1.02	0.83
TKNout	4.28	4.33	0.57	23.85	0.04	4.85	3.71

SD, Standard deviation; SE, $1.96 \times$ Standard Error; Max., Maximum of observation; Min., Minimum of observation; UB, Upper bound of 95% confidence interval; LB, Lower bound of 95% confidence interval.

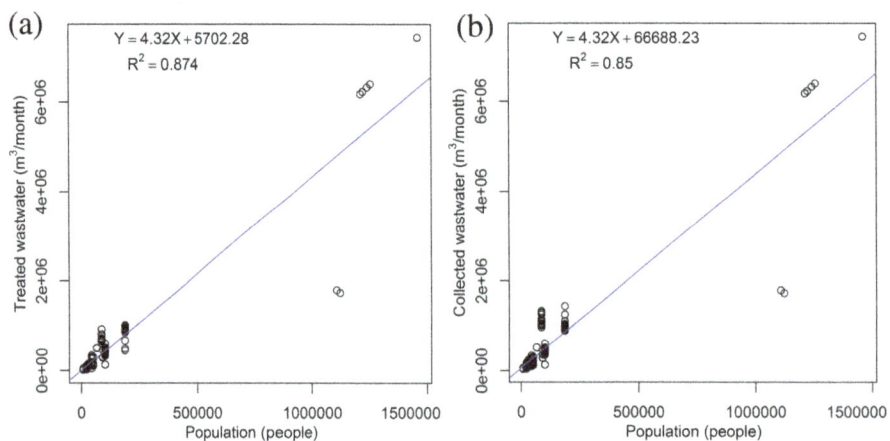

Figure 1 The flow of treated (a) and collected (b) wastewater versus population.

Table 3 Changes of Tehran wastewater quality contents from 1984 to 2013

Year	Parameters							Ref.
	pH	BOD₅	COD	TSS	TDS	P	TKN	
1984	n.a.	288.33	n.a.	353.33	495	25	n.a.	[17]
1993	7.43	184.47	261.87	256.03	n.a.	n.a.	45.56	[18]
1995	7.39	170	237.8	226.3	n.a.	9.9	39.37	[19]
1997	7.8	129.8	225	189.7	n.a.	2.7	38.3	[20]
2013	7.96	171.85	259.6	198.65	494.88	11.24	35.16	This study

n.a: not available.

in 1997), they were found GPCD values of 36, 62, 52, 10.5 and 0.74 g/d.cap for BOD_5, COD, TSS, TKN and P, respectively. Although their estimated values are not representative of all treatment plants, comparison of them with the GPCD estimated in our study show that the GPCD of BOD_5, TKN, and P were not considerably changed, but it was decreased by 71% and 79% of estimated values (in 1997) for TSS and COD parameters, respectively. This result is consistent with the pattern of TSS changes in Table 3.

Table 4 The estimated values of per capita parameters

Parameter	Mean	SD	SE	Max.	Min.	UB	LB
$GCPD_{Slu\_E}$	0.11	0.14	0.02	0.89	0.00	0.12	0.09
$GCPD_{Slu\_P}$	0.16	0.23	0.02	1.07	0.00	0.19	0.14
$GCPD_{TS}$	128.96	56.39	6.69	395.04	40.07	135.65	122.27
$GCPD_{OrgN}$	2.07	1.18	0.39	5.76	0.17	2.45	1.68
$GCPD_{TDS}$	92.23	43.03	5.68	303.75	27.98	97.90	86.55
$GCPD_P$	1.96	0.91	0.11	5.60	0.53	2.07	1.85
$GCPD_{TKN}$	6.77	3.88	0.53	31.14	2.19	7.30	6.24
$GCPD_{TSS}$	37.31	20.76	2.44	160.44	3.20	39.76	34.87
$GCPD_{COD}$	49.25	21.25	2.49	140.06	5.76	51.73	46.76
$GCPD_{BOD5}$	32.96	16.41	1.91	104.07	2.82	34.86	31.05
$(BOD_5/COD)_{out}$	0.41	1.06	0.12	7.25	0.95	2.51	2.27
$(BOD_5/COD)_{in}$	0.61	1.20	0.13	21.87	0.23	1.75	1.49
$Tret_{PerP}$	136.72	51.53	5.43	360.75	33.61	142.15	131.28
$Coll_{PerP}$	186.06	74.42	7.85	522.71	44.83	193.91	178.22
PDC_{TS}	89.49	38.92	4.63	283.05	0.39	94.12	84.86
PDC_{OrgN}	1.82	0.92	0.32	4.03	0.17	2.14	1.49
PDC_{TDS}	87.32	36.07	4.77	279.52	29.75	92.09	82.55
PDC_P	0.76	0.29	0.04	1.60	0.25	0.80	0.73
PDC_{TKN}	0.73	0.84	0.12	4.88	0.01	0.85	0.61
PDC_{TSS}	2.47	2.33	0.28	21.54	0.26	2.75	2.20
PDC_{COD}	4.46	2.93	0.34	25.51	0.84	4.81	4.12
PDC_{BOD}	2.15	1.60	0.19	15.31	0.36	2.34	1.96

SD, Standard deviation; SE, 1.96 × Standard Error; Max., Maximum of observation; Min., Minimum of observation; UB, Upper bound of 95% confidence interval; LB, Lower bound of 95% confidence interval.

For evaluating the biodegradability of Tehran's domestic wastewater the BOD_5/COD ratio, is called Biodegradability Index (BI), was determined. The BI Index varies from 0.4 to 0.8 for domestic wastewaters. If BOD_5/COD is > 0.6 then the waste is fairly biodegradable and can be effectively treated biologically. If BOD_5/COD ratio is between 0.3 and 0.6, then seeding is required to treat it biologically. If BOD_5/COD is < 0.3 then it cannot be treated biologically [21,22]. From data in the literature (Table 3) and the results of this study, the BI index was obtained to be 0.70, 0.71, 0.57 and 0.61 for the years of 1993, 1995, 1997 and 2013, respectively. Although these values reveal that the wastewater may be classified as highly biodegradable, BI values for later two years are almost 15% lower than those for previous years. This may be resulted from this fact that over the past 15-20 years a wide range of diverse industrial synthetic detergents which are mostly made from petroleum products and alcohols, has been produced and extensively used for cleaning and disinfection purposes and then discharged into sewage systems [23]. Many detergent compounds was found to be resistant to biodegradation by bacteria [24]. This may be the main reason of deceasing in the biodegradation potential of Tehran's wastewater according to the BI index.

Table 5 summarizes the mathematical simple relationships among the estimated GCPD parameters. The strongest relationship was found between $GCPD_{COD}$ and $GCPD_{BOD}$ ($R^2 = 0.74$). The next relationship with high correlation ($R^2 = 0.66$) was found between $GCPD_{BOD}$ and $GCPD_{TSS}$. These relationships may be useful for designers to help them to estimate the GPCD values from each other.

The per capita collected wastewater ($Coll_{PerP}$) was determined to be 186.06 ± 7.85 while the $Coll_{PerP}$ of 199.67 L/d.cap was estimated from the study's results of Sharifi Sistani [25]. This finding shows that this

Table 5 The main relationships between Tehran wastewater quality parameters

X	Y	Equation	R^2
$GCPD_{COD}$	$GCPD_{BOD}$	$Y = 0.66X + 0.16$	0.74
$GCPD_{BOD}$	$GCPD_{TSS}$	$Y = 1.03X + 3.61$	0.66
$GCPD_{COD}$	$GCPD_{TSS}$	$Y = 0.76X - 0.15$	0.60
$GCPD_{BOD}$	$GCPD_{TDS}$	$Y = 2.03X + 25.88$	0.58
$GCPD_{COD}$	$GCPD_{TDS}$	$Y = 1.55X + 16.55$	0.56
$GCPD_{BOD}$	$GCPD_{TKN}$	$Y = 0.17X + 1.17$	0.52
$GCPD_{COD}$	$GCPD_{TKN}$	$Y = 0.12X + 0.62$	0.45
$GCPD_{TSS}$	$GCPD_{TKN}$	$Y = 0.12X + 2.31$	0.42
$GCPD_{COD}$	$GCPD_P$	$Y = 0.03X + 0.72$	0.35
$GCPD_{BOD}$	$GCPD_P$	$Y = 0.03X + 0.92$	0.33
$GCPD_{TSS}$	$GCPD_P$	$Y = 0.02X + 1.16$	0.30

parameter has been decreased in 2013 by 93% of its value in 2000. This seven percent decrease during recent thirteen years may be results from some reasons such as increasing the level of public awareness on water saving tips, using of treated water for construction purposes and cars or yards washing and probably increased use of fast foods instead of cooking meals at home.

Table 6 compares the per capita values of wastewater quality parameters estimated for Iran's domestic wastewater treatment plants in 2001. These values are drawn from a report of Department of Energy [26]. As shown in this table, different cities in Iran have different values for per capita loadings.

As shown in Table 6 the GCPD values of BOD_5, TSS, TKN and TP in 2001, according only to one wastewater treatment plant data (Zargandeh), were more than those for 2013 obtained to be 28.9, 32.3, 6.1 and 1.6 g/d.cap, respectively. The result indicates and confirms that the GCPD of TSS considerably reduced during the recent decade.

According to the technical criteria standard N. 3-129 [27], a new sewage treatment plant in Iran should be designed at least based on a per capita average of 40 to 50 g/d.cap BOD_5 and 50 to 60 g/d.cap of total suspended solids (TSS). These values have been universally used in the design of wastewater treatment systems and unchallenged since the publication of this standard in Iran. The recommended values of 40-50 g/d.cap for BOD_5 and 50-60 g/d.cap for TSS are more than the values of 31.05-34.86 and 34.87-39.76 g/d.cap estimated in this study, respectively. In another word, the estimated values of GCPD for BOD_5 and TSS for Tehran's wastewater are 27% and 32% lower than the mean of recommended values by Department of Energy, respectively. Using the guideline values recommended by Department of Energy for the design of wastewater treatment plant for the city of Tehran, with different living habits in different parts of the city, may results in considerable overestimations in the design process of new treatment plants. Thus, it is recommended that our estimated GPCDs be used in the design of new wastewater treatment plants in Tehran instead of recommended values by Department of Energy.

In Table 7, the GCPD values for main wastewater parameters in different countries were compared with the estimated GCPD values for Tehran's wastewater. The estimated values for Tehran are close to the per capita values obtained for countries such as Turkey, India, and Egypt. In these countries, the GCPD values are lower than that for European countries and United States. Low consumption of toilet papers in this countries and especially in Tehran may be one of the reasons for the low values of GCPD for BOD_5 and TSS [20].

It is estimated that over 90% of the treated wastewater effluent from treatment plants across Iran country is reused in some way; however, much of it is mixed with freshwater before further use, particularly in the suburban areas [29]. The direct use of untreated wastewater from sewage outlet, directly used for crop production is not a common scene in Iran. However, treated or partially treated wastewater used directly for irrigation without being mixed or diluted is more common. This is practiced in many treatment plants and there is no exact estimate about the amount used by this method to irrigate fodders, cereals, fruit trees, and vegetables eaten cooked or uncooked [29]. Thus the pollutants discharge per capita (PDC; g/d.cap) with the wastewater treatment systems is important parameter should be considered to be estimated.

The PDC with the wastewater treatment systems in Tehran was defined and determined in terms of GCPD and pollutant removal efficiency (PRE; %). The pollutants removal efficiency of wastewater treatment systems in Tehran ranged 92.11–93.41% for BOD_5, 89.65–90.97% for COD, 91.48–93.18% for TSS, 28.65–31.72% for TS, 1.73-2.76% for organic nitrogen (ON), 58.52-61.24% for P and 84.62-88.26% for TKN.

The ranges of estimated PDC summarized in Table 4, were compared with conventional domestic wastewater treatment systems in the United States [30] (Table 8). The generation per capita of BOD_5, COD, ON and TP of the United States wastewater treatment systems were almost thrice, four times, nine times and twice as those in Tehran, respectively. Conventional and nutrient removal activated sludge treatment systems are regarded as representative wastewater treatment processes in urban area of developed

Table 6 Basic characteristics of raw sewage intended for the design of wastewater treatment plants in Iran [26]

Paramater	Per capita parameters as g/d.cap for different plant or cities					
	Toiserkan	Zargandeh (Tehran)	Parkandabad (Mashhad)	Hovaizeh	Arak	Zabol
BOD_5	40	40	50	54	57	57
TSS	50	83	46	50	60	50
N(TKN)	-	14	-	-	5	-
TP	-	1	-	-	0.45	

Table 7 Comparison of per capita loadings (g/d.cap) of wastewater parameters in different countries with estimated values in Tehran

Parameter	BOD5	TSS	TKN	TP
Turkey[a]	27-50	41-68	8-14	0.4-2
India[a]	27-41	n.a.	n.a.	n.a.
Japan[a]	40-45	n.a.	1-3	0.2-0.4
Egypt[a]	27-41	41-68	8-14	0.4-0.6
Uganda[a]	55-68	41-55	8-14	0.4-0.6
Italy[a]	49-60	55-82	8-14	0.6-1
Germany[a]	55-68	82-96	11-16	1.2-1.6
Denmark[a]	55-68	82-96	14-19	1.5-2
Sweden[a]	68-82	82-96	11-16	0.8-1.2
Brazil[a]	55-68	55-68	8-14	0.6-1
United States[a]	50-120	60-150	9-22	2.7-4.5
Tehran (This study)	31-34	35-40	6.2-7.3	1.8-2

a: Adapted from *Wastewater engineering: treatment, disposal, and reuse* [28].

countries such as USA. PDCs with these treatment systems were estimated as 16.1 g-BOD_5/d.cap, 31.6 g-COD/d.cap, 4.9 g-ON/d.cap, and 3.2 g-TP/d.cap [30] while PDCs with Tehran's wastewater treatment systems were estimated as 2.15 ± 0.19 g-BOD_5/d.cap, 4.46 ± 0.34 g-COD/d.cap, 1.82 ± 0.32 g-ON/d.cap, and 0.76 ± 0.04 g-TP/d.cap. Deployment of larger removal efficiency treatment systems will decrease PDC with increase of PRE. When judging only from pollutant discharge reduction function of treatment systems, countries or cities with smaller GCPD and larger PDC-BOD_5 should be prioritized for investments on wastewater treatment measurements including assistances in order to improve ambient water quality [31].

The estimated PDC of some regions studied by Tsuzuki [15] were compared with that of Tehran and summarized in Table 9. When the criteria are applied, the group of countries in the Region of South China Sea and Tehran were found to be with higher and lower priority of BOD_5 discharge reduction from domestic wastewater, respectively (Table 9). The PDC of ROP-ME[a] sea area region, Red sea region and Gulf of Aden region is 2.5 g/d.cap and comparable with that obtained for Tehran (2.15 g/d.cap).

The comparatively low BOD_5-PDC in sewage treatment plants of Tehran may results from the fact that the per capita loading of BOD_5 in Tehran is lower than that for other countries and on the other hand the removal efficiency of BOD_5 in most of treatment plants are more than 90%.

Tajrishi [29] estimated that less than 40% of the total domestic sludge in Iran's wastewater treatment plants is being treated completely. In other words; of more than 200,000 cubic meters of daily produced sludge (2000 tons/d dry solids) of total fecal, septic and waste excrements sludge, only about 80,000 cubic meters (800 tons) is being digested and/or stabilized daily by different treatment methods. According to the finding of our study, the per capita values of 0.16 ± 0.02 and 0.11 ± 0.02 L sludge/d were determined for the produced and waste excrements sludge in Tehran's WWTPs, respectively. According to these values and the population covered by Tehran's WWTPs (3,150,000 persons/d), it is estimated that 504 m^3/d and 346 m^3/d of sludge will be produced and waste as excrement raw sludge, respectively, in Tehran. This high volume of sludge needs to be managed and discarded properly. The most common method of treatment for these sludge is digestion (aerobically and an aerobically). The lagooning, composting, and landfilling are the next methods of treatment. Mechanical dewatering may also be implemented as final treatment to reduce the volume of the stabilized sludge.

Conclusions

In this study, the basic characteristic data regarding to the nine main domestic wastewater treatment plants in Tehran was assessed and analyzed. The BOD_5 and TSS parameters of 32.96 ± 1.91, 37.31 ± 2.44 g/d.cap obtained in this study are considerably lower than the values of 40 to 50 and 50 to 60 g/d.cap, respectively, recommended by Department of Energy of Iran for the design of new treatment plants. The per capita loading of wastewater were used by designers for the design of a wastewater treatment plant in Tehran, was estimated to be 215 L/d.cap while the actual per capita collected wastewater according to the real data of treatment plants was determined to be 186.06 ± 7.85 L/d.cap. This disparity has interesting implication when these

Table 8 Comparison of GDPCs and PDCs of conventional activated sludge wastewater treatment systems in the USA [30] with those in Tehran

Country	Parameters	BOD5	COD	ON	TP
United States (Qasim, 1998)	GDPC (g/d.cap)	95	180	18	4
	Removal efficiency (%)	80-85	80-85	60-85	10-25
	PDC (g/d.cap)	16.1	30.6	4.9	3.2
Tehran (This study)	GDPC (g/d.cap)	31.0-34.8	46.7-51.7	1.6-2.4	1.8-2.0
	Removal efficiency (%)	92.1-93.4	89.6-90.9	1.7-2.7	58.5-61.2
	PDC ((g/d.cap)	1.96-2.3	4.1-4.8	1.4-2.1	0.7-0.8

Table 9 Comparison of PDC of other regions from 1998 to 2002 [15] with that of Tehran

Region/country	PDC (g/d.cap)	
	BOD	TP
South China Sea Region	43	n.a
Caspian Sea Region	24	n.a
Eastern African Region	11	n.a
Pacific Island Region	8.5	0.51
West and Central African (WACAF) Region	4.6	1.8
ROPME[a] sea area and Red sea and gulf of Aden region	2.50	1.10
Tehran (this study)	2.15	0.76

a: Regional Organization for the Protection of the Marine Environment.

estimated values are applied to the design of new wastewater treatment plant and calculation of an equivalent population for an industrial waste. The design of new treatment plant in Tehran with a per capita of 186.06 ± 7.85 L/d.cap may lower the required capacity for the treatment of wastewater up to 40%. The results of this study reveal that the Tehran's wastewater may be classified as highly degradable, but during recent decades the Biodegradability Index has been reduced up to 15%. This may be resulted from this fact that over the past 15-20 years, a wide range of diverse industrial synthetic detergents have been produced and extensively used for cleaning and disinfection purposes and then discharged into sewage systems. According to the PDC of BOD_5, Tehran was found to be with lower priority of BOD_5 discharge reduction from domestic wastewater. It is estimated that 504 m^3/d and 346 m^3/d of sludge will be produced and waste as excrement raw sludge, respectively, in Tehran.

In conclusion, the use of new revised per capita parameters obtained in this study could help in designing more efficient treatment systems and generating more reliable data for operational control of wastewater treatment process in Tehran. However, further research on wastewater quality and quantity assessment and adequate monitoring measures are required for existing and future treatment facilities in Tehran to ensure that they comply with safe operational and environmental standards.

Endnote

[a]Regional Organization for the Protection of the Marine Environment.

Nomenclature

Nomin; the nominal capacity of WWTP (m^3/month)
Collect; the collected volume of wastewater (m^3/month)
Treat; the treated volume of wastewater (m^3/month)
ProSlu; the produced sludge (m^3/month)
ExcSlu; the excess sludge (m^3/month)
CurentPop; the current covered population
Tin; the Influent temperature (C°)

pHin; the Influent pH
BODin; the Influent BOD_5 (mg/L)
CODin; the influent COD (mg/L)
TSin; the influent total solids (mg/L)
TSSin; the influent suspended solids (mg/L)
TDSin; the influent dissolved solids (mg/L)
Pin; the influent phosphorous (mg/L)
OrNin; the influent organic nitrogen (mg/L)
TKNin; the influent total Kjeldahl nitrogen (mg/L)
Tout; the effluent temperature (C°)
pHout; the effluent pH
BODout; the effluent BOD_5 (mg/L)
CODout; the effluent COD (mg/L)
TSout; the effluent total solids (mg/L)
TSSout; the effluent suspended solids (mg/L)
TDSout; the effluent dissolved solids (mg/L)
Pout; the effluent phosphorous (mg/L)
OrNout; the effluent organic nitrogen (mg/L)
TKNout; the effluent total Kjeldahl nitrogen (mg/L)
$GCPD_{Slu\_E}$; the generation per capita per day of excess sludge (L sudge/d.cap)
$GCPD_{Slu\_P}$; the generation per capita per day of produced sludge (L sludge /d.cap)
$GCPD_{TS}$; the generation per capita per day of total solids (g /d.cap)
$GCPD_{OrgN}$; the generation per capita per day of organic nitrigen (g /d.cap)
$GCPD_{TDS}$; the generation per capita per day of total dissolved solids (g /d.cap)
$GCPD_P$; the generation per capita per day of total phosphorous (g /d.cap)
$GCPD_{TKN}$; the generation per capita per day of total kjeldahl nitrogen (g /d.cap)
$GCPD_{TSS}$; the generation per capita per day of total suspended solids (g /d.cap)
$GCPD_{COD}$; the generation per capita per day of chemical oxygen demand (g /d.cap)
$GCPD_{BOD5}$; the generation per capita per day of biological oxygen demand (g /d.cap)
$(BOD_5/COD)_{out}$; the biodegradability index of effluent wastewater
$(BOD_5/COD)_{in}$; the biodegradability index of influent wastewater
$Tret_{PerP}$; the per capita per day of treated wastewater (L/d.cap)
$Coll_{PerP}$; the per capita per day of collected wastewater (L/d.cap)

Competing interests
The authors declare that they have no competing interests.

Authors' contributions
AM, SN and AHM contributed to drafting and editing the manuscript. HRT participated in raw data provision. MH contributed in the design of study, data analysis and drafting the manuscript. All authors read and approved the final manuscript.

Acknowledgements
The authors are grateful to the Center for Water Quality Research (CWQR) at the Institute for Environmental Research (IER) of Tehran University of Medical Sciences for providing facilities and supports for this research.

Author details
[1]Center for Water Quality Research (CWQR), Institute for Environmental Research (IER), Tehran University of Medical Sciences, Tehran, Iran. [2]Department of Environmental Health Engineering, Faculty of Health, Tehran University of Medical Sciences, Tehran, Iran. [3]Center for Solid Waste Research (CSWR), Institute for Environmental Research (IER), Tehran University of Medical Sciences, Tehran, Iran. [4]Department of Environmental Health Engineering, School of Public Health, Islamic Azad University-Tehran Medical Branch, Tehran, Iran.

References
1. Zanoni A, Rutkowski R. Per capita loadings of domestic wastewater. J Water Pollut Control Fed. 1972;44:1756–62.
2. Calvert C, Parks EH. The population equivalent of certain industrial wastes. Sewage Work J. 1934;6:1159–64.
3. Goldstein SN, Moberg WJ. Wastewater treatment systems for rural communities. DC: Washington; 1973.
4. Hadi M, Shokoohi R, Ebrahimzadeh Namvar A, Karimi M, Solaimany Aminabad M. Antibiotic resistance of isolated bacteria from urban and hospital wastewaters in Hamadan City. Iran J Health and Environ. 2011;4:105–14.
5. Hadi M, Samarghandi MR, McKay G. Simplified fixed bed design models for the adsorption of acid dyes on novel pine cone derived activated carbon. Water Air Soil Pollut. 2011;218:197–212.
6. Mahvi A. Sequencing batch reactor: a promising technology in wastewater treatment. Iran J Environ Health Sci Eng. 2008;5:79–90.
7. Mahvi A, Nabizadeh R, Pishrafti M. Evaluation of single stage USBF in removal of nitrogen and phosphorus from wastewater. Eur J Sci Res. 2008;23:204-211
8. Naghizadeh A, Mahvi A, Mesdaghinia A, Alimohammadi M. Application of MBR Technology in Municipal Wastewater Treatment. Arab J Sci Eng. 2011;36:3–10.
9. Naghizadeh A, Mahvi A, Vaezi F, Naddafi K. Evaluation of hollow fiber membrane bioreactor efficiency for municipal wastewater treatment. Iran J Environ Health Sci Eng. 2008;5:257–68.
10. Karagozoglu B, Altin A. Flow-rate and pollution characteristics of domestic wastewater. Int J Environ Pollut. 2003;19:259–70.
11. Tajrishy M, Cities S, Abdolghafoorian A, Abrishamchi A. Water reuse and wastewater recycling: Solutions to Tehran's growing water crisis. In: Quentin Grafton R, Wyrwoll P, White C, Allendes D, editors. Global Water: Issues and Insights. Australia: ANU Press; 2014. p. 223.
12. APHA, AWWA, WEF: *Standard methods for the examination of water and wastewater*. 22 edn: American Public Health Association (APHA), American Water Works Association (AWWA) & Water Environment Federation (WEF); 2012.
13. Teng CM. Correcting Noisy Data. In: Proceedings of 16th International Conference on Machine Learning; San Francisco. 1999. p. 239–48.
14. Masters T. Practical Neural Network Recipes in C++. San Diego: Academic Press; 1993.
15. Tsuzuki Y. Comparison of pollutant discharge per capita (PDC) and its relationships with economic development: An indicator for ambient water quality improvement as well as the Millennium Development Goals (MDGs) sanitation indicator. Ecol Indic. 2009;9:971–81.
16. Statistics and basic operational items of Tehran's wastewater company [http://ts.tpww.ir/abfa_content/media/image/2015/03/33358_orig.pdf]
17. Seyed Morteza H. Wastewater Quality and Quantity. Hoseinian: Tehran; 1974.
18. Sazeh Consultants Engineering & Construction Co.: Assessment the results of wastewater quality analyses of Tehran's Jonoob wastewater treatment plant. Sazeh Consultants Engineering & Construction Co. Tehran; 1993
19. Afshar J. Assessment of the quality and quantity of Tehran's wastewater. Master of Sciences Thesis: Tehran University of Medical Sciences, Department of Environmental Health Engineering; 1995.
20. Azimi AA, Ameri M. Determination of per capita flow rate and wastewater pollutants of saheb-gharanieh treatment plant in Tehran. J Environ Stud. 2002;28:93–100.
21. Rim-Rukeh A, Agbozu L. Impact of partially treated sewage effluent on the water quality of recipient Epie Creek Niger Delta, Nigeria using Malaysian Water Quality Index (WQI). J Appl Sci Environ Manag. 2013;17:5–12.
22. Srinivas T. Environmental biotechnology. New Delhi: New Age International Publishers; 2008
23. Matthijs E, Debaere G, Itrich N, Masscheleyn P, Rottiers A, Stalmans M, et al. The fate of detergent surfactants in sewer systems. Water Sci Technol. 1995;31:321–8.
24. Scott MJ, Jones MN. The biodegradation of surfactants in the environment. Biochim Biophys Acta Biomembr. 2000;1508:235–51.
25. Sharifi Sistani M. Wastewater treatment in Iran, past, present and future. Water Environ J. 2000;38:25–32.
26. Department of Energy: In Specialized workshop proceedings on assessment of wastewater treatment plants challenges; Shiraz. Water and Wastewater Company; 2001
27. Department of Energy. Technical criteria for reviewing and approval of urban sewage treatment projects, Standards No. 3-129. Tehran: Plan and Budget Organization, Office of Research and Technical Criteria; 1993.
28. Metcalf L, Eddy H, Tchobanoglous G: Wastewater engineering: treatment, disposal, and reuse. New York: McGraw-Hill; 2003
29. Tajrishy M. Wastewater Treatment and Reuse in Iran: Situation Analysis. Tehran: Departement of Civil Engineering, Sharif University of Technology, Environment and Water Research Center (EWRC); 2010.
30. Qasim SR: Wastewater treatment plants: planning, design, and operation. New York: CRC Press; 1998.
31. Tsuzuki Y. Relationships between pollutant discharges per capita (PDC) of domestic wastewater and the economic development indicators. J Environ Syst Eng. 2007;63:224–32.

4

Application of Reverse Transcriptase –PCR (RT-PCR) for rapid detection of viable *Escherichia coli* in drinking water samples

Neda Molaee[1], Hamid Abtahi[2], Mohammad Javad Ghannadzadeh[3], Masoude Karimi[4] and Ehsanollah Ghaznavi-Rad[2,5]*

Abstract

Background: Polymerase chain reaction (PCR) is preferred to other methods for detecting *Escherichia coli* (*E. coli*) in water in terms of speed, accuracy and efficiency. False positive result is considered as the major disadvantages of PCR. For this reason, reverse transcriptase-polymerase chain reaction (RT-PCR) can be used to solve this problem. The aim of present study was to determine the efficiency of RT-PCR for rapid detection of viable *Escherichia coli* in drinking water samples and enhance its sensitivity through application of different filter membranes.

Materials and methods: Specific primers were designed for *16S rRNA* and elongation Factor II genes. Different concentrations of bacteria were passed through FHLP and HAWP filters. Then, RT-PCR was performed using 16srRNA and *EF –Tu* primers. Contamination of 10 wells was determined by RT-PCR in Arak city. To evaluate RT-PCR efficiency, the results were compared with most probable number (MPN) method.

Results: RT-PCR is able to detect bacteria in different concentrations. Application of EF II primers reduced false positive results compared to *16S rRNA* primers. The FHLP hydrophobic filters have higher ability to absorb bacteria compared with HAWB hydrophilic filters. So the use of hydrophobic filters will increase the sensitivity of RT-PCR.

Conclusion: RT-PCR shows a higher sensitivity compared to conventional water contamination detection method. Unlike PCR, RT-PCR does not lead to false positive results. The use of *EF-Tu* primers can reduce the incidence of false positive results. Furthermore, hydrophobic filters have a higher ability to absorb bacteria compared to hydrophilic filters.

Keyword: Coliforms, Reverse transcriptase, Drinking water

Background

The presence of pathogens in drinking water is a major health problem. Therefore, detection and removal of pathogens are considered as main health issues for drinking water. In diagnostic workup, presence of *E.coli* is considered as an indicator of water pollution by wastewater. The conventional methods for detecting pathogens in water such as (most probable number) MPN and water culture are usually costly and time consuming and presumably lack of sufficient accuracy. This is why new methods are expected to detect pathogens in drinking water [1].

Polymerase chain reaction (PCR) could be a good alternative to conventional assays. PCR is a sensitive, accurate and rapid method that its effectiveness has been demonstrated in numerous studies [2]. PCR is able to detect even one bacterium per 100 ml of water [3]. Polymerase chain reaction is a sensitive and accurate method that not only detects pollution, but determines the type of bacteria [4]. However, due to relatively high stability of DNA in the culture medium, presence of dead cells will produce positive result. Some studies showed that the PCR result for bacteria killed by boiling or UV radiation becomes positive. Studies on *E. coli* and *Listeria monocyte genes* showed that the PCR result for autoclaved bacteria is still positive [5].

Due to incidence of false positive results in DNA-based detection procedures, other methods such as initial

---
* Correspondence: e.ghaznavirad@arakmu.ac.ir
[2]Molecular and Medicine Research Center, Arak University of Medical Sciences, Arak, Iran
[5]Department of Medical Microbiology and Immunology, Faculty of Medicine, Arak University of Medical Sciences, Arak, Iran
Full list of author information is available at the end of the article

Table 1 The sequence and position of the oligonucleotide primers used in the study

Gene	Forward	Reverse	Size (bp)	Access number
16S rRNA	5'CGA GTG GCG GAC GGG TGA GT3' (FROM 81)	5' TCG ACA TCG TTT ACG CGC TGG A3' (FROM 786)	723	EF620925
EF-Tu	5'CGCTGGAAGGCGACGCAGAG 3' (FROM 1253)	5'CGGAAGTAGAACTGCGGACGGTAG3' (FROM 1698)	470	X57091

enrichment before PCR testing are recommended. Although this method increases the detection efficiency of the test for live and dead samples, the duration of the assay will be increased. In addition, in the case where the number of dead bacteria is high, it may lead to false positive results [6].

Due to the low stability of RNA outside the cell, the use of reverse transcriptase polymerase chain reaction (RT-PCR) for detecting RNA is considered as a suitable marker for tracking the living cells. At the same time, the stability of different RNAs varies outside the cell. Davis *et al.,* found that of the four species of nucleic acid, mRNA is the most promising candidate as an indicator of viability in bacteria [7]. Klein and Juneja showed a good correlation between the presence of mRNA and the viability of *L. monocytogenes* when comparing growing cells with those killed by autoclaving [8]. Therefore, this study designed to examine the efficiency of RT-PCR by preparing various concentrations of *E. coli* in water. In addition, the impact of bacterial filters (hydrophilic and hydrophobic) on concentrated water samples and RT-PCR results are studied.

Materials and methods

PCR primers targeting *16S rRNA* and *EF-Tu* regions of the *E. coli* genome were designed using Primer Select (DNAstar, Inc., Madison, WI) software package. Two sets of primer pairs were designed and tested. The nucleotide sequences of the primers have been shown in Table 1.

Water dilutions

To prepare water dilutions, sequential dilutions of *E. coli* bacteria were prepared in 100 ml of sterile water. The McFarland Solution No.1 was used to prepare different bacterial concentrations. For this purpose, the bacterial suspension is adjusted to match the one McFarland turbidity standard.

Then, the suspension was used to prepare sequential dilutions of 8/100 to 1/1600 as listed in Table 2 (two samples for each dilution). To verify the number of bacteria in various concentrations, the bacterial sediment

was obtained by centrifuging the whole water at 5000 rpm for 5 min and cultured by pour plate method in nutrient agar medium. Two concentrations were prepared in two series.

After preparing water samples with various bacterial concentrations, a series of dilutions were passed through FHLP filter (Millipore, pore diameter of 0.5 microns) and the second series were passed through HAWP filter (Millipore, pore diameter of 0.45 microns). Then, the filters were placed inside sterile micro tubes and 50 ml water containing 0.1% diethylpyrocarbonate (DEPC water) was added. The micro tubes were vortexed vigorously to release bacteria from the filter surface to the fluid.

RT-PCR

The bacterial RNA was extracted and purified using the RNX-plus kit (Sinagene, Iran). Deoxyribonuclease I enzyme was used to remove DNA contamination of the purified RNA. To remove the enzyme, 1 µl ethylene diamine tetraacetic acid (EDTA, 25 Mm) was added for 10 min at 65°C. RT-PCR was performed for *16 s rRNA* and *EF-Tu* under the circumstances described elsewhere [9]. The method is described briefly in Figure 1.

The first phase of PCR is composed of 35 cycles, each consists of three steps including 1-denaturation (at 94°C for 1 min), 2- binding the primers to the DNA template (at 59°C, 1 min for EF-Tu and 1 min for 16SrRNA) and 3- target gene amplification (at 72° for 1 min). After the end of the cycles, the final extension was performed for one cycle at 72 ° C for 5 min.

To evaluate the PCR products, the products were electrophoresed on 1% agarose gel in Tris base- boric acid-EDTA (TBE, pH = 8) buffer solution. To analyze the electrophoresis results, the samples were stained with ethidium bromide solution and visualized by an UV transilluminator apparatus.

To confirm the difference between PCR and RT-PCR in diagnosis of live and death bacteria, control for living and dead bacteria through RT-PCR method and a control of PCR for dead bacteria have been performed in this experiment (Figures 2).

Table 2 Coliforms serial dilutions prepared in laboratory

Dilutions	1600/1	1/800	1/400	1/200	1/100	2/100	4/100	8/100
Number of bacteria	1	1	1	1	1	2	4	8
The volume of water(ml)	1600	800	400	200	100	100	100	100

Serial dilutions of *E.coli*

Live bacteria ←→ Killed bacteria

↓

filtration

↙ ↓ ↘

FHLP   HAWP

Cell collected on membrane

↓

Extracted RNA(RNX-PLUS)
Chloroform
Isopropyl alcohol

↓

Supernatant

↓

RNA PELLET

DNAse1

↓

RT-PCR

↙ ↘

16SrRNA   EF-Tu

↘ ↙

↓

CDNA

↓

PCR

↓

Analyzed on agarose gel 1%

Figure 1 Flow chart illustrates the RT-PCR procedure for detecting *16S rRNA* and *EF-Tu* mRNA of *E.coli*.

Figure 2 16SrRNA gene RT-PCR and PCR results : Lane 1: Marker 100 bp, Lane2, a live bacteria RT-PCR, Lane3, death bacteria RT-PCR, Lane 4, a live bacteria PCR, Lane 5, death bacteria PCR.

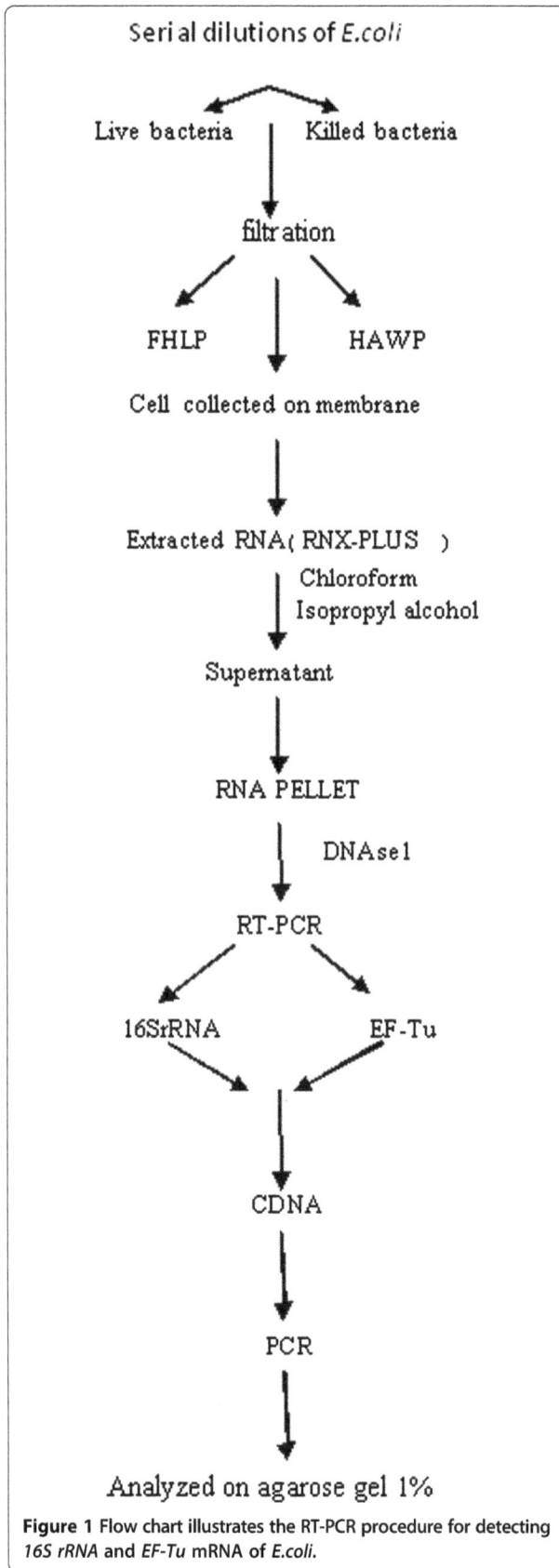

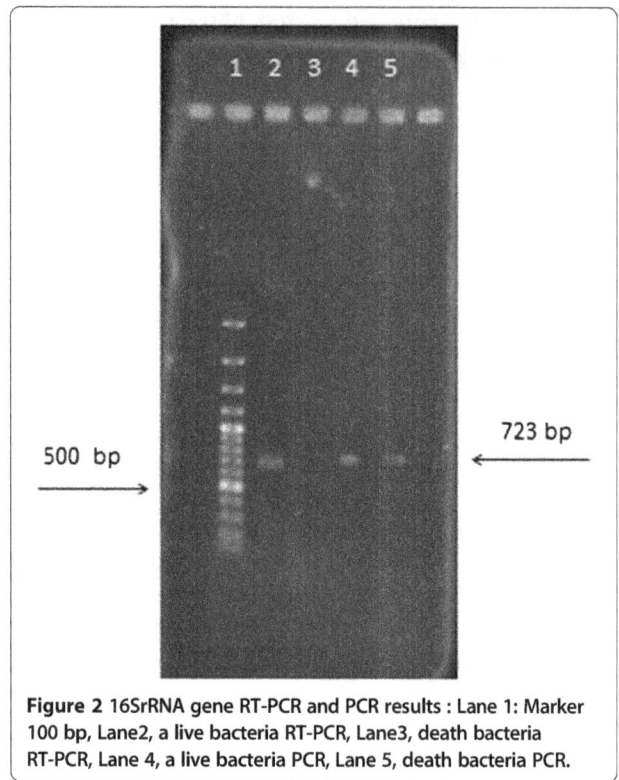

The stability of *EF-Tu* and *16S rRNA*: mRNA detection in live bacteria

To investigate the stability of RNA in the medium, the bacterial dilutions were killed by chlorine. Then, RNA was extracted and purified at different times (0, 0.5, 1, 2, 3, 4, 6, 10 and 16 h after bacterial death).

To prepare the chlorine solution, 0.1 g of the powdered calcium hypochlorite granules 70% was dissolved in 100 ml water. Then, 30 ml chlorine solution was added to the bacterial dilutions so that the concentration of the residual chlorine was 0.2 to 0.4 mg/l after 30 minutes. After chlorination, the water dilutions were cultured in nutrient agar medium by pour plate method to ensure the death of all bacteria.

After chlorination, series of bacterial dilutions passed through both filters. RT-PCR was performed for genes, *EF-Tu* and *16srRNA* as described before.

Examination of well water samples by RT-PCR and MPN

To achieve this purpose of the study 10 wells were selected and 250 ml water was collected in sterile containers from each well in Arak city according to standard procedure.

The MPN method was used to examine the consistency of RT-PCR results. The presence of coliform in the samples was evaluated using both methods. In the RT-PCR method, 100 ml water was filtered using HAWP and FHLP filters and then checked for *Escherichia coli*. The MPN test was performed by three-tube method and the

Table 3 The number of colony obtained on in agar nutrient medium by pour plate method

Dilution	1/1600	1/800	1/400	1/200	1/100	2/100	4/100	8/100
Colony number	1	1	1	1	1	2	4	8

number of *E. coli* bacteria was determined based on gas production.

Results

Water dilutions

The number of bacteria was confirmed by culturing the water dilutions in agar nutrient medium by pour plate method. The result has been shown in Table 3.

Water dilutions filtration and RT-PCR

The result of RT-PCR on water dilutions showed that the RT-PCR results become positive only after filtration with hydrophobic filters (FHLP). When FHLP filters were used, the presence of even one bacterium in 1600 ml water can be detected by RT-PCR. The RT-PCR results for the genes *16S rRNA* (Figures 3 and 4) and *EF-Tu* (Figures 5 and 6) demonstrate the role of FHLP and HAWP filters in the detection of low numbers of bacteria in water samples. The RT-PCR results for all dilutions passed through FHLP showed a band as observed by UV transilluminator.

The stability of *16SrRNA* and *EF-Tu* after chlorination

No bacteria were found in cultures after chlorination of water dilution samples. Therefore, it was confirmed that bacteria were killed.

Table 4 shows the stability of *16SrRNA* and *EF-Tu* RNAs after the death of bacteria by chlorine. As shown, *EF-Tu* and 16S rRNA remained stable up to 4 h and over 16 h, respectively.

Comparison of molecular and MPN methods

Table 5 shows the results of MPN and RT-PCR tests for 10 drinking water wells. Comparison of RT-PCR results demonstrates the higher efficiency of RT-PCR method for bacteria detection compared to MPN method. The results of this comparison are given in Table 5.

Discussion

Due to the importance of water borne diseases, in particular those transmitted by drinking water, the evaluation of water quality in terms of pathogenic microorganisms has a major influence on public health program. The presence of coliforms (especially *E. coli*) in water should be considered as a pollution indicator. Various methods have been designed to detect pathogens in water samples. To approve the water quality in terms of pathogens, the water samples should be exactly checked for the presence of coliforms, especially *Escherichia coli*. Therefore, the presence of coliforms after treatment indicate the water pollution, needs further action. In the case of water contamination with coliforms, the water samples may be contaminated with other pathogens, parasites, viruses, and etc. [10].

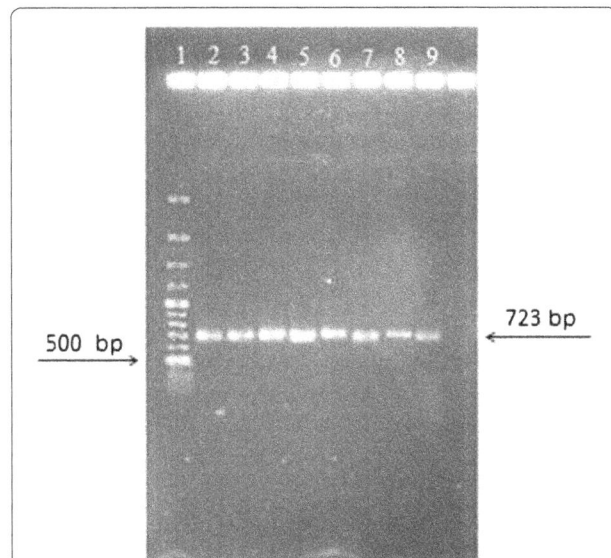

Figure 3 Dilution of bacterial *16SrRNA* gene RT-PCR results filtered with FHLP, Lane1, Marker 100 bp, Lane2, 1/1600Dilution, Lane3, 1/800Dilution, Lane 4, 1/400Dilution, Lane 5, 1/200Dilution, Lane 6, 1/100Dilution, Lane 7, 2/100Dilution, Lane 8, 4/100Dilution, Lane 9, 8/100Dilution.

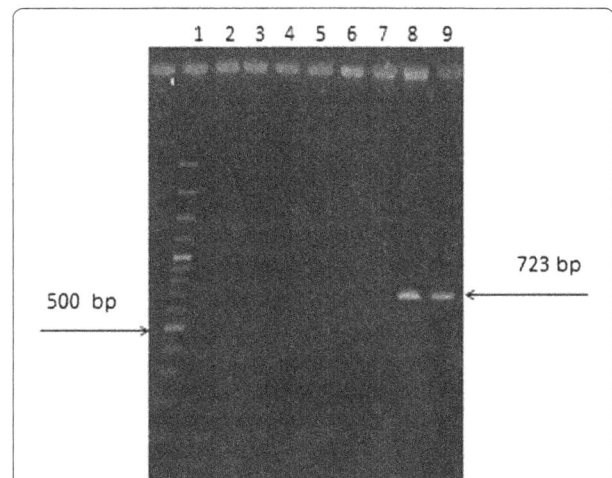

Figure 4 Dilution of bacterial *16SrRNA* gene RT-PCR results filtered with HAWP, Lane 1, Marker 100 bp, Lane2,1/1600Dilution, Lane3, 1/800Dilution, Lane 4, 1/400Dilution, Lane 5, 1/200Dilution, Lane 6, 1/100Dilution, Lane 7, 2/100 Dilution, Lane 8, 4/100Dilution, Lane 9, 8/100Dilution.

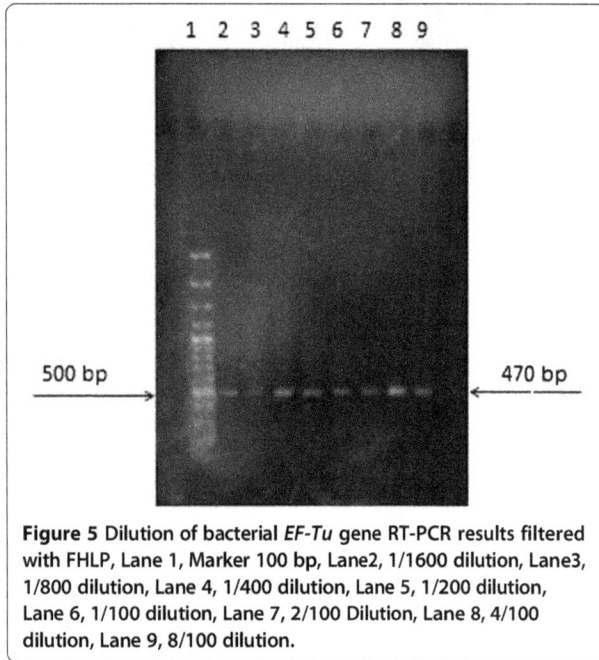

Figure 5 Dilution of bacterial *EF-Tu* gene RT-PCR results filtered with FHLP, Lane 1, Marker 100 bp, Lane2, 1/1600 dilution, Lane3, 1/800 dilution, Lane 4, 1/400 dilution, Lane 5, 1/200 dilution, Lane 6, 1/100 dilution, Lane 7, 2/100 Dilution, Lane 8, 4/100 dilution, Lane 9, 8/100 dilution.

Table 4 Stability of the mRNA (*EF-Tu* and *16S rRNA*) in different times was shown

Target	0 min	30 min	1 h	2 h	3 h	4 h	6 h	10 h	16 h
16SrRNA	Y	Y	Y	Y	Y	Y	Y	Y	Y
EF-Tu	Y	Y	Y	Y	Y	Y	N	N	N

Y, Positive RT-PCR. N, négative RT-PCR.NP not performed.

There are several problems with viable culture methods (MPN) used for routine monitoring of the bacteriological safety of water supplies, including maintaining the viability of bacteria between the time of collection and enumeration, lack of growth of viable but nonculturable bacteria. Therefore, the use of an ideal method with advantages such as high speed, sensitivity and accuracy, ease of doing and low cost to evaluate a high volume of water samples can be effective in preventing the spread of infectious

diseases through water. In addition, detection, separation of most pathogens in contaminated water is difficult, time consuming and often associated with high costs [11]. Although the use of molecular techniques for detecting pathogens is preferred in terms of speed and accuracy, techniques like PCR have some drawbacks. The major disadvantages is that this method is not able to discriminate dead and living bacteria, hence presence of dead bacteria leading to false positive result [6].

Methods that only detect live bacteria must be used to fix this problem. Techniques like RT-PCR can be used to detect live bacteria. In general, RNA-based techniques can prove the presence of living cells only when the target molecule has little stability after cell death. In fact, mRNA molecules are constantly reproduced and increased inside live cells. However, RNA unlike the DNA has a very short half-life, so that RNA content decreases rapidly after the cell death. Therefore, the presence of RNA molecules is a good marker to prove the existence of living cells. Vaitilingom *et al.*, used RT-PCR method and found RNA molecules only in living cells [3]. However, some types of RNA molecules are more stable. For instance, the results of study conducted by Sheridan *et al.*, showed that *16S rRNA* genes remain stable for about 16 h. Therefore, less stable RNA molecules should be sought [12].

The results of the present study showed that the molecular RT-PCR method is more efficient for detecting less stable *EF-Tu*, can reduce the detection time to 4 hours.

Another important point is the higher efficiency of methods like RT-PCR. Various studies suggest the lower sensitivity of RT-PCR than PCR. In this regard, Liu et al. indicated that RT-PCR is not capable of detecting bacteria in suspension with 1 CFU/ml [13]. Thus, hydrophobic filters can be used to concentrate bacteria and increase the sensitivity of RT-PCR to prevent loss of RNA molecules. Bej *et al.*, found that hydrophobic filters have a higher ability to absorb bacteria and DNA

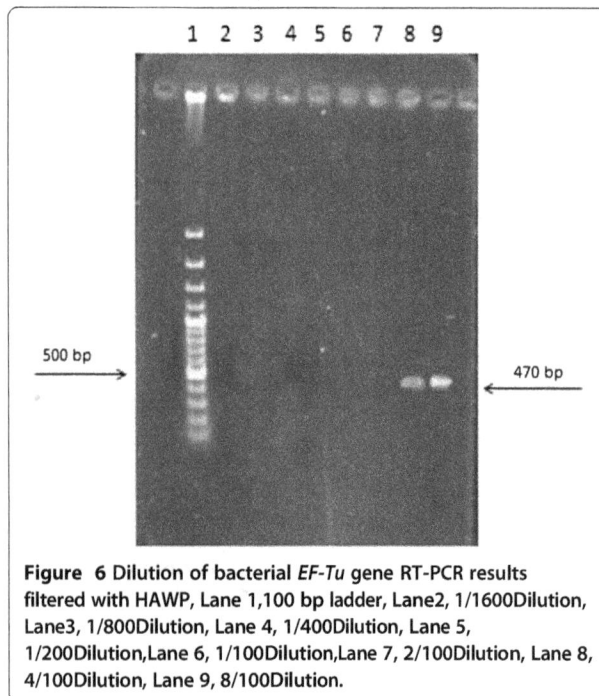

Figure 6 Dilution of bacterial *EF-Tu* gene RT-PCR results filtered with HAWP, Lane 1,100 bp ladder, Lane2, 1/1600Dilution, Lane3, 1/800Dilution, Lane 4, 1/400Dilution, Lane 5, 1/200Dilution,Lane 6, 1/100Dilution,Lane 7, 2/100Dilution, Lane 8, 4/100Dilution, Lane 9, 8/100Dilution.

Table 5 Comparison of RT-PCR result for *EF-Tu* and *16S rRNA* gene with the MPN method of 10 wells

Number of wells	1	2	3	4	5	6	7	8	9	10
RT-PCR(*16S rRNA*)	-	+	+	+	-	+	+	+	+	+
RT-PCR(*EF-Tu*)	-	+	+	+		+	+	+	+	+
MPN		-	-	+	-	-	-	-	+	+	-

particles than hydrophilic filters [11]. Our results indicating that the RT-PCR efficiency increases using hydrophobic filter even at low bacterial concentrations. The enhanced efficiency can be attributed to the lack of bacteria loss and thus the presence of bacteria nucleic acids. Although the pore diameter of hydrophilic filter (e.g. HAWP) is less than FHLP filters, the hydrophilic filters are less able to absorb bacteria. HAWP filters only able to isolate the bacteria at dilutions of 4 bacteria in 100 ml water. In other words, RT-PCR is not able to detect bacteria at dilutions less than 4 bacteria using hydrophilic filters.

Beside the ability to detect one bacterium in 100 ml water, the results also indicating that RT-PCR is able to detect a bacterium in higher volumes (up to 1600 ml water) using hydrophobic filters. In addition to the high accuracy and efficiency of the RT-PCR method in water pollution detection, it reduces the time of bacterial detection. The routine MPN test lasts at least 24 h, while this time is reduced to 6 to 7 h using RT-PCR method. Therefore, the test could be started within four hours after sampling to avoid detection of possible contamination with dead bacteria.

The results of polymerase chain reaction (PCR) test in water purification and disinfection systems become false positive due to the stability of DNA molecule. Accordingly, molecular methods like PCR cannot reflect the impact of disinfectants on water pollution index. To resolve this problem, other cell molecules (e.g. RNAs) with relatively low stability such as *EF-Tu* with the stability period of 4 h can be used. Reverse transcriptase polymerase chain reaction (RT-PCR) must be used to determine the presence of RNA in the medium. The detection performance can be enhanced using hydrophobic filters for relatively high volumes of water with low bacterial concentrations.

Conclusions

Reverse transcriptase polymerase chain reaction (RT-PCR) is an efficient method for detecting the bacterial contamination of water. The detection performance can be enhanced using hydrophobic filters. The detection time is significantly reduced using less stable ribonucleic acids such as *EF-Tu*.

Competing interests
The authors declare that they have no competing interests.

Authors' contribution
All authors have equal contribution. All authors read and approved the final manuscript.

Acknowledgement
This project was supported by Arak University of Medical Sciences, Arak, Iran.

Author details
[1]Department of Microbiology and Immunology, Arak University of Medical sciences, Arak, Iran. [2]Molecular and Medicine Research Center, Arak University of Medical Sciences, Arak, Iran. [3]Department of Environmental Health, Faculty of Health, Arak University of Medical Sciences, Arak, Iran. [4]Department of Medical Microbiology and Immunology, Arak University of Medical sciences, Arak, Iran. [5]Department of Medical Microbiology and Immunology, Faculty of Medicine, Arak University of Medical Sciences, Arak, Iran.

References
1. Toze S. PCR and the detection of microbial pathogens in water and waste water. Wat Res. 1999;33(17):3545–56.
2. Procop GW. Molecular diagnostics for the detection and characterization of microbial pathogens. Clin Infect Dis. 2007;45 Suppl 2:S99–111.
3. Vaitilingom M, Gendre F, Brignon P. Direct detection of viable bacteria, molds, and yeasts by reverse transcriptase PCR in contaminated milk samples after heat treatment. Appl Environ Microbiol. 1998;64(3):1157–60.
4. Horakova K, Mlejnkova H, Mlejnek P. Direct detection of bacterial faecal indicators in water samples using PCR. Water Sci Technol J Int Assoc Water Pollut Res. 2006;54(3):135–40.
5. Kobayashi H, Oethinger M, Tuohy MJ, Procop GW, Hall GS, Bauer TW. Limiting false-positive polymerase chain reaction results: detection of DNA and mRNA to differentiate viable from dead bacteria. Diagn Microbiol Infect Dis. 2009;64(4):445–7.
6. Sabat G, Rose P, Hickey WJ, Harkin JM. Selective and sensitive method for PCR amplification of Escherichia coli 16S rRNA genes in soil. Appl Environ Microbiol. 2000;66(2):844–9.
7. Davis BD, Luger SM, Tai PC. Role of ribosome degradation in the death of starved Escherichia coli cells. J Bacteriol. 1986;166(2):439–45.
8. Klein PG, Juneja VK. Sensitive detection of viable Listeria monocytogenes by reverse transcription-PCR. Appl Environ Microbiol. 1997;63(11):4441–8.
9. Morin NJ, Gong Z, Li XF. Reverse transcription-multiplex PCR assay for simultaneous detection of Escherichia coli O157:H7, Vibrio cholerae O1, and Salmonella Typhi. Clin Chem. 2004;50(11):2037–44.
10. Pommepuy M, Le Guyader F. Molecular approaches to measuring microbial marine pollution. Curr Opin Biotechnol. 1998;9(3):292–9.
11. Bej AK, Mahbubani MH, Dicesare JL, Atlas RM. Polymerase chain reaction-gene probe detection of microorganisms by using filter-concentrated samples. Appl Environ Microbiol. 1991;57(12):3529–34.
12. Sheridan GE, Masters CI, Shallcross JA, MacKey BM. Detection of mRNA by reverse transcription-PCR as an indicator of viability in Escherichia coli cells. Appl Environ Microbiol. 1998;64(4):1313–8.
13. Liu Y, Gilchrist A, Zhang J, Li XF. Detection of viable but nonculturable Escherichia coli O157:H7 bacteria in drinking water and river water. Appl Environ Microbiol. 2008;74(5):1502–7.

5

Investigation on up-flow anaerobic sludge fixed film (UASFF) reactor for treating low-strength bilge water of Caspian Sea ships

Seyyed Mohammad Emadian[1], Mostafa Rahimnejad[2*], Morteza Hosseini[2] and Behnam Khoshandam[1]

Abstract

Background: In order to meet the International Maritime Organization (IMO) objectives, the main purpose of this study was using the cheap and practical wastewater treatment system for low-strength bilge water of Caspian Sea ships; therefore, the low-strength bilge water of the Caspian Sea ships has been treated by up-flow anaerobic sludge fixed film (UASFF) reactor at the ambient temperature.

Results: The reactor operated at two hydraulic retention times (HRTs) of 10 h and 8 h. The organic loading rates (OLR) ranged (0.12-0.6) g chemical oxygen demand (COD)/l.day. At the beginning of the experimental procedure, the sludge was immobilized on the surface of the support materials. After 10 days of batch feeding of the reactor with the wastewater as an acclimation period (with COD removal of 59%), the reactor operated continuously. At the end of the experiment, with the HRT of 8 h and OLR of 0.6 g COD/l.day, the COD and total suspended solid (TSS) removal efficiencies reached the amounts of 75% and 99%, respectively. In addition to the good features of the reactor in removing COD and TSS, the effluent oil concentration was significantly lower than the standard value (15 ppm) which has been laid down for the discharge of the bilge water from ships by the IMO.

Conclusions: The obtained data demonstrated that UASFF reactor is an appropriate system for treatment of a low-strength bilge water.

Keywords: Anaerobic treatment, UASFF reactor, COD, pH, TSS, Oil content

Background

Three kinds of wastewater exist which are produced on ships: black water, grey water and bilge water. Bilge water is the mixture of water, oily fluids, lubricants, cleaning fluids and other similar wastes that accumulate in the lowest part of a ship. The International Maritime Organization (IMO) regulations necessitate that any oil and oil residue discharged in wastewater streams must contain less than 15 mg/l of oil [1]. The common technology is used in ships for treating bilge water is oil water separator (OWS) using the buoyancy difference of oil and water for separation. Cleaning agents in bilge water can create an emulsion of oil in water. When emulsification takes place, buoyancy difference of oil and water is too small to be treated properly via the existing OWS technology.

Other techniques have been studied in order to treat bilge water including membrane technology [2,3], electrocoagulation [4,5], UF/photocatalytic oxidation [6]. Some disadvantages were reported associated with the application of membrane in treatment of bilge water such as: their relatively high cost of production because of the expensive raw materials, fouling which has a number of negative effects such as the reduction in membrane flux, additional capital and maintenance cost due to membrane replacement and regeneration [2,7]. Karakulski et al. reported a promising usage of laboratory-scale ultrafiltration pilot plant with tubular membranes for the treatment of bilge water. However, the use of additional photocatalytic oxidation stage was necessary to eliminate the residual oil [6]. Rincon et al. concluded that the electrocoagulation process was an effective method in destabilization of oil in water emulsions and

* Correspondence: rahimnejad@nit.ac.ir
[2]Department of Chemical Engineering, Babol University of Technology, Babol, Iran
Full list of author information is available at the end of the article

removing of heavy metals. However, the electricity consumption and the use of additional flotation method should be considered for improving the treatment efficiency [5].

Anaerobic treatment is a well-established technology for treatment of wastes and wastewaters because it is technologically simple for low energy consumption and it is an efficient, economical and environmentally-friendly method. The final product of anaerobic digestion is biogas which is a mixture of methane and carbon dioxide. These produced components can be applied for heating and upgrading natural gas quality or co-generation [8]. One of the most notable developments in anaerobic treatment process technology is the up-flow anaerobic sludge blanket (UASB) reactor. The UASB reactor has some positive features, such as short hydraulic retention time that allows high organic loadings. Furthermore, it has a low energy demand and area requirement [9,10]. A major problem of UASB reactor is the long period (several months) required for the formation of granule sludge in the reactor [11]. Although formation of granule in UASB reactors has some advantages, successful treatment of wastewaters with flocculent sludge UASB reactors have been reported [12,13]. The up-flow anaerobic sludge fixed film (UASFF) reactor configuration has combined the advantages of both UASB and Up-flow anaerobic fixed film (UAFF) reactors. This kind of reactor is efficient in the treatment of dilute to high strength wastewaters at low to high Organic Loading Rates [14,15]. The packing medium in the hybrid reactor plays an important role in giving a better performance to the UASB reactor such as increasing solids retention by dampening

short circuiting, improving gas/liquid/solid separation, and providing surface for biomass attachment.

Bilge water is classified as the low strength group of wastewater [14]. Although anaerobic process is used for the treatment of medium and high strength wastewaters, it has already been applied successfully for a number of waste streams including low strength wastewaters [16-18].

In this study, the efficiency of UASFF reactor (on the basis of COD, TSS, oil removal and biogas production) has been studied in treatment of low-strength bilge water under different low organic loading rates at the ambient temperature.

Methods
Experimental system
The schematic diagram of the laboratory-scale UASFF reactor used in this study is presented in Figure 1. The fabricated Plexiglas bioreactor column had an internal diameter of 4.4 cm and a liquid height of 194 cm. The column consisted of three sections including bottom, middle and top. The bottom part of the column, with a volume of 1823 ml operated as a UASB reactor whereas the middle part of the column with a volume of 855 ml was used as a fixed film reactor. The top part of the bioreactor with a volume of 273 ml was an unpacked column prior to the effluent overflow. The fixed film section of the column was randomly packed with 270 billowy pieces of PVC rings with diameter of 15 mm and the height of 13 mm (150 m^2/m^3 specific surface areas for each one). The media in the reactor were stabled by using a plastic mesh. The wastewater as a substrate was continuously fed to the

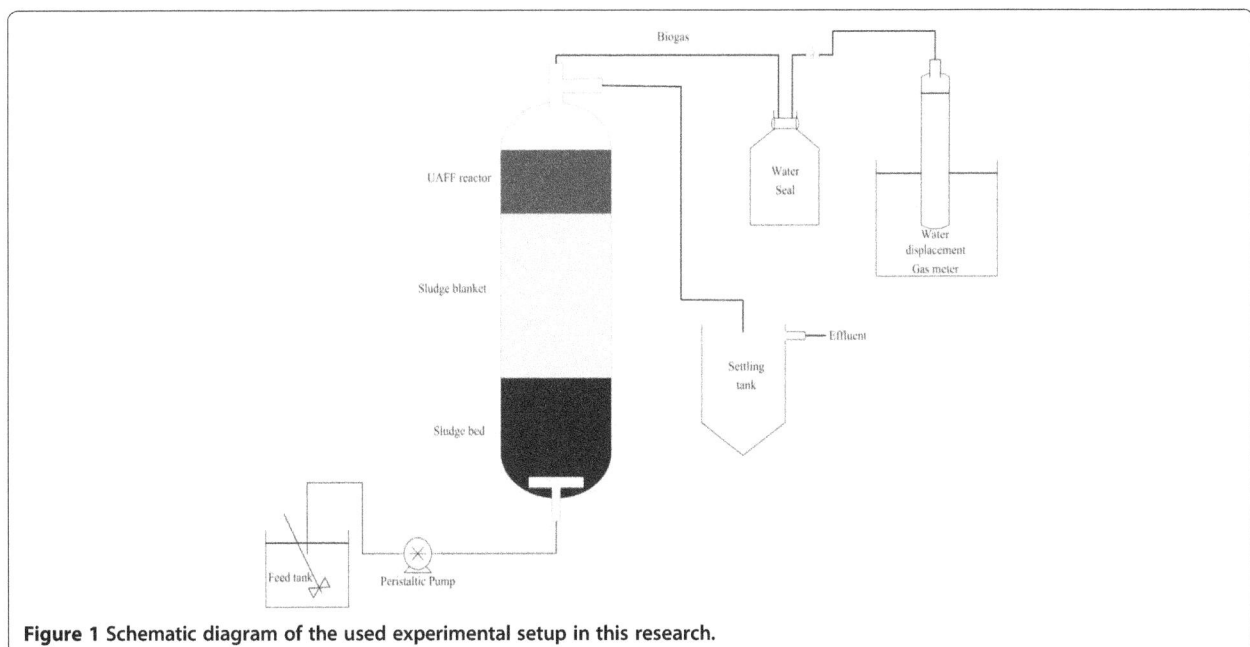

Figure 1 Schematic diagram of the used experimental setup in this research.

base of the reactor, under the bed of active sludge, through a T-inlet connected to a peristaltic pump. An outlet was provided at the top of the reactor that was connected to a 1 liter funnel shaped settling compartment served as a sedimentation part where the final effluent was collected from the top of this tank. The effluent tube was connected to a gas tank for gas collection by water displacement whenever wanted to measure the produced biogas volume. The reactor operated at ambient temperature (15–25)°C.

Wastewater characteristics

The bilge water was collected from a tank with which anchored cargo ships typically discharged their bilge water to it at Amirabad port, Behshahr, Mazandaran, Iran. The samples were collected from the top, middle and the bottom of the tank in order to provide a uniform sample from all parts of the tank. The UASFF reactor was fed with bilge water pre-settled for 10 min. the characteristics of pre-settled bilge water are summarized in Table. 1. The pH of the feed was adjusted to 6.8 to 7.2 by adding diluted HCl. The only supplementary nutrient, $MgNO_3$ as a nitrogen supply, was added to yield a COD: N ratio of 250:5.

Inoculum (seed sludge)

The reactor was seeded with a mixture of activated sludge from the aerobic wastewater treatment of the Mazandaran pulp and paper industry and a non-granular sludge obtained from an up-flow anaerobic sludge blanket reactor operating with cheese whey wastewater from the Gela food industry of Amol, Mazandaran, Iran. The TSS of the mixture was 13 g/l. The non-granular sludge was methanogenically active as the biogas bubbles were apparently observed stripping from the sample surface which was collected in a closed bottle.

Analytical methods

Several monitoring parameters were evaluated during the entire operation, including COD, TSS and oil concentrations, as well as pH, temperature and biogas production volume rate. For COD analysis, HACH's Method 8000, a combination of reactor digestion method and colorimetric method, was used [19]. This method is equivalent to

Table 1 characteristics of pre-settled bilge water; TN and TP were measured in COD = 50 mg/l

Parameter	Value
PH	8 – 9
COD (mg/l)	20 – 200
TS (mg/l)	800 – 2400
TSS (mg/l)	220 – 1760
TN (µg/l)	836
TP (µg/l)	211

standard method 5220D: closed reflux, colorimetric method [20]. Analytical determination of TSS was carried out in agreement with the standard methods for the examination of water and wastewater [20]. Analysis of oil was determined according to USEPA Method 1664, N-Hexane gravimetric method. Temperature and pH were measured using a pH/temperature probe (HANNA, PH212, Germany) with automatic temperature compensation. The method used in pH measurement was generally in compliance with standard method 4500B [20]. Biogas was collected by water displacement and the volume was read from a calibrated gas collection cylinder.

Start-up and operation scheme

Start-up period usually takes a long time. In order to decrease this time, the immobilization of biomass on the support material was done. So, the mentioned mixture of sludge was used by means of a technique described by Zaiat et al. [21]. The support material in combination with the sludge was stored in 1.5 l closed bottle and homogenized for the period of a week by using a shaker so as to secure steadier immobilization of bio-particles in the supporting material. It is noticeable that this initial immobilization of biomass in the support materials has never been done by the other authors. After this stage, the packing material was filled in its place in the UASFF reactor.

The reactor was inoculated with 500 ml of the same sludge mixture. In order to acclimatize the sludge with bilge water, the reactor was daily batch feed with the bilge water (50 mg/l) for 10 days. After each feed, the liquid content of the reactor was continuously circulated for 1 day (until the next feed). The acclimation period permitted oxygen level decrease to prevent inhibition of anaerobic bacteria as well as the bacteria population to adjust with the feed wastewater. The TSS concentration of the sludge after the 10-day batch-fed period was 16.5 g/l. A COD removal of about 59% was achieved at the end of this acclimation period.

The purpose of the start-up of anaerobic bioreactors is to grow, build up and retain a sufficient concentration of active and well balanced biomass. The start-up was carried out by using stepped organic loading to produce the most rapid biomass development. The start-up stage of the process was began by continuous feeding of the reactor with an initial influent COD concentration of 50 mg/l, HRT of 10 h and consequently organic loading rate of 0.12 g COD/l day which is remarkably a low value. This influent COD concentration was applied for 21 days. After that, it is increased to 100 mg/l from the day of 21 to 49. The HRT of 10 h was kept constant throughout the start-up duration. The reactor was allowed to reach steady state condition before each OLR change. When effluent COD reached a relatively constant value, the steady state condition was achieved and then influent OLR can be raised [22]. The experimental procedure is illustrated in Figure 2.

During the experiment, COD reduction, pH and biogas production were monitored daily. The TSS reduction was usually measured every other day. Also oil reduction was checked 2 times throughout the experiment. The first check was after the end of the start-up period and the second check was after the completion of the whole experiment.

Results and discussion
Bioreactor start-up
pH
Changes in acidity (pH) of the effluent from the UASFF reactor during the start-up stage is shown in Figure 3. The pH was comparatively stable (varying from 8.3 to 8.78), which was suitable for efficient methanogenesis, indicating that the system had sufficient alkalinity to neutralize organic acids coming from the hydrolysis and fermentation stage [23]. After 22 days, a sudden decrease in pH from 8.68 to 8.3 took place which is attributed to the accumulation of the produced VFA (Volatile Fatty Acid) because of enhancement of the OLR. Accumulation of VFA in the reactor did not sour the reactor. The similar result was reported by Van Haandel and Lettinga about the treatment of domestic wastewater [24].

COD removal efficiency
The bioreactor performance during the start-up is shown in Figure 4. The reactor was fed with an influent COD of 50 mg/l and the COD removal efficiency was increased from 40% to 68% in the first 21 days. Subsequently, the influent COD concentration was enhanced to 100 mg/l for the remaining 28 days of the start-up period. As the graph shows, increase in influent COD

from 50 mg/l to 100 mg/l caused a decline in the COD removal efficiency from 68% to fewer than 42% which can be attributed to the fact that the system was put under stress. This phenomenon can be due to the increase in VFAs concentration which is recognizable from sudden decrease in effluent pH at day 22. Similar observation was reported by other authors [23,25,26]. The system recovered shortly and adapted to the new condition with time. Though, in terms of removal efficiency, the increase in influent COD from 50 mg/l to 100 mg/l led to an increment in terms of the COD removal efficiency from 68% to 77%, implying that the sludge was acclimated appropriately to the bilge water. A comparison between Figures 2 and 3 shows a similar trend between effluent pH and COD removal efficiency which concurs with the results obtained by Zhang et al. [23].

Biogas production rate
The biogas production rate along the start-up is shown in Figure 5. As the profile shows, the biogas volume rate increased from 0.06 l/day to 0.37 l/day in the first 21 days. The introduction of higher COD to the reactor was caused a sudden decrease in biogas production at day 22. The excess VFA which was produced at this time inhibited the methanogenic bacteria from their efficient performance and as a consequence, the biogas production decreased [27,28]. However, the biogas production increased again from day 26 and this indicated that the microorganisms acclimated to the new condition. At the end of the start-up period, the biogas production reached an amount of 0.48 l/day. During the start-up, the biogas production raised like the COD removal efficiency which was in agreement with another author's result [28]. The reason for the increased biogas production is due to proper anaerobic population development [26].

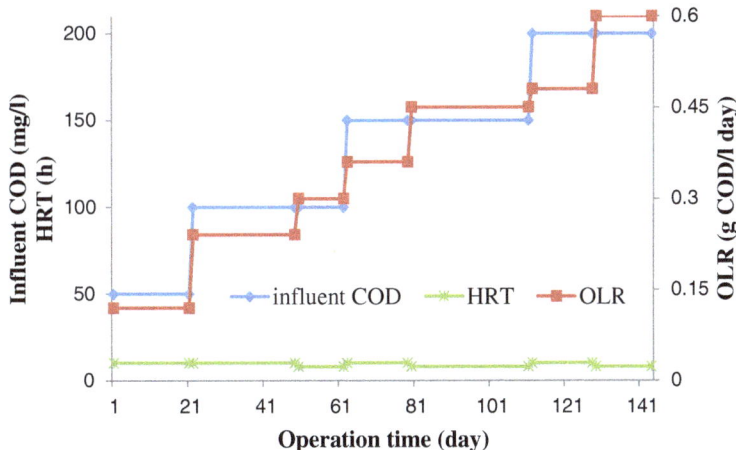

Figure 2 Start-up and operation scheme for UASFF reactor.

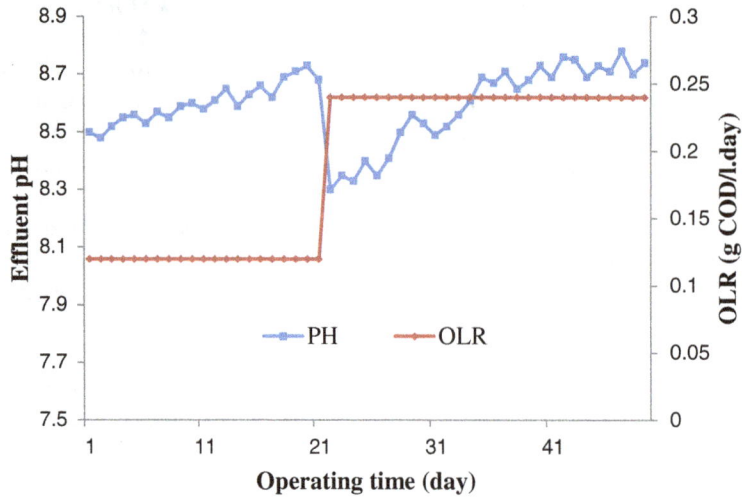

Figure 3 Change of PH during start-up.

The overall performance of the reactor during the startup was satisfactory. It is known that the selection of seed material plays a crucial role in minimizing the time required for start-up duration [26]. In addition, it is clearly understood that the initial immobilization of microorganisms on the surface of the support materials had a key role in progressing the start-up procedure.

Later operation stage

After a 49-days startup period, the reactor was operated at HRTs of 10 h and 8 h with three different influent COD concentrations (from 100 mg/l to 200 mg/l) to evaluate the effect of low organic loadings on the reactor performance.

pH

Figure 6 shows the variation of effluent pH during the operation. As it shows, the pH of the treated wastewater was in the range of 8.04-8.61 which is indicative of the buffering capacity of the reactor. There was a sudden decrease in pH from 8.52 at the day of 96 to 8.1 at the day of 111 because at this stage of the operation, the effect of the nutrient was tested. For testing this effect, from the day of 96 to 101, the addition of the nutrient was ceased and after that the new nutrient, NH_4Cl, was added to the reactor till the day of 111. Decrease in pH in this period of the operation proved that more VFA was accumulated in the reactor due to the lower activity of the methanogenic bacteria which was responsible for

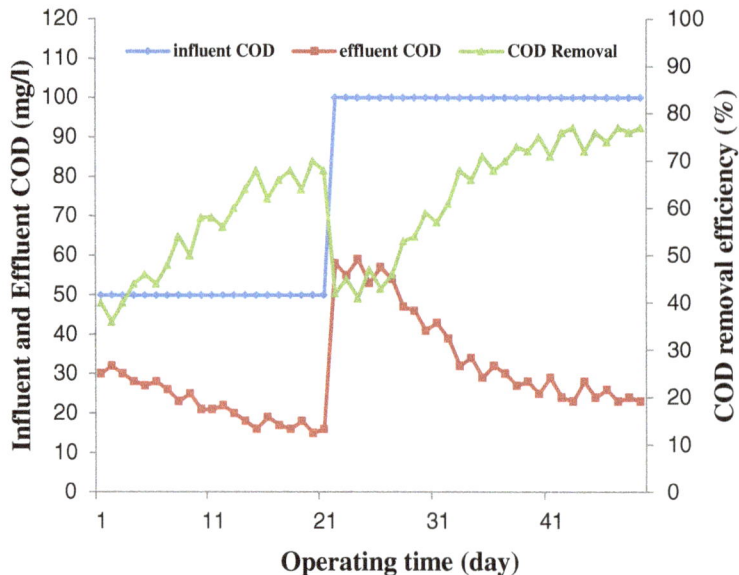

Figure 4 Bioreactor performance during start-up period.

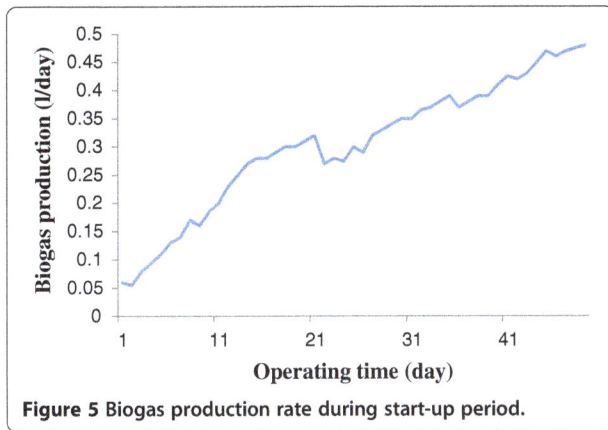

Figure 5 Biogas production rate during start-up period.

consuming of the VFA. However, the reactor recovered itself because of introducing $MgNO_3$ as the nutrient to the reactor and pH increased again which is indicative of the increase in methanogenic bacteria functionality.

COD and TSS removal efficiencies

The performance of UASFF reactor based on COD and TSS removal efficiencies during the operation period is shown in Figure 7 and Figure 8, respectively. As illustrated in the Figure 7, the COD removal efficiency went through an increasing trend from a low amount of 59% to a maximum of 77% during the first 46 days except at the beginning of each OLR increment, there was a corresponding decrease in COD removal efficiency but the system recovered shortly and adapted to the new conditions with time like the start-up period [14,26]. As Figure 8 illustrates, the influent TSS concentration is unstable because of the poor agitation that was provided in feed tank. As the graph shows, the effluent TSS concentration was very low which is indicative of the good performance of the reactor in eliminating the suspended solids. As it was mentioned before, the effect of the

nutrient on the performance of the reactor was tested during the days of 96 to 111. According to Figure 7, the COD removal efficiency decreased from 77% to 42% during the days 96 to 101, the period that the addition of $MgNO_3$ was ceased to the reactor. After that, by addition of new nutrient (NH_4Cl) to the reactor, the COD removal efficiency increased a little and reached an amount of 50% at day 111. The obtained data demonstrated that $MgNO_3$ was a better choice than NH_4Cl in the present study. Therefore, $MgNO_3$ was introduced to the reactor again from day 111. Although the COD removal efficiency decreased considerably during the days of 96 to 111, the TSS removal efficiency was still as high as the other days of the operation (see Figure 8). For instance at day 98, the COD removal efficiency declined to the amount of 40% while the TSS removal efficiency was 97%. This phenomenon indicates that most of the COD removal during this study was due to the reduction of the soluble COD and not the suspended COD. By increasing the COD influent and introduction of $MgNO_3$ as the nutrient to the reactor at the day of 111, the COD removal efficiency raised again and it reached an amount of the 75% at the end of the study. The obtained result is comparable with the COD removal efficiency achieved by Sun et al. in which they reached the percentages of 59% in treatment of synthetic bilge water by using an aerobic moving bed bio-reactor (MBBR) [29]. In addition, the reactor achieved TSS removal efficiency of 99% at the end of the experiment which was a noticeable result. The good performance of the reactor in the eliminating of the TSS content of the wastewater during the operation suggests that most proportion of the TSS removal is due to the entrapment and adsorption of the suspended solids at sludge bed and fixed film [30]. The TSS removal efficiency throughout the experiment did not differ significantly which is in agreement with Ligero et al. results who reported the TSS removal efficiency of

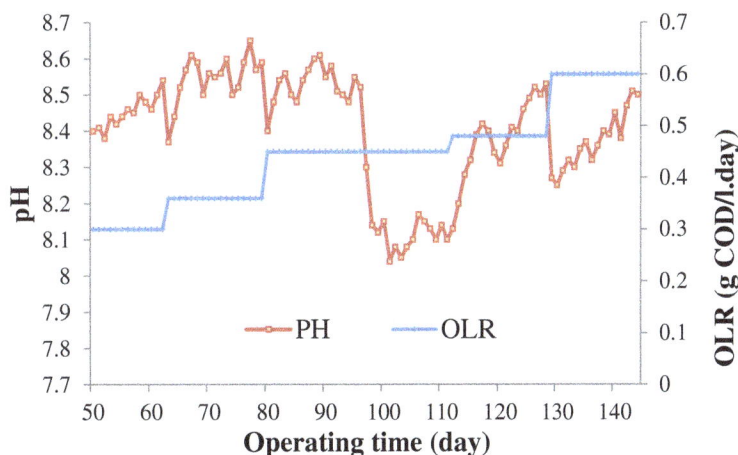

Figure 6 Change of pH during later operation.

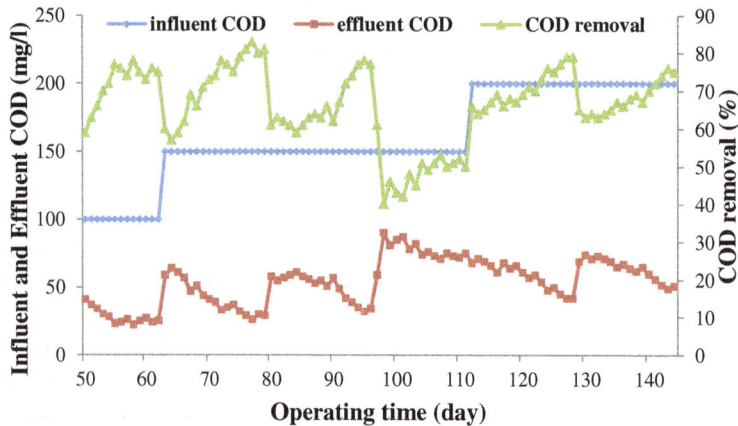

Figure 7 Bioreactor performance during the later operation stage.

an UASB reactor for all values of HRT was not very different [31].

Biogas production rate

As Figure 9 shows, the biogas production rate increased from 0.52 l/day at the day of 50 to 0.85 l/day at the day of 96. Lettinga reported that the reduction of BOD and COD contributed to the gas production [32]. One can see in Figure 9, the biogas production decreased from 0.85 l/day at the day of 96 to 0.41 l/day at day of 101 (without nutrient addition) and then it reached to the amount of 0.53 at the day of 111 (with addition of NH_4Cl as the nutrient) which can explain that the activity of methanogenic bacteria decreased at this stage of the operation. By increasing the influent COD and the addition of $MgNO_3$ as the nutrient at the day of 111, the biogas production increased again and it continued to the amount of 0.93 l/day at the end of the study.

Oil content

The reduction of oil content of the wastewater at the end of start-up and operation of the reactor is shown in

Figure 10. Presence of oil in wastewaters leads to the accumulation of it on the surface of the sludge which causes foaming and scum formation which eventually lowers the digestion efficiency [10]. There was no sign of foam and scum in the reactor which was indicative of the good performance of the reactor. As Figure 10 shows, either at the end of the start-up or at the end of the operation, the oil effluent concentration was below 15 mg/l which is IMO standard level for discharging the wastewater from ships [1]. The obtained result was so promising in comparison with the outcome of the Sun et al. [29]. They reported that the effluent oil content from the MBBR at the HRT of 8 h was about 30 mg/l which was about double times higher than the standard level of discharging [29].

Sludge

The TSS concentration of the sludge in the reactor increased from 16.5 g/l at the beginning of the start-up to 67 g/l at the end of the study. This sludge production in the reactor may be attributed to (1) flocculation and entrapment of the non-biodegradable influent TSS, forming

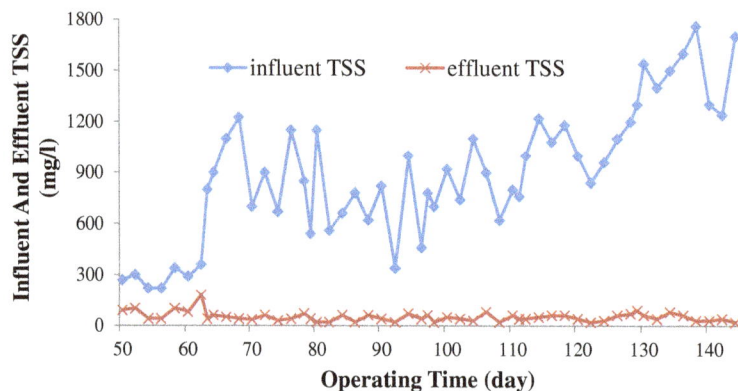

Figure 8 TSS removal during operation.

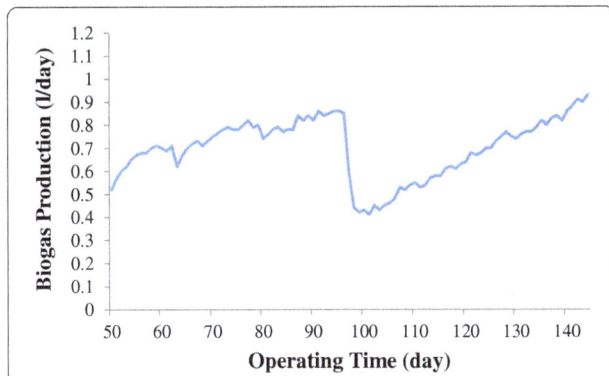

Figure 9 Biogas production during later stage.

Figure 11 Influent and effluent of the reactor at the end of the operation.

the inert sludge mass fraction and (2) the biological sludge mass that is generated as a result of anaerobic conversion in the hybrid reactor but because of the mentioned reasons in COD and TSS removal section, the entrapment of the suspended solids in the sludge seems to have more effect on increasing the TSS content of the reactor sludge. So, the sludge acted as a filter for removing the suspended solids from the wastewater [33]. Therefore, the UASB reactor had a noticeable effect on removing the TSS content of the wastewater [34-36]. At the end of this study, a flocculent sludge was observed without any granule formation in it. As the other authors reported, low strength wastewater can lead to substrate transfer limitation and cause inhibition of granulation or can make it difficult to maintain granules [37,38].

Wastewater appearance

Figure 11 illustrates the apparent difference between influent and effluent of the reactor at the end of the operation. As the Figure 11 shows, the reactor had a good performance in decolorizing of the wastewater which can be as another advantage of the reactor.

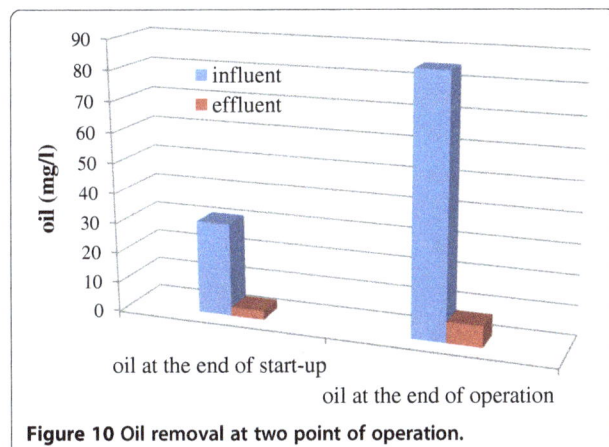

Figure 10 Oil removal at two point of operation.

Conclusions

In this study, anaerobic treatment of dilute bilge water was performed by using UASFF reactor at ambient temperature. After a good resulted immobilization of sludge in the support materials and start-up period, the COD and TSS removal efficiencies reached the amounts of 75% and 99% at the end of the operation, respectively. The results showed that the sludge blanket acted as a filter for removing the suspended solids from the wastewater and the major proportion of COD removal was due to the soluble and not suspended COD. The biogas production rate reached an amount of 0.93 l/day at the end of the experiment and effluent oil concentration is remarkably below the standard amount which has been set by the IMO (15 ppm). The good performance of the bioreactor on appearance of the wastewater can be considered as another advantage of this type of the UASFF reactor. The immobilization of the biomass in the support materials had an important role in reducing the influent COD because they created a good media for methanogenic bacteria on their surface. According to the obtained results, it can be concluded that the UASFF reactor is a very promising option for the treatment of the low-strength bilge water, produced from the ships in Caspian Sea, at the ambient temperatures for implementation on the ships in a large scale.

Competing interests
The authors declare that they have no competing interests.

Authors' contributions
SME participated in fabricated of the set up plant and contributed to obtained data. MR participated in the fabrication of set up, carried out the experiments and optimization and drafted the manuscript. MH participated in modeling process and helped to draft manuscript. BK participated in the experiments and helped to draft manuscript. All authors have contribution in this research Also all of them read and approved the final format of this manuscript.

Acknowledgments
The authors wish to acknowledge Biofuel & Renewable Energy Research Center, Noshirvani University of Technology (Babol, Iran) for the facilities provided to accomplish the present research.

Author details
[1]Department of Chemical Engineering, Semnan University, Semnan, Iran.
[2]Department of Chemical Engineering, Babol University of Technology, Babol, Iran.

References
1. MARPOL:International Convention for the Prevention of Pollution from Ships, 1973, as modified by the protocol of 1978 relating thereto (MARPOL 73/78). in: (IMO) I.M.O., ed; 1973

2. Ghidossi R, Veyret D, Scotto JL, Jalabert T, Moulin P. Ferry oily wastewater treatment. Sep Purif Technol. 2009;64:296–303.

3. Peng H, Tremblay AY, Veinot DE. The use of backflushed coalescing microfilteration as pretreatment for the ultrafilteration of bilge water. Desalination. 2005;181:109–20.

4. Korbahti BK, Artut K. Electrochemical oil/water demulisification and purification of bilge water using Pt/Ir electrodes. Desalination. 2010;258:219–28.

5. Rincon GJ, La Motta EJ. Simultaneous removal of oil and grease, and heavy metals from artificial bilge water using electrocoagulation/flotation. J Environ Manage. 2014;144:42–50.

6. Karakulski K, Morawski WA, Grzechulska J. Purification of bilge water by hybrid ultrafiltration and photocatalytic processes. Sep Purif Technol. 1998;14:163–73.

7. Benito JM, Sanchez MJ, Pena P, Rodriguez MA. Development of a new high porosity ceramic membrane for the treatment of bilge water. Desalination. 2007;214:91–101.

8. Weiland P. Biogas production: current state and perspectives. Appl Microbiol Biotechnol. 2010;85:849–60.

9. Kivaisi AK. The potential for constructed wetlands for wastewater treatment and reuse in developing countries: a review. Ecol Eng. 2001;16:545–60.

10. Tchobanoglous G, Burton FL, Stensel HD. Wastewater Engineering: Treatment and Reuse. New York: McGraw-Hill Education; 2003.

11. Liu Y, Tay JH. State of the art of biogranulation technology for wastewater treatment. Biotechnol Adv. 2004;22:533–63.

12. Goodwin JAS, Finlayson JM, Low EW. A further study of the anaerobic biotreatment of malt whisky distillery pot ale using an UASB system. Bioresour Technol. 2001;78:155–60.

13. Sabry T. Application of the UASB inoculated with flocculent and granular sludge in treating sewage at different hydraulic shock loads. Bioresour Technol. 2008;99:4073–7.

14. Kumar A, Yadav AK, Sreekrishnan TR, Satya S, Kaushik CP. Treatment of low strength industrial cluster wastewater by anaerobic hybrid reactor. Bioresour Technol. 2008;99:3123–9.

15. Chan YJ, Chong MF, Law CL, Hassell DG. A review on anaerobic–aerobic treatment of industrial and municipal wastewater. Cheml Eng J. 2009;155:1–18.

16. Aiyuk S, Amoako J, Raskin L, van Haandel A, Verstraete W. Removal of carbon and nutrients from domestic wastewater using a low investment, integrated treatment concept. Water Res. 2004;38:3031–42.

17. Gomec CY. High-rate anaerobic treatment of domestic wastewater at ambient operating temperatures: A review on benefits and drawbacks. J Environ Sci Health A Tox Hazard Subst Environ Eng. 2010;45:1169–84.

18. Leitao RC, Santaellla ST, van Haandel AC, Zeeman G, Lettinga G. The effect of operational conditions on the hydrodynamic characteristics of the sludge bed in UASB reactors. Water Sci Technol. 2011;64:1935–41.

19. DR/890:colorimeter, Procedures Manual, Method 8000. in: Hach Company L., CO, ed; 2009

20. APHA. Standard methods for the examination of Water and Wastewater. Washington, DC: American Public Health Association/American Water Work Association/Water Environmental Federation; 2008.

21. Zaiat M, Cabral AKA, Foresti E. Horizontal-flow anaerobic immobilized sludge reactor for wastewater treatment: conception and performance evaluation. Revista Brasileira de Engenharia. 1994;11:33–42.

22. Sunil Kumar G, Gupta SK, Singh G. Biodegradation of distillery spent wash in anaerobic hybrid reactor. Water Res. 2007;41:721–30.

23. Zhang Y, Yan L, Chi L, Long X, Mei Z, Zhang Z. Startup and operation of anaerobic EGSB reactor treating palm oil mill effluent. J Environ Sci. 2008;20:658–63.

24. Van Haandel AC, Lettinga G. Anaerobic sewage treatment- a practical guide for regions with a hot climate. England: John Wiley & Sons; 1994.

25. Najafpour GD, Zinatizadeh AAL, Mohamed AR, Hasnain Isa M, Nasrollahzadeh H. High-rate anaerobic digestion of palm oil mill effluent in an upflow anaerobic sludge-fixed film bioreactor. Process Biochem. 2006;41:370–9.

26. Selvamurugan M, Doraisamy P, Maheswari M. An integrated treatment system for coffee processing wastewater using anaerobic and aerobic process. Ecol Eng. 2010;36:1686–90.

27. Buyukkamaci N, Filibeli A. Volatile fatty acid formation in an anaerobic hybrid reactor. Process Biochem. 2004;39:1491–4.

28. Chan YJ, Chong MF, Law CL. An integrated anaerobic–aerobic bioreactor (IAAB) for the treatment of palm oil mill effluent (POME): Start-up and steady state performance. Process Biochem. 2012;47:485–95.

29. Sun C, Leiknes T, Weitzenbock J, Thorstensen B. Development of an integrated shipboard wastewater treatment system using biofilm-MBR. Sep Purif Technol. 2010;75:22–31.

30. Tawfik A, Sobhey M, Badawy M. Treatment of a combined dairy and domestic wastewater in an up-flow anaerobic sludge blanket (UASB) reactor followed by activated sludge (AS system). Desalination. 2008;227:167–77.

31. Ligero P, de Vega A, Soto M. Influence of HRT (hydraulic retention time) and SRT (solid retention time) on the hydrolytic pre-treatment of urban wastewater. Water Sci Technol. 2001;44:7–14.

32. Lettinga G. Anaerobic digestion and wastewater treatment systems. Antonie Van Leeuwenhoek. 1995;67:3–28.

33. Nadais H, Capela I, Arroja L, Duarte A. Optimum cycle time for intermittent UASB reactors treating dairy wastewater. Water Res. 2005;39:1511–8.

34. Álvarez JA, Ruíz I, Soto M. Anaerobic digesters as a pretreatment for constructed wetlands. Ecol Eng. 2008;33:54–67.

35. Green M, Shaul N, Beliavski M, Sabbah I, Ghattas B, Tarre S. Minimizing land requirement and evaporation in small wastewater treatment systems. Ecol Eng. 2006;26:266–71.

36. Ruiz I, Díaz MA, Crujeiras B, García J, Soto M. Solids hydrolysis and accumulation in a hybrid anaerobic digester-constructed wetlands system. Ecol Eng. 2010;36:1007–16.

37. Aiyuk S, Verstraete W. Sedimentological evolution in an UASB treating SYNTHES, a new representative synthetic sewage, at low loading rates. Bioresour Technol. 2004;93:269–78.

38. Aiyuk S, Xu H, van Haandel A, Verstraete W, Verstraete W. Removal of ammonium nitrogen from pretreated domestic sewage using a natural ion exchanger. Environ Technol. 2004;25:1321–30.

6

Optimizing photo-mineralization of aqueous methyl orange by nano-ZnO catalyst under simulated natural conditions

Ahed Zyoud[1*], Amani Zu'bi[1], Muath H. S. Helal[2], DaeHoon Park[3], Guy Campet[4] and Hikmat S. Hilal[1]

Abstract

Background: Photo-degradation of organic contaminants into non-hazardous mineral compounds is emerging as a strategy to purify water and environment. Tremendous research is being done using direct solar light for these purposes. In this paper we report on optimum conditions for complete mineralization of aqueous methyl orange using lab-prepared ZnO nanopowder catalyst under simulated solar light.

Results: Nano-scale ZnO powder was prepared in the lab by standard methods, and then characterized using electronic absorption spectra, photolumenscence emission (PL) spectra, XRD, and SEM. The powder involved a wurtzite structure with ~19 nm particles living in agglomerates. Photo-degradation progressed faster under neutral or slightly acidic conditions which resemble natural waters. Increasing catalyst concentration increased photodegradation rate to a certain limit. Values of catalyst turn over number and degradation percentage increased under higher light intensity, whereas the quantum yield values decreased. The photocatalytic efficiency of nano-ZnO powders in methyl orange photodegradation in water with solar light has been affected by changing the working conditions. More importantly, the process may be used under natural water conditions with pH normally less than 7, with no need to use high concentrations of catalyst or contaminant. The results also highlight the negative impact of possible high concentrations of CO_2 on water purification processes. Effects of other added gaseous flows to the reaction mixture are also discussed.

Conclusion: ZnO nano-particles are useful catalyst for complete mineralization of organic contaminants in water. Photo-degradation of organic contaminants with ZnO nano-particles, methyl orange being an example, should be considered for future large scale water purification processes under natural conditions.

Keywords: Methyl orange, Contaminant mineralization, Solar simulated light, ZnO nanopowder

Introduction

Purification of water from hazardous chemicals is an important research area. Organic contaminants, such as industrial dyes, halocarbons and phenol derivatives, are among the main contaminants that demand complete safe removal [1]. Different strategies are being investigated for water remediation, including biological treatment [2, 3], ultra-filtration [4], adsorption methods [5] and others. Such methods may not be favored as they may not cause complete mineralization of the organic contaminant. They simply transfer the pollutant from one phase to another [6]. Advanced Oxidation Processes (AOP) have been proposed as alternative routes for water purification. Among those, oxidation via ozone or hydrogen peroxide has been reported as an effective technique [7–11]. Unfortunately, such methods may be costly, as ozonation demands artificial UV radiations, and hydrogen peroxide is not available free of charge. Contaminant complete mineralization with natural solar light seems to be the most practical process for future water purification. A semiconductor photo-catalyst speeds up the action of light by first absorbing photon and producing electrons and holes [12]. With the abundance of cost-less solar radiations, a low cost catalyst may thus be useful. Different semiconducting materials, in the powder form, have been assessed as photo-catalysts [13, 14]. TiO_2 in its anataze form is the most widely used effective photo-catalyst for its high efficiency, photochemical

* Correspondence: ahedzyoud@najah.edu
[1]SSERL, Department of Chemistry, An-Najah National University, Nablus, Palestine
Full list of author information is available at the end of the article

stability, non-toxic nature and low cost. It has been described for degradation of a wide range of organic contaminants [15–22]. Zinc oxide ZnO is a semiconductor with a comparable band gap ~3.2 eV (with wavelength shorter than 400 nm), but has been investigated to a lesser extent in water purification. ZnO is evaluated in many advanced applications such as field-effect transistors, lasers, photodiodes, chemical and biological sensors and solar cells, but to a lesser extent in photo-degradation catalysis [23–26]. One main advantage for ZnO is that it absorbs a larger fraction of solar spectrum, than TiO_2 does [27]. The performance of ZnO in degrading a number of organic contaminants has been reported [28, 29]. The quantum efficiency of ZnO nano-particles in photo-degrading organic contaminants process is higher than that of TiO_2 [30, 31], due to its higher absorptivity in waves shorter than 400 nm, which accounts to about 5 % of the reaching solar light.

In this communication the photo-catalytic activity of ZnO powder in complete mineralization of organic contaminants is revisited, focusing on finding optimum conditions that yield highest contaminant removal, for the first time. For application purposes, it is necessary to emulate natural conditions of contaminated waters in terms of pH, low contaminant concentration, low allowable catalyst amount and moderate water temperatures. Influence of CO_2 and other gas flows on methyl orange photo-degradation process will also be assessed here for the first time. All such reaction parameters will be studied in order to assess feasibility of using ZnO activated photo-degradation of methyl orange in water under simulated natural conditions. To assess feasibility of ZnO catalyst to function on its own in simulated natural waters, multiple use of the added ZnO to photo-degrade fresh methyl orange samples will also be investigated. All such studies are being investigated for the first time.

Methyl orange, with molecular structure shown in Fig. 1, is a dye that is believed to be mutagenic [32]. It slightly dissolves in water. Its color changes with pH, from yellow (at pH higher than 4.4) to red (at lower pH values), and therefore it is used as an indicator [33]. Methyl orange is also used as a dye in textile industry [34]. It is an example of the widely spread azo dies [35], which are resistant to complete biodegradation [36]. For these reasons, methyl orange is commonly used as a model dye to study in environmental cleanup, and this work is no exception.

In earlier study [37, 38], we reported on using commercial ZnO powders as catalysts in photo-degradation of methyl orange. As mentioned above, this work is intended to find optimum conditions for using ZnO nano-particles in methyl orange photo-degradation under natural water conditions, where ZnO is added to contaminated waters and allowed to function on its own under direct solar light.

Experimental

Chemicals

Hydrochloric acid, sodium hydroxide, and methyl orange were purchased from Merck. ZnO powder was prepared in the lab to obtain small particle sizes, as described earlier [39, 40]. A $ZnCl_2$ solution (250 mL, 0.25 M) was drop-wise added (within 40 min) to NaOH solution (200 mL, 0.90 M) with continuous stirring. The system was then left to settle, and the supernatant was decanted. The resulting precipitate was washed with water many times to remove any remaining ions. Enough amount of distilled water was added to convert the precipitate into slurry. The slurry was centrifuged at 6000 rpm for 10 min, and the supernatant was carefully decanted leaving the solid catalyst which was dried at 120 °C.

The CO_2 gas was prepared by adding concentrated HCl solution (5 M) drop-wise to Na_2CO_3 solid in a stoppered flask with only one outlet. The outlet was connected with a glass tube bubble the CO_2 through the reaction mixture at a flow rate 90 mL/L.

Equipment

A 400 W Osram Tungsten Halogen lamp was used as a source for solar simulator light. The lamp spectrum is a

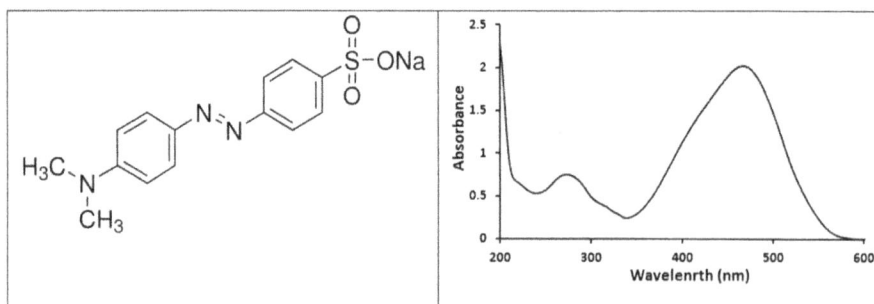

Fig. 1 Molecular structure and UV–vis absorption spectrum of methyl orange

bell curve typical with little (~5 %) in the UV region, just like natural solar light that reaches earth. A light meter (Model lx-102) from Lutron was used to measure the radiation intensity at the reaction mixture surface. A Shimadzu UV-1601 spectrophotometer was used to measure remaining methyl orange concentration using calibration curves, pre-prepared at different pH values, as methyl orange spectra may change with pH value.

The electronic absorption spectra were measured on a Shimadzu UV-1601 spectrophotometer for ZnO powders as suspensions in minimal water amounts. Photoluminescence (PL) Emission Spectra were measured for aqueous suspensions of ZnO powder on a Perkin-Elmer LS50 Luminescence Spectrometer. Excitation wavelength 325 nm was used. XRD patterns were measured on a Philips XRD XPERTPRO diffracto-meter with Cu K_α radiation ($\lambda =$ 1.5418 Å) located in the labs of Dansuk Industrial Co., LTD., South Korea. Field Emission-Scanning Electron Micrographs (FE-SEM) were measured on a Jeol Model JSM-6700 F microscope, in the labs of Dansuk Industrial Co., LTD. South Korea. Atomic absorption spectra (AAS) were used to measure zinc ions resulting from possible degradation of ZnO. The AAS results were measured on an ICE3000 Thermoscientific Atomic Absorption Spectrophotometer equipped with a zinc lamp.

Catalytic experiment

Photo-catalytic experiments were conducted under direct irradiation from the solar simulator lamp. Water samples pre-contaminated with known concentrations of methyl orange were placed inside a 250 mL beaker. Known nominal amounts of the catalyst ZnO powder were added. The mixture was magnetically stirred with a magnetic bar for 15 min in the dark, to allow adsorption equilibrium and to assess amounts of adsorbed methyl orange on the solid catalyst. The pH of the reaction mixture was controlled by adding drops of dilute HCl or NaOH solutions. The solar simulator lamp, vertically clamped above the solution with an adjustable stand, was then switched on with continuous stirring. The desired irradiation intensity on the mixture surface was achieved by controlling the lamp distance. The reaction time was calculated the time the lamp was switched on. Certain experiments were conducted in duplicate to check the reproducibility of the process.

The reaction progress was followed by measuring the amount of remaining methyl orange with time. This was performed by syringing out small aliquots of reaction mixture at certain times. The aliquots were then centrifuged at high speed (5000 rpm) for 5 min. The liquid phase was then carefully syringed out and analyzed spectrophotometrically at 480 nm.

Results and discussion

Zinc oxide powder characterization

Solid state electronic absorption spectra were measured for the prepared ZnO nano-powder as aqueous suspension. ZnO showed absorption with a maximum at ~368 nm (equivalent to 3.37 eV), Fig. 2. The wide band gap is presumably due to small average particle size, since smaller particles exhibit wider band gap values [41].

The photoluminescence emission spectrum was measured for lab-prepared ZnO powder as dispersion in water, using excitation wavelength 325 nm. The ZnO suspension showed an emission peak at ~385 nm (~3.2 eV), Fig. 3. The other two emission peaks at ~445 nm and ~483 nm are attributed to the presence of oxygen vacancies which cause crystal imperfections, as reported earlier [42]. The value of the band gap is not far different from that measured by electronic absorption spectra discussed above.

Figure 4 shows the XRD patterns for ZnO nano-powder. The ZnO clearly involves a wurtzite form, based on comparison with earlier literature results [43]. Based on Scherrer equation approximations, the average particle size was calculated to be ~19 nm.

SEM was used to study the surface morphology and to further estimate the particle size of prepared ZnO powders, Fig. 5. The micrograph showed elongated nano-rods (with rice-shape) of ZnO agglomerates. The agglomerates involved nano-particles of less than 20 nm size.

Methyl orange photo-degradation

Exposure of aqueous solutions of methyl orange to solar simulator lamp, in the presence of ZnO nano-powder, caused appreciable de-colorization of methyl orange solution in soundly short times. Control experiments conducted in the dark, using catalyst while keeping other conditions the same, showed no detectable de-colorization. This means that no degradation occurred in the dark, and that the ZnO powder adsorbs only little fraction of the contaminant. Control experiments conducted with irradiation in the absence of ZnO did not show any noticeable de-colorization with prolonged exposure. This indicates the necessity of the ZnO particles to activate the methyl orange

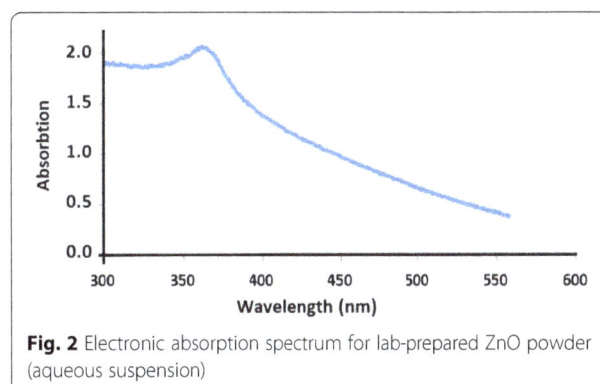

Fig. 2 Electronic absorption spectrum for lab-prepared ZnO powder (aqueous suspension)

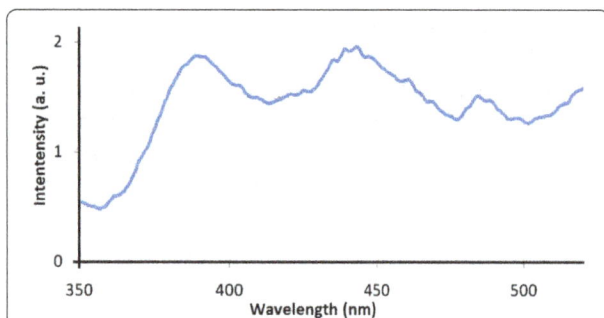

Fig. 3 Photoluminescence emission spectrum measured for ZnO powder (aqueous suspension)

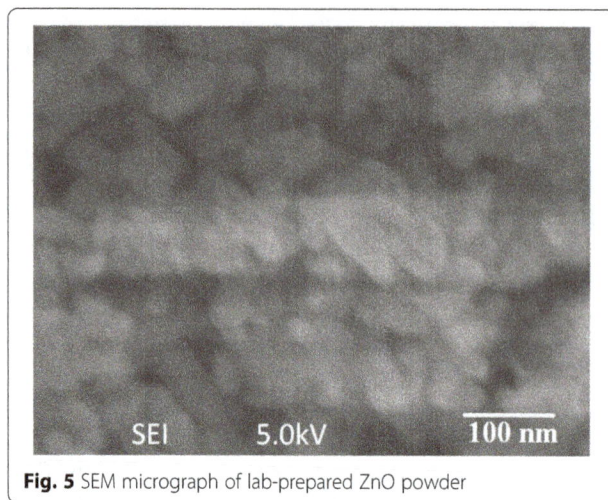

Fig. 5 SEM micrograph of lab-prepared ZnO powder

degradation. Using a cut-off filter (that blocks light 400 nm and shorter) caused severe lowering in methyl orange degradation. Collectively, these results indicate that light waves shorter than 400 nm are the driving force for the degradation of methyl orange, and that the ZnO particles are needed to observe the degradation process. Degradation is thus due to the shorter wave length tail available in the lamp light. With a band gap more than 3.2 eV, ZnO powder catalyst employs photons with wavelength shorter than 400 nm in the photo-degradation process.

Therefore, methyl orange de-colorization here, even in case of incomplete process, is due to complete mineralization of the degraded molecules. This was evident from the disappearance of absorbance bands in the range 200–300 nm (characteristic for the phenyl group) and the range 400–500 nm (characteristic for the azo group). When left for enough time, complete removal and complete mineralization were observed, as shown in Fig. 6 below. Therefore, the reacted phenyl group is believed to be completely degraded leaving no organic products. The reacted azo group is also believed to escape as N_2 gas [44, 45]. Complete photo-mineralization of methyl orange molecules lost under the working conditions is well documented [37, 46–54].

In order to assess the ability of ZnO to photo-degrade aqueous methyl orange under simulated natural conditions

different parameters were investigated here. Effects of different reaction parameters, such as pH, catalyst concentration and contaminant concentration, on rate of photodegradation were reported earlier [55]. The parameters (initial nominal pH, methyl orange concentration and ZnO amount) have been revisited here together with other new parameters (aqueous CO_2 concentration, temperature, light intensity and catalyst reuse).

The remaining methyl orange concentrations were plotted with exposure time. The catalytic efficiency is better understood in terms of turnover number, T.N. (degraded molecules/Zn atoms) and quantum yield, Q.Y., (degraded molecules/total incident photons) measured after 30 min exposure to radiation.

Effect of pH on photo-degradation process

Because the amphoteric nature of ZnO, it is necessary to study the effect of the pH value on the methyl orange photo-degradation process. This is also necessary as

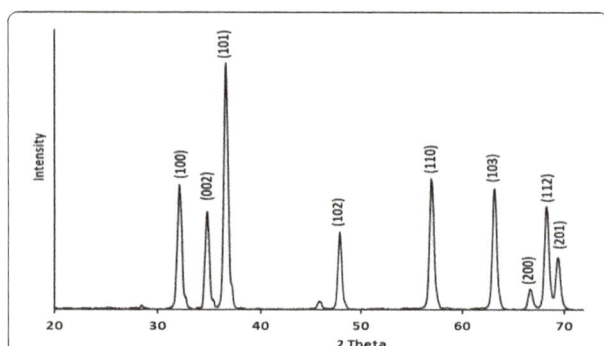

Fig. 4 XRD pattern measured for lab-prepared ZnO powder

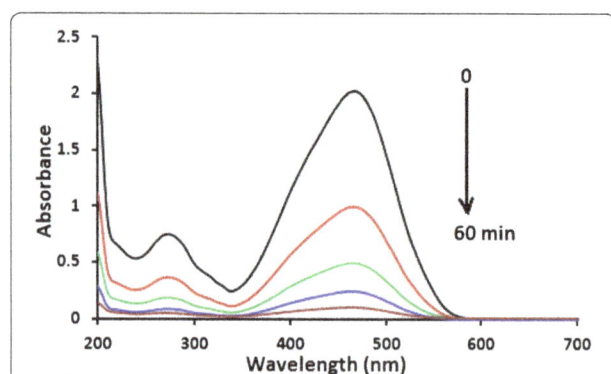

Fig. 6 Electronic absorption spectra showing continued mineralization of methyl orange under photo-degradation conditions. Reaction was conducted using methyl orange solution (100 mL, 10 ppm), at 20 °C, under 19.0 mW/cm^2 total radiation intensity, pH ~7, and ZnO (0.1 g)

natural waters normally have pH values in the range 5–8 [56, 57].

The pH value affects the ZnO surface OH groups [58], the methylene orange and the aqueous solution species. The pH value affects the generation of the oxidizing species ($\cdot$OH, $O_2\cdot^-$, H_2O_2 and $HO_2\cdot$) that result in the reaction system [59, 60]. The nature of methyl orange molecule varies with pH value, as stated above. ZnO has a point of zero charge at pH 9.0, above which the ZnO surface is predominantly negatively charged [13]. The electrical properties of the ZnO surface may thus vary with the pH.

Experiments were carried out at pH values of (2.5, 5, 7, 9 and 11), Fig. 7. The slightly acidic solution (pH ~5) showed highest T.N. and Q.Y. values, followed by neutral solution. Both solutions with pH 5 and 7 showed complete removal of methyl orange after 60 min of irradiation. As discussed above removal of methyl orange involves complete mineralization, which shows the practicality of using the photo-degradation in natural systems. More acidic or basic solutions (less 2.5 or higher than 9) showed T.N. and Q.Y. values nearly half of those under mild solutions. Table 1 summarizes these results.

In basic media, ZnO becomes $Zn(OH)_2$ form with lower semiconducting properties [61]. Under lower pH conditions the lowering in removal efficiency is possibly due to the dissolution of ZnO into Zn^{2+} ions [62], and in highly basic media the ZnO yields zincate ion ZnO_2^{-2} [63]. The results show that the optimum photo-degradation is close to natural water conditions (neutral to slightly acidic), which adds to the credibility of using ZnO catalyst system. Therefore, unless otherwise stated, all results described here-in-after were obtained under initial nominal pH 7.

Table 1 Effect of pH on photo-degradation of aqueous methyl orange. Reactions were conducted using methyl orange solutions (100 mL, 10 ppm) and ZnO (0.1 g) at 20 °C under total radiation intensity 19.0 mW/cm²

pH	~7	~5	~2.5	~9	~11
T.N.	1.62×10^{-3}	1.94×10^{-3}	1.17×10^{-3}	8.42×10^{-4}	6.71×10^{-4}
Q.Y.	4.36×10^{-4}	4.28×10^{-4}	2.92×10^{-4}	2.12×10^{-4}	1.74×10^{-4}
% Degradation	100 %	100 %	73 %	60 %	48 %

Effect of temperature

At temperatures in the range 20–40 °C, the ZnO catalyst showed sound activity, reaching up to 85 % methyl orange removal within 60 min. At lower temperatures, ~ 10 °C, the reaction went slower, reaching up to 65 % removal within 60 min. Figure 8 shows effect of working temperature on the methyl orange phtodegradation reaction profiles. If left for longer time, methyl orange is expected to continue even at lower temperatures. Table 2 summarizes values of TN, QY and degradation% measured at different temperatures. The results show the applicability of using the ZnO system in removing methyl orange with simulated solar light. In this work, unless otherwise stated, all study was performed at 20 °C.

If higher temperatures (above 45 °C) are used, the reaction goes slower. This is due to possible escape of the oxygen molecules dissolved inside oxygen. The oxygen molecules are involved with the mechanism of photodegradation reaction. Similar results were observed in earlier reports [53].

Effect of CO_2 and other gas flows

The effect of CO_2 flow on methyl orange photo-degradation was investigated by measuring the reaction profiles for the process while passing a flow of CO_2 gas. The stream of CO_2 (90 mL/min), while keeping all other

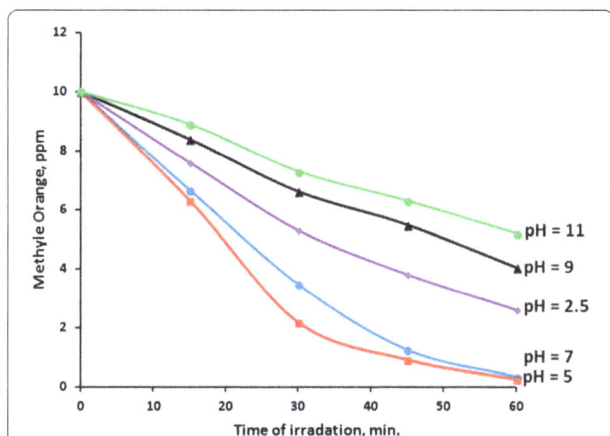

Fig. 7 Effect of pH on photo-degradation of aqueous methyl orange. Reactions were conducted using methyl orange solutions (100 mL, 10 ppm) and ZnO (0.1 g) at 20 °C under total radiation intensity 19.0 mW/cm²

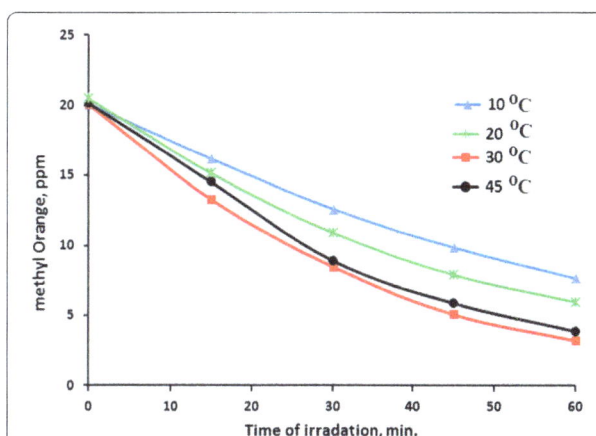

Fig. 8 Effect of temperature on methyl orange photo-degradation. Reactions were conducted using methyl orange solution (100 mL, 20 ppm) at different temperatures, using ZnO catalyst (0.1 g) under total irradiation intensity of 19.0 mW/cm² at pH ~7

Table 2 Effect of temperature on methyl orange photo-degradation. Reactions were conducted using methyl orange solution (100 mL, 20 ppm) at different temperatures, using ZnO catalyst (0.1 g) under total irradiation intensity of 19.0 mW/cm^2 at pH ~7

Temp.	10 °C	20 °C	30 °C	45 °C
T.N.	1.86×10^{-3}	2.38×10^{-3}	2.87×10^{-3}	2.80×10^{-3}
Q.Y.	3.52×10^{-4}	4.28×10^{-4}	5.44×10^{-4}	5.29×10^{-4}
% Degradation	62 %	70 %	84 %	81 %

Table 3 Effect of gas streams on methyl orange photo-degradation. Reactions were conducted using methyl orange solution (100 mL, 20 ppm) under 19.0 mW/cm^2 irradiation using ZnO (0.1 g) with continuous stirring at 20 °C

	Exposed to air only	Air flow	N$_2$ flow open system	CO$_2$ and air flows together	CO$_2$ flow
T.N.	2.87×10^{-3}	1.72×10^{-3}	1.44×10^{-3}	0.75×10^{-3}	1.03×10^{-3}
Q.Y.	5.44×10^{-4}	3.25×10^{-4}	2.72×10^{-4}	1.42×10^{-4}	1.95×10^{-4}
% Degradation	84 %	52 %	50 %	25 %	19 %

reaction parameters the same, significantly slowed down the removal of the methyl orange, as shown in Fig. 9. Table 3 summarizes values of TN, QY and degradation% while using CO$_2$ gas stream. The mode of action of CO$_2$ on lowering the reaction rate is not due to lowering the solution pH, as the pH was lowered only slightly, from 6.8 (the nominal used pH) to 6.4, after 60 min. As discussed above the pH in the range 5–8 does not inhibit methyl orange removal process. Alternatively, the mode of action of CO$_2$ is thus due to its ability to react with the radicals formed during the photo-degradation process. As reported earlier, *"CO$_2$ may interact with some free radical species and may either propagate or inhibit free radical chain reactions"* [64]. In this work the CO$_2$ clearly inhibits photo-degradation of methyl orange by interacting with the free radicals believed to be formed during the reaction process. The results reflect a warning signal about the negative impact of possible

CO$_2$ higher concentrations in natural waters, which would result from increased atmospheric CO$_2$ concentration. Having natural waters with higher CO$_2$ concentrations may negatively affect future water purification processes.

CO$_2$ gas stream may arguably affect photodegradation rate by removing oxygen dissolved in the reaction mixture. This was investigated using two continuous streams of CO$_2$ and air together. Figure 9 shows that using the air (1000 mL/min) stream with the CO$_2$ stream (90 mL/min) did not show significant enhancement in photo-degradation reaction rate. This result further confirms the discussion above, where CO$_2$ captures the free radicals necessary for the photodegradation to occur.

Adding a stream of air alone did not increase the reaction rate. Figure 9 shows that the air stream lowered the reaction rate by about 25 % compared to experiments conducted under normal air. In a well known mechanism [40], oxygen molecule is assumed to abstract one electron from the excited ZnO particle leading to O$_2{}^{\bullet-}$ radical anion species as explained in the equation [e$^-$ + O$_2 \rightarrow$ O$_2{}^{\bullet-}$]. The O$_2{}^{\bullet-}$ species is believed to react with the other H$_2$O$_2$ resulting species to yield the chemically active OH$^{\cdot}$ radical as in the equation [H$_2$O$_2$ + O$_2{}^{\bullet-} \rightarrow$ OH$^{\bullet}$ + OH$^-$ + O$_2$]. This radical is assumed to react with an organic contaminant molecule and degrade it.

Therefore, the presence of O$_2$ in water is necessary. However, as the results here show, excess of O$_2$ in water inhibits the photodegradation reaction. This is due to the adverse effect of excess O$_2$ which blocks OH$^{\bullet}$ radical as shown in equation [H$_2$O$_2$ + O$_2{}^{\bullet-} \rightarrow$ OH$^{\bullet}$ + OH$^-$ + O$_2$].

Adding nitrogen to the reaction mixture together with CO$_2$ practically stopped the reaction progress (Fig. 9). The effect is perhaps dual in nature, where the CO$_2$ behaves as scavenger, as discussed above, while the nitrogen lowers concentration of oxygen that is necessary for the reaction to proceed. Adding a nitrogen flow (240 mL/min) to the reaction mixture, while exposed to atmospheric air, slowed down the process, but did not stop it completely. Nitrogen lowers concentrations of oxygen in the reaction but traces are left therein which are enough for the reaction to occur. All reactions above

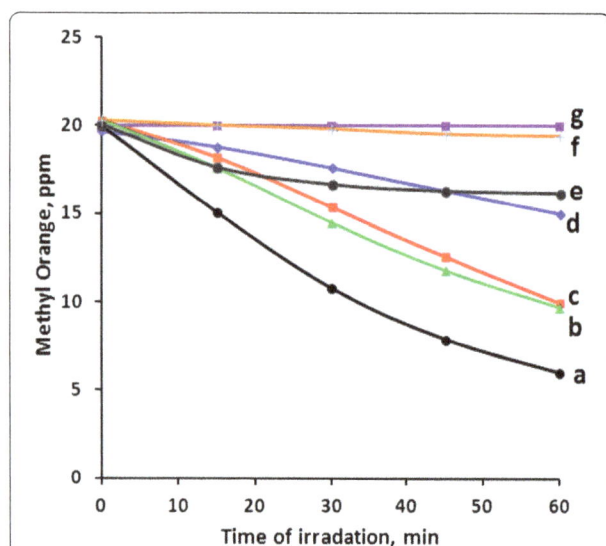

Fig. 9 Effect of gas streams on methyl orange photo-degradation. Reactions were conducted using methyl orange solution (100 mL, 20 ppm) under 19.0 mW/cm^2 irradiation using ZnO (0.1 g) with continuous stirring at 20 °C with different gas flows: (**a**) exposed to air only, (**b**) air flow, (**c**) N$_2$ flow open system, (**d**) CO$_2$ and air flows together, (**e**) CO$_2$ flow, (**f**) closed system with nitrogen flow, (**g**) N$_2$ and CO$_2$ flows together

were conducted under exposure to atmospheric air. This is evident because complete coverage from air while under nitrogen stream caused complete reaction inhibition.

Effect of contaminant concentration

The effect of methyl orange concentration (10, 20, 30 and 40 ppm) on its photodegradation process was studied, Fig. 10. Values of percent methyl orange removal, after 60 min, were 100 % for 10 ppm, 70 % for 20 ppm, 66 % for 30 ppm, 45 % for 40 ppm, as summarized in Table 4. Despite the lowering in removal percentage with higher contaminant concentration, the average reaction rate increased. Values of T.N. and Q.Y. show comparable values for the 20, 30, and 40 ppm concentrations. Lower values were observed for the 10 ppm concentration, which is due to the fact that all added methyl orange molecules were mineralized with no more left. This naturally yields lower TN and QY values at lower contaminant concentrations. The results indicate the suitability of ZnO catalyst to function over a relatively wide range of contaminant concentrations. More importantly, the catalyst completely activates mineralization of low contaminant concentrations, which are more likely to occur in nature [37, 53].

Effect of catalyst nominal amount on photo-degradation process

Catalyst amount may affect photo-degradation processes. The effect of catalyst nominal amount on the photodegradation of methyl orange was investigated using different amounts of ZnO (0.05, 0.10, 0.20 or 0.30 g) in 100 mL solution of 10 ppm methyl orange, under 19.0 mW/cm^2 irradiation intensity. Figure 11 shows that the average reaction rate was unchanged with

Table 4 Effect of contaminant concentration on its photo-degradation. Reactions were conducted using different contaminant concentrations in aqueous solution (100 mL), pH ~7, at 20 °C with ZnO (0.1 g) under 19.0 mW/cm^2 irradiation intensity

	10 ppm	20 ppm	30 ppm	40 ppm
T.N.	1.73×10^{-3}	2.38×10^{-3}	2.37×10^{-3}	2.29×10^{-3}
Q.Y.	4.36×10^{-4}	4.28×10^{-4}	4.48×10^{-4}	4.17×10^{-4}
% Degradation	100 %	70 %	66 %	45 %

catalyst nominal amount. The T.N. value decreased with increasing ZnO amount, as shown in Table 5. This means that the relative efficiency of the catalyst is lowered by increasing catalyst loading. Similar results were observed in other photo-degradation processes [37, 53, 54, 65, 66]. This is attributed to the increased blocking of light with higher catalyst loading. With more catalyst particles, the short wave tail photons are not able to enter the reaction mixture, and the ZnO particles in the reaction mixture do not receive photons. Similar behaviors have been observed in earlier studies [14, 64, 65].

The QY value showed an increase with increasing nominal catalyst amount at the beginning, but then slightly decreased. Again the lowering in QY is due to the blocking of light by the abundant ZnO particles at the mixture surface which prevent photons from reaching other catalyst sites [14, 37, 53, 54, 65–67]. Agglomeration of smaller ZnO particles into larger ones, in case of higher concentrations, may also play a role, as the total number of surface active sites may decrease [67]. The results indicate that using smaller amounts of catalyst enhanced catalyst efficiency without lowering the average reaction rate or the removal percentage. This is

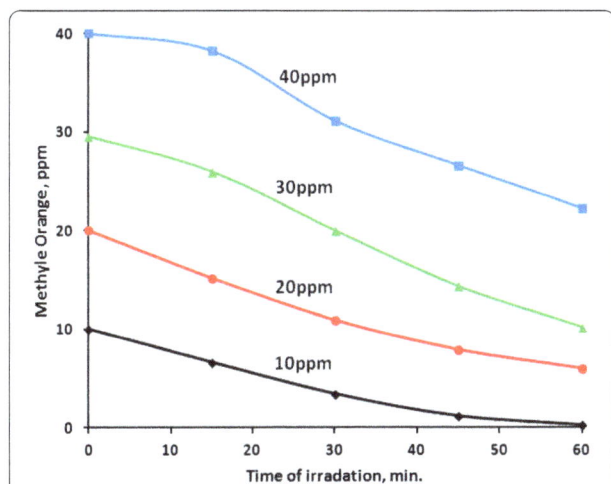

Fig. 10 Effect of contaminant concentration on its photo-degradation. Reactions were conducted using different contaminant concentrations in aqueous solution (100 mL), pH ~7, at 20 °C with ZnO (0.1 g) under 19.0 mW/cm^2 irradiation intensity

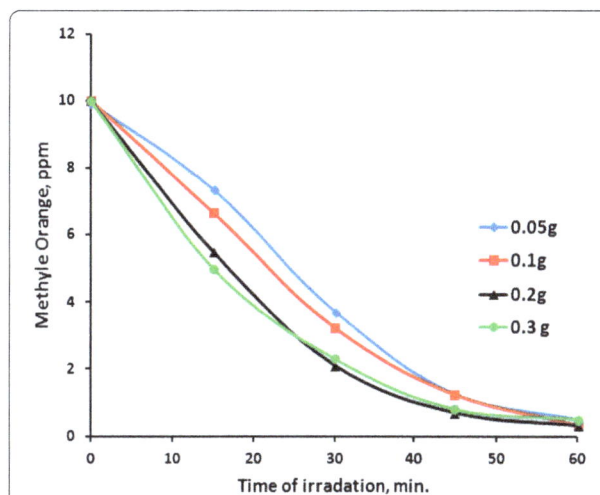

Fig. 11 Effect of catalyst amount on methyl orange photo-degradation. Reactions were conducted using methyl orange solution (100 mL, 10 ppm) at 20 °C pH ~7, under 19.0 mW/cm^2

Table 5 Effect of catalyst amount on methyl orange photo-degradation. Reactions were conducted using methyl orange solution (100 mL, 10 ppm) at 20 °C pH ~7, under 19.0 mW/cm^2

	0.05 g	0.10 g	0.20 g	0.3 g
T.N.	3.10×10^{-3}	1.68×10^{-3}	9.82×10^{-4}	9.57×10^{-4}
Q.Y.	2.93×10^{-4}	3.19×10^{-4}	3.72×10^{-4}	3.62×19^{-4}
% Degradation	95 %	100 %	100 %	100 %

a positive feature of the ZnO catalyst described here, as in case of treating natural waters, smaller amounts of catalyst will be highly favored.

Effect of light intensity on the photodegraation process

The effect of incident light intensity on photodegradation rate of methyl orange was investigated, using solutions (100 mL each) with two different concentrations (10 and 20 ppm). Experiments were conducted under different light intensities using 0.10 g of ZnO lab prepared powder. In each contaminant concentration, four different light intensities (1.90, 5.12, 8.78 and 19.0 mW/cm^2) were used. This range includes the average reported daylight intensity of 120 W/m^2 (12 mW/cm^2) [68]. Figure 12 shows how methyl orange removal percentage changed with illumination intensity in a period of 30 min for each contaminant concentration. For the 10 ppm concentration case, up to 100 % removal was observed for the higher radiation intensities (5.12, 8.78, and 19 mW/cm^2), and up to 50 % removal for the lower intensity (1.9 mW/cm^2), in 30 min. For the 20 ppm concentration case, up to 70 % contaminant removal was observed in 30 min (at the higher radiation intensities, and up to 35 % at the lower irradiation intensity). The results indicate the applicability of using ZnO in

Table 6 Effect of light intensity on methyl orange photo-degradation. Reactions were conducted using two different methyl orange concentrations in 100 mL solution at pH ~7 with ZnO (0.1 g) at 20 °C

10 ppm

	1.90 mW/cm^2	5.12 mW/cm^2	8.78 mW/cm^2	19.0 mW/cm^2
T.N.	6.96×10^{-4}	1.5×10^{-3}	1.58×10^{-3}	1.73×10^{-3}
Q.Y.	1.42×10^{-3}	1.35×10^{-3}	6.57×10^{-4}	4.36×10^{-4}
% Degradation	54 %	95 %	96 %	100 %

20 ppm

	1.90 mW/cm^2	5.12 mW/cm^2	8.78 mW/cm^2	19.0 mW/cm^2
T.N.	7.96×10^{-4}	1.86×10^{-3}	1.99×10^{-3}	2.38×10^{-3}
Q.Y.	1.62×10^{-3}	1.41×10^{-3}	8.76×10^{-4}	4.28×10^{-4}
% Degradation	34 %	58 %	65 %	70 %

removing methyl orange from contaminated water even under radiation intensities lower than normal day light.

Values of TN and QY calculated after 30 min for each case are shown in Table 6. In case of 10 ppm, the lower irradiation intensity (1.9 mW/cm^2) showed lower TN value (about 50 %) than its higher intensity counterparts (5.12, 8.78, and 19 mW/cm^2). From the Table and the Figure, it can be seen that under 5.12 mW/cm^2 or higher, the removal percentage reaches approximately constant value. With lower intensity, lower TN value should be expected, as less photons are available for the catalyst sites. However, as radiation intensity is increased above 5.12 mW/cm^2, efficiency of light harvesting becomes less. This is evident from values of QY, which decreased when using higher irradiation intensities.

For the 20 ppm methyl orange concentration, similar behavior occurred, but the irradiation intensity 5.12 mW/cm^2 showed highest QY. Collectively the results suggest

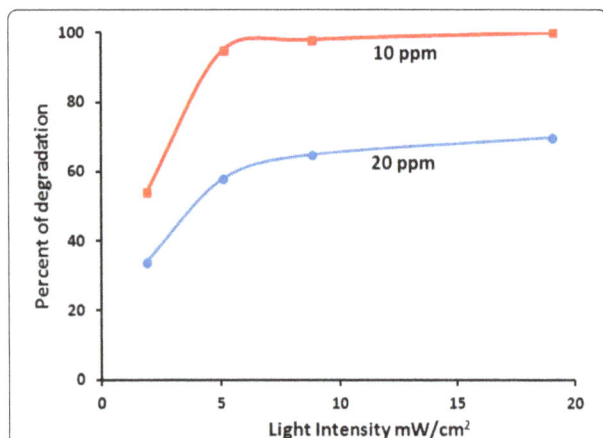

Fig. 12 Effect of light intensity on methyl orange photo-degradation. Reactions were conducted using two different methyl orange concentrations in 100 mL solution at pH ~7 with ZnO (0.1 g) at 20 °C

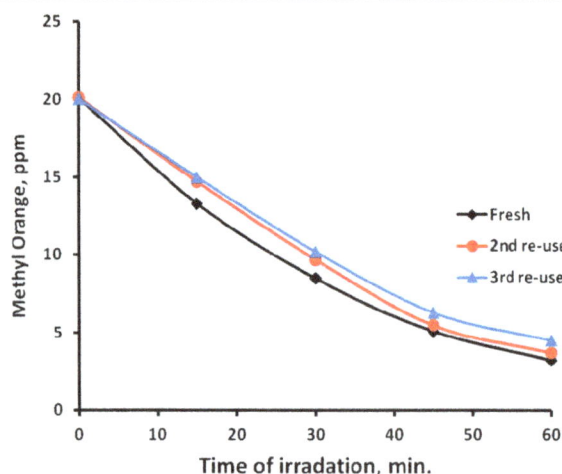

Fig. 13 Effect of catalyst reuse on methyl orange photo-degradation. Reactions were conducted using fresh methyl orange solutions (100 mL, 20 ppm) in multiple use of the ZnO catalyst (0.1 g) at 20 °C, pH 7 under 19.0 mW/cm^2 irradiation intensity

that it is not necessary to use high irradiation intensities to remove methyl orange from water. This adds to the applicability of using ZnO in natural water purification processes in different environments with different light intensities.

Catalyst re-use experiments

Reusability of the ZnO catalyst for methyl orange photo-degradation in water was studied, by adding fresh amount of methyl orange to the stirred solution after earlier reaction cessation. Figure 13 shows reaction profiles for three time reuse experiments. The Figure shows that at the beginning of each reuse experiment, the measured amount of the methyl orange was 20 ppm. This is due to complete removal of the earlier methyl orange contaminant in the preceding study, as the reaction was left for enough time. In each experiment, after 60 min time of exposure, more than 80 % removal was achieved, showing only low loss of efficiency with reuse.

The amount of Zn^{2+} ions resulting from dissolution of the used ZnO catalyst (with different amounts 0.05–0.3 g per 100 mL solution) during photo-degradation experiments in neutral media was measured by AAS, and was found to be 6 ppm, when the mixture was left overnight. In case of more acidic media the amount was higher, up to 8 ppm. This indicates that only a small fraction of ZnO dissolved under the working conditions. Based on literature [69] the value for solubility of Zn ions resulting from nano-scale ZnO is about 7 ppm. The WHO recommended upper limit for Zn ions is ~5 ppm [70]. The Zn^{2+} ions dissolved in this work is not far from the recommended WHO threshold limits. The results add to the credibility of using ZnO in purification of natural waters.

Conclusion

Nano-scale ZnO particles can be effectively used as catalysts for complete mineralization of methyl orange in water with solar simulated light. The catalyst can be effectively used under different working conditions (including temperature and pH) that resemble natural waters, and can thus be investigated at larger scale in natural water purification. Adding streams of air, CO_2 gas, and/or N_2 gases may affect the reaction progress and may inhibit the reaction.

Competing interests

The authors declare that they have no competing interests.

Authors' contributions

AZ participated in supervising research, catalytic experimental design, experimental work, data analysis and writing up. AZ performed catalytic experiments. DP and GC conducted advanced characterization. MH participated with writing up, library search and research ideas. The preparations and catalytic study were performed in H. S. H. laboratory. He participated in writing up, provided technical solutions and new ideas. All authors read and approved the final manuscript.

Acknowledgement

The thrust of this work has been done at SSERL, Department of Chemistry, ANU. The authors wish to thank the technical staff at ANU for help. XRD and SEM measurements were performed in the laboratories of Dansuk Industrial Co., LTD., South Korea.

Author details

[1]SSERL, Department of Chemistry, An-Najah National University, Nablus, Palestine. [2]College of Pharmacy and Nutrition, University of Saskatchewan, 116 Thorvaldson Building, Saskatoon S7N 5C9, Canada. [3]Dansuk Industrial Co, LTD. #1239-5, Jeongwang-Dong, Shiheung-Si, Kyonggi-Do 429-913, South Korea. [4]Institut de Chimie de la Matie're Condense'ie de Bordeaux (ICMCB), 87 Avenue du Dr. A Schweitzer, Pessac 33608, France.

References

1. Bianco Prevot A, Baiocchi C, Brussino MC, Pramauro E, Savarino P, Augugliaro V, et al. Photocatalytic degradation of acid blue 80 in aqueous solutions containing TiO_2 suspensions. Environ Sci Technol. 2001;35:971–6.
2. McMullan G, Meehan C, Conneely A, Kirby N, Robinson T, Nigam P, et al. Microbial decolourisation and degradation of textile dyes. Appl Microbiol Biotechnol. 2001;56:81–7.
3. Pearce C, Lloyd J, Guthrie J. The removal of colour from textile wastewater using whole bacterial cells: a review. Dyes and Pigments. 2003;58:179–96.
4. Marcucci M, Nosenzo G, Capannelli G, Ciabatti I, Corrieri D, Ciardelli G. Treatment and reuse of textile effluents based on new ultrafiltration and other membrane technologies. Desalination. 2001;138:75–82.
5. Robinson T, McMullan G, Marchant R, Nigam P. Remediation of dyes in textile effluent: a critical review on current treatment technologies with a proposed alternative. Bioresour Technol. 2001;77:247–55.
6. Tahir S, Rauf N. Removal of a cationic dye from aqueous solutions by adsorption onto bentonite clay. Chemosphere. 2006;63:1842–8.
7. Arslan I, Akmehmet Balcioğlu I, Tuhkanen T. Oxidative treatment of simulated dyehouse effluent by UV and near-UV light assisted Fenton's reagent. Chemosphere. 1999;39:2767–83.
8. Kuo W. Decolorizing dye wastewater with Fenton's reagent. Water Res. 1992;26:881–6.
9. Rice RG. Applications of ozone for industrial wastewater treatment—a review. Ozone Sci Eng. 1996;18:477–515.
10. Dodd MC, Buffle M-O, Von Gunten U. Oxidation of antibacterial molecules by aqueous ozone: moiety-specific reaction kinetics and application to ozone-based wastewater treatment. Environ Sci Technol. 2006;40:1969–77.
11. Glaze WH, Kang J-W, Chapin DH. The chemistry of water treatment processes involving ozone, hydrogen peroxide and ultraviolet radiation. Environ Sci Technol. 1987;39(10):3409–20.
12. Herrmann J-M. Heterogeneous photocatalysis: fundamentals and applications to the removal of various types of aqueous pollutants. Catal Today. 1999;53:115–29.
13. Daneshvar N, Salari D, Khataee A. Photocatalytic degradation of azo dye acid red 14 in water on ZnO as an alternative catalyst to TiO_2. J Photochem Photobiol A Chem. 2004;162:317–22.
14. Mai F, Chen C, Chen J, Liu S. Photodegradation of methyl green using visible irradiation in ZnO suspensions: determination of the reaction pathway and identification of intermediates by a high-performance liquid chromatography–photodiode array-electrospray ionization-mass spectrometry method. J Chromatogr A. 2008;1189:355–65.
15. Turchi CS, Ollis DF. Photocatalytic degradation of organic water contaminants: mechanisms involving hydroxyl radical attack. J Catal. 1990;122:178–92.
16. Umebayashi T, Yamaki T, Tanaka S, Asai K. Visible light-induced degradation of methylene blue on S-doped TiO2. Chem Lett. 2003;32:330–1.
17. Tayade RJ, Surolia PK, Kulkarni RG, Jasra RV. Photocatalytic degradation of dyes and organic contaminants in water using nanocrystalline anatase and rutile TiO_2. Sci Technol Adv Mater. 2007;8:455–62.
18. Ding Z, Zhu H, Lu G, Greenfield P. Photocatalytic properties of Titania pillared clays by different drying methods. J Colloid Interface Sci. 1999;209:193–9.
19. Fox MA, Dulay MT. Heterogeneous photocatalysis. Chem Rev. 1993;93:341–57.

20. Hagfeldt A, Graetzel M. Light-induced redox reactions in nanocrystalline systems. Chem Rev. 1995;95:49–68.

21. Hoffmann MR, Martin ST, Choi W, Bahnemann DW. Environmental applications of semiconductor photocatalysis. Chem Rev. 1995;95:69–96.

22. Panayotov DA, Yates JT. Spectroscopic detection of hydrogen atom spillover from Au nanoparticles supported on TiO_2: use of conduction band electrons. J Phys Chem C. 2007;111:2959–64.

23. Hong W-K, Sohn JI, Hwang D-K, Kwon S-S, Jo G, Song S, et al. Tunable electronic transport characteristics of surface-architecture-controlled ZnO nanowire field effect transistors. Nano Lett. 2008;8:950–6.

24. Dorfman A, Kumar N, Hahm J-i. Nanoscale ZnO-enhanced fluorescence detection of protein interactions. Adv Mater. 2006;18:2685–90.

25. Bao J, Zimmler MA, Capasso F, Wang X, Ren Z. Broadband ZnO single-nanowire light-emitting diode. Nano Lett. 2006;6:1719–22.

26. Law M, Greene LE, Johnson JC, Saykally R, Yang P. Nanowire dye-sensitized solar cells. Nat Mater. 2005;4:455–9.

27. Sakthivel S, Neppolian B, Shankar M, Arabindoo B, Palanichamy M, Murugesan V. Solar photocatalytic degradation of azo dye: comparison of photocatalytic efficiency of ZnO and TiO_2. Sol Energy Mater Sol Cells. 2003;77:65–82.

28. Gouvea CA, Wypych F, Moraes SG, Duran N, Nagata N, Peralta-Zamora P. Semiconductor-assisted photocatalytic degradation of reactive dyes in aqueous solution. Chemosphere. 2000;40:433–40.

29. Lizama C, Freer J, Baeza J, Mansilla HD. Optimized photodegradation of reactive blue 19 on TiO2 and ZnO suspensions. Catal Today. 2002;76:235–46.

30. Lachheb H, Puzenat E, Houas A, Ksibi M, Elaloui E, Guillard C, et al. Photocatalytic degradation of various types of dyes (Alizarin S, Crocein Orange G, Methyl Red, Congo Red, Methylene Blue) in water by UV-irradiated titania. Appl Catal Environ. 2002;39:75–90.

31. Kandavelu V, Kastien H, Thampi KR. Photocatalytic degradation of isothiazolin-3-ones in water and emulsion paints containing nanocrystalline TiO_2 and ZnO catalysts. Appl Catal Environ. 2004;48:101–11.

32. MSDS. Material Safety Data Sheet,Methyl orange MSDS, Fisher Scientific, 1 Reagent Lane, Fair Lawn, NJ 07410. 2000.

33. Sandberg RG, Henderson GH, White RD, Eyring EM. Kinetics of acid dissociation-ion recombination of aqueous methyl orange. J Phys Chem. 1972;76:4023–5.

34. Windholz M, Budavari S. The merck index. Merck and Co. Rahway; 1983.

35. Maynard C. Riegel's Handbook of Industrial Chemistry. In Book Riegel's Handbook of Industrial Chemistry (Editor ed.^eds.). City: JA Kent, ed; 1983.

36. Shih Y-H, Tso C-P, Tung L-Y. Rapid degradation of methyl orange with nanoscale zerovalent iron particles. Nanotechnology. 2010;7:16–7.

37. Hilal HS, Al-Nour GY, Zyoud A, Helal MH, Saadeddin I. Pristine and supported ZnO-based catalysts for phenazopyridine degradation with direct solar light. Solid State Sci. 2010;12:578–86.

38. Hilal SH, Nour G, Zyoud A. Photo-degradation of Methyl Orange with Direct Solar Light Using ZnO and Activated Carbon-supported ZnO. In: Water purification. New York: Nova Science Publishers, Inc; 2009. p. 227–46.

39. Hejjawi S. TiO_2 and ZnO photocatalysts for degradation of widespread pharmaceutical wastes: Effect of particle size and support. MSc Thesis, An-Najah N. University. 2013.

40. Amer H. ZnO nano-particle catalysts in contaminant degradation processes with solar light naked and supported systems. MSc Thesis, An-Najah N. University. 2012.

41. Gupta A, Bhatti H, Kumar D, Verma N, Tandon R. Nano and bulk crystals of ZnO: synthesis and characterization. Digest Int J Nanomat Biostruct. 2006;1:1–9.

42. Lopez-Romero S. Growth and characterization of ZnO cross-like structures by hydrothermal method. Matéria (Rio de Janeiro). 2009;14:977–82.

43. Akhtar MJ, Ahamed M, Kumar S, Khan MM, Ahmad J, Alrokayan SA. Zinc oxide nanoparticles selectively induce apoptosis in human cancer cells through reactive oxygen species. Int J Nanomedicine. 2012;7:845.

44. Dai K, Chen H, Peng T, Ke D, Yi H. Photocatalytic degradation of methyl orange in aqueous suspension of mesoporous Titania nanoparticles. Chemosphere. 2007;69:1361–7.

45. Attia AJ, Kadhim SH, Hussein FH. Photocatalytic degradation of textile dyeing wastewater using titanium dioxide and zinc oxide. J Chem. 2008;5:219–23.

46. Chen C, Wang Z, Ruan S, Zou B, Zhao M, Wu F. Photocatalytic degradation of CI Acid Orange 52 in the presence of Zn-doped TiO_2 prepared by a stearic acid gel method. Dyes Pigments. 2008;77:204–9.

47. Zhou G, Deng J. Preparation and photocatalytic performance of Ag/ZnO nano-composites. Mater Sci Semicond Process. 2007;10:90–6.

48. Wang H, Xie C, Zhang W, Cai S, Yang Z, Gui Y. Comparison of dye degradation efficiency using ZnO powders with various size scales. J Hazard Mater. 2007;141:645–52.

49. Yu D, Cai R, Liu Z. Studies on the photodegradation of rhodamine dyes on nanometer-sized zinc oxide. Spectrochim Acta A Mol Biomol Spectrosc. 2004;60:1617–24.

50. Daneshvar N, Aber S, Dorraji MS, Khataee A, Rasoulifard M. Preparation and investigation of photocatalytic properties of ZnO nanocrystals: effect of operational parameters and kinetic study. Evaluation. 2008;900:6.

51. Shvalagin V, Stroyuk A, Kotenko I, Kuchmii SY. Photocatalytic formation of porous CdS/ZnO nanospheres and CdS nanotubes. Theor Exp Chem. 2007;43:229–34.

52. Byrappa K, Subramani A, Ananda S, Rai KL, Sunitha M, Basavalingu B, et al. Impregnation of ZnO onto activated carbon under hydrothermal conditions and its photocatalytic properties. J Mater Sci. 2006;41:1355–62.

53. Zyoud AH, Zaatar N, Saadeddin I, Ali C, Park D, Campet G, et al. CdS-sensitized TiO_2 in phenazopyridine photo-degradation: Catalyst efficiency, stability and feasibility assessment. J Hazard Mater. 2010;173:318–25.

54. Zyoud A, Zaatar N, Saadeddin I, Helal MH, Campet G, Hakim M, et al. Alternative natural dyes in water purification: anthocyanin as TiO_2-sensitizer in methyl orange photo-degradation. Solid State Sci. 2011;13:1268–75.

55. Hilal SH, Nour G, Zyoud A. Photo-degradation of methyl orange with direct solar light using ZnO and activated carbon-supported ZnO. Chapter Book Water Purif. 2009;Chapter 6:227–46.

56. Neal C, Hill T, Alexander S, Reynolds B, Hill S, Dixon AJ, et al. Stream water quality in acid sensitive UK upland areas; an example of potential water quality remediation based on groundwater manipulation. Hydrol Earth Syst Sci Discuss. 1997;1:185–96.

57. Hem JD. Study and interpretation of the chemical characteristics of natural water. Water Supply Paper 2254, Department of the Interior, US Geological Survey; Alexandria, 1985.

58. Zhang F, Zhao J, Shen T, Hidaka H, Pelizzetti E, Serpone N. TiO_2-assisted photodegradation of dye pollutants II. Adsorption and degradation kinetics of eosin in TiO_2 dispersions under visible light irradiation. Appl Catal Environ. 1998;15:147–56.

59. Chang C, Hsieh Y, Cheng K, Hsieh L, Cheng T, Yao K. Effect of pH on Fenton process using estimation of hydroxyl radical with salicylic acid as trapping reagent. 2008.

60. Zepp RG, Faust BC, Hoigne J. Hydroxyl radical formation in aqueous reactions (pH 3–8) of iron (II) with hydrogen peroxide: the photo-Fenton reaction. Environ Sci Technol. 1992;26:313–9.

61. Velmurugan R, Swaminathan M. An efficient nanostructured ZnO for dye sensitized degradation of reactive red 120 dye under solar light. Sol Energy Mater Sol Cells. 2011;95:942–50.

62. Marci G, Augugliaro V, Lopez-Munoz MJ, Martin C, Palmisano L, Rives V, et al. Preparation characterization and photocatalytic activity of polycrystalline ZnO/TiO2 systems. 2. Surface, bulk characterization, and 4-nitrophenol photodegradation in liquid–solid regime. J Phys Chem B. 2001;105:1033–40.

63. Laudise ER, Kolb E, Laboralories BT, Hiti IM, Jeriey N. The solubility of zincite in basic hydrothermal solvents. Amer Mineral. 1963;48:642.

64. Vesela A, Wilhelm J. The role of carbon dioxide in free radical reactions in organism. Physiol Res. 2002;51:335–40.

65. Zyoud AH, Hilal HS. Silica-supported CdS-sensitized TiO_2 particles in photo-driven water purification: Assessment of efficiency, stability and recovery future perspectives. Chapter in a book, Water Purification, Novascience Pub, NY (in Press, 2008) 2009.

66. Hilal HS, Nour GY, Zyoud A. Photodegradation of Methyl orange and phenazopyridine HCl with direct solar light using ZnO and activated carbonsupported ZnO, Water Purification Novascience Publ, NY. 2009. p. 227e246.

67. Chen C-C. Degradation pathways of ethyl violet by photocatalytic reaction with ZnO dispersions. J Mol Catal A Chem. 2007;264:82–92.

68. Guide W. Guide to Meteorological Instruments and Methods of Observation. In: Book Guide to Meteorological Instruments and Methods of Observation. Genf, Schweiz: Secretariat of the WMO; 2006.

69. Reed RB, Ladner DA, Higgins CP, Westerhoff P, Ranville JF. Solubility of nano-zinc oxide in environmentally and biologically important matrices. Environ Toxicol Chem. 2012;31:93–9.

70. Krenkel P. Water quality management. Elsevier; Burlington, e-Book 2012.

7

Ultrafiltration of natural organic matter from water by vertically aligned carbon nanotube membrane

Ali Jafari[1], Amir Hossein Mahvi[2,1,3*], Simin Nasseri[2,1], Alimorad Rashidi[4], Ramin Nabizadeh[1] and Reza Rezaee[5]

Abstract

In this study vertically aligned carbon nanotubes (VA-CNT) was grown on anodized aluminum oxide (AAO) substrate. The synthesized AAO-CNT membrane was characterized using Raman spectroscopy, field emission scanning electron microscopy (FESEM), contact angle and BET. The pure water flux, humic acid (HA) (as representative of natural organic matters) rejection and fouling mechanism were also evaluated. The fabricated membrane has pore density of 1.3×10^{10} pores per cm^2, average pore size of 20 ± 3 nm and contact angle of $85 \pm 8°$. A significant pure water flux of 3600 ± 100 $L/m^2.h$ was obtained at 1 bar of pressure by this membrane due to the frictionless structure of CNTs. High contact angle exhibited the hydrophobic property of the membrane. It was revealed that HA is primarily rejected by adsorption in the membrane pores due to hydrophobic interactions with HA. Flux decline occurred rapidly through both cross flow and dead end filtration of the HA. Based on the blocking laws, internal pore constriction is dominant fouling mechanism in which HA adsorbs in membrane pores results in pores blockage and flux decline.

Keywords: Membrane, Carbon nanotube, Fouling, Natural organic matter, Ultrafiltration

Introduction

Natural organic matters (NOMs) are known as problematic substances in environment and health. Nowadays their various direct and indirect effects are well understood. They own different characteristics in terms of reactivity, structure and they enter to water bodies through various natural and man-made sources [1]. NOMs are present to different extents in waters [2, 3]. Typical ranges of 0.1–0.2 mg/L and 1–20 mg/L of NOM based on total organic carbon (TOC) have been reported for ground and surface waters, respectively. However, the TOC concentrations can be very higher (100–200 mg/L) in colored waters of swamps and marshes [4, 5].

These substances form hazardous disinfection byproducts (DBPs) such as trihalomethanes (THMs) in reaction with chlorine. NOM is often made of two fractions, namely hydrophobic and hydrophilic. Hydrophobic and hydrophilic fractions have the potential of THMs and haloacetic acids (HAAs) formation, respectively. Hydrophobic components are humic acid and fulvic acid and the hydrophilic components are proteins, amino acids and carbohydrates [4, 5].

Hydrophobic compounds have the greatest effect on DBPs formation. Furthermore, some hydrophobic substances may intrinsically be much more toxic than the chlorinated components [6].

The presence of NOMs have other problems such as negative effect on the water quality and treatment process, increasing the coagulants and disinfectants demand, biological growth in distribution network, decomposition of organic matter within the network and creating a slimy layer on the pipes [5, 7]. In addition, NOM adversely affects the membrane performance in water purification [8].

Different methods with varying efficiencies such as chemical coagulation and precipitation, ion exchange, adsorption, electrocoagulation, advanced oxidation and membrane process have been applied for NOM mitigation [5, 9–14]. Although in some cases enhanced coagulation has been proposed for NOMs removal, but large application of coagulants, pH modification problems and large amount of produced sludge are the main related obstacles of this method.

* Correspondence: ahmahvi@yahoo.com
[2]Center for Water Quality Research (CWQR), Institute for Environmental Research (IER), Tehran University of Medical Sciences, Tehran, Iran
[1]Department of Environmental Health Engineering, School of Public Health, Tehran University of Medical Sciences, Tehran, Iran
Full list of author information is available at the end of the article

Furthermore, adsorption as a known process in water treatment, have been studied and applied for NOM removal using various materials. In addition to activated carbon materials, new nano adsorbents such as CNTs and zero valent iron nanoparticles have also been studied as promising adsorbents [15, 16], but the questions related to release of nanomaterials and regeneration costs are the main drawback of adsorption application.

Membranes are used for NOMs fractions removal from aqueous solution as one of the important processes in water treatment. Membrane process is of interest due to no changes in the structure of pollutants, no intermediates addition to water, no adverse environmental effects, no need of chemicals and easy navigation [17].

The major obstacles associated with the membrane are energy consuming and fouling problem. Researchers are attempting to change the structure of conventional polymeric membranes or developed new membranes with higher permeability and higher pollutant rejection.

In this regard, carbon nanotubes (CNTs) have been considered for membrane synthesis due to their exceptional properties and high adsorption capabilities [18, 19]. Due to porous structure of the tubes and high surface area, a wide range of contaminants have been effectively removed by CNTs [20].

Promising results show that CNTs in membranes structure have higher flux, higher performance, less fouling, less required cleaning, higher thermal stability, higher consistency and lower energy requirement than conventional polymeric membranes [21–23].

Carbon nanotube membranes can be synthesized by various methods. One of these techniques is template carbonization to synthesize CNTs with desired diameter and high purity. In this method, the situation is prepared in such a way that the CNTs grow inside the channels of anodic aluminum oxide (AAO) template [23]. Thus, the membrane is created through growing CNT arrays and known as vertically aligned carbon nanotubes (VA-CNT) membrane.

One of the most important advantages of the VA-CNT membrane is high flux of water through the CNTs due to low length to the high density of nanotubes [21]. The first plan of VA-CNT membranes was developed by Hinds research team that CNTs were grown on iron as a catalyst using chemical vapor deposition (CVD) method [21]. It has been reported that hydrophobic channel of CNT is smooth and frictionless that facilitate the rapid movement of water [21]. Accordingly, in recent years, CNTs have been mixed with polymers in order to improve the performance of polymeric membranes with higher flux and less fouling [24–29].

To our knowledge, relatively few studies have been conducted in the development of vertically aligned CNT membranes for water purification. With regard to promising results related to CNTs application for membrane synthesis, present work is going to synthesize and characterize the VA-CNT membrane through AAO technique and investigate for NOMs rejection from water.

Materials and methods
Membrane experimental set up
The schematic diagram of the experimental setup is shown in Fig. 1. It consists of a 2-L feed tank, membrane module with the effective area of 1.4 cm^2, feeding pump, cooling system, valves, gauges and flow meters.

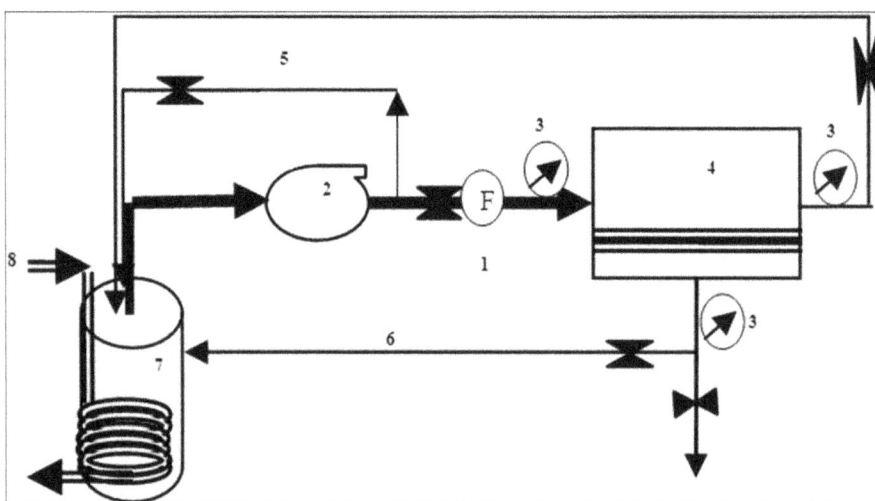

Fig. 1 Simplified schematic of experimental set-up unit. 1. Flow meter 2. Low pressure pump 3. Gauges 4. Membrane module 5. Recirculation (bypass) line 6. Concentrate line 7. Feed tank 8. Cooling circuit

Synthesis, preparation and characterization of membrane

All the chemicals reagents were of reagent grade and no further purification was done. The fabricated membrane was synthesized via anodic aluminum oxide (AAO) method through a two-step process (anodizing and growth of CNTs in the porous AAO).

In this study, a similar procedure as Gilani et al. was used for membrane synthesis and preparation [22, 30].

High purity (99.99) aluminum (Al) foil of 300 μm thickness was cut in small rectangular pieces. The foils were sonicated in acetone solution and then rinsed with double distilled water in order to degrease and subsequently dried in room temperature. Then, Al plates were electropolished in ethanol and perchloric acid (60 %) mixture solution (4/1 v/v) under constant cell voltage of 20 V for 2 min. The back surface of the plates was protected by an insulting tape.

The anodizing process started using oxalate acid (0.3 M) as electrolyte solution and under voltage of 40 volts for 2 h. Separation of the oxide layer was conducted using a mixture of phosphoric acid (6 % wt) and chromic acid (1.8 wt %) for 2 h at 65 °C.

Aluminum foil was re-anodized for 58 h under identical manner used for first anodization step. Then the unoxidized part was removed by putting in saturated mercuric chloride. The barrier layer was removed by soaking the template in phosphoric acid (5 wt %) at temperatures of 50 °C for 3 h [22, 30]. Deposition of CNTs onto the interior walls of the template was conducted by placing the AAO in CVD furnace. The temperature of the furnace was gradually increased to 650 °C at a rate of 5 °C/min. Meanwhile, a controlled argon flow (200 ml/min) was induced to the furnace. A mixture of acetylene and argon was inserted into the furnace as the carbon precursor and carrier gas, respectively with a ratio of 0.01 for 12 h. After the carbon deposition, the acetylene flow was turned off, under the condition of argon flow the reactor was allowed to cool to room temperature for 12 h. Finally, the synthesized membrane was washed in ethanol and dried in vacuum oven at 60 °C [22, 30].

Membrane characteristics were performed using Raman spectroscopy, FESEM, contact angle, and BET. Field emission scanning electron microscopy (FESEM) (Hitachi-S4160) was used to characterize the uniformity and morphology of the AAO–CNT membrane. The specific surface area of the AAO–CNTs was determined using the Brunauer–Emmett–Teller (BET) method by ASAP 2010 (Micromertics Inc., Norcross, GA).

Raman spectrum was used to observe the uniformity of graphite stracture of the growing CNTs in the membrane using a Raman spectrometer (Almega Thermo Nicolet Dispersive). Contact angle as an important factor in membrane characterization was measured by a contact angle analyzer (OCA 15 plus, dataphysics Instruments, Germany) using the sessile drop technique.

Solutions and analytical measurements

A laboratory grade humic acid (Acros Organics Company, NJ – USA) was used in this study as NOM model to evaluate the performance and removal mechanism by synthesized VA-CNT-AAO membrane. A known amount of HA powder was dissolved in distilled water and pH was adjusted around 7 for all experiments. The concentration of HA was reported as the term of TOC, as a surrogate measure using a TOC analyzer (TOC-VCPH, Shimadzu, Japan). For this purpose different samples (feed, permeate and concentrate) were taken at defined interval times and analyzed for TOC concentration.

Generally, in membrane process, materials may be removed from water by adsorption and/or repulsion mechanisms. The portion of desorbed or adsorbed solutes on the membrane surface is important in fouling and flux analysis.

From a practical standpoint, to investigate the removal mechanism, a certain volume of synthetic solution was placed in the feed tank (Fig. 1). Then, the system was operated in closed loop condition where permeate and concentrate were recycled back to the feed tank at constant transmembrane pressure (TMP) of 1 bar. A cooling circuit was applied to feed tank to maintain the temperature of the feed solution at a constant value of 25 ± 0.5 °C. Sampling was performed at specified intervals from feed tank, permeate and retentate flows for further analysis. All experiments were conducted in duplicate. The observed percentage of HA rejection (%) was calculated as Eq. 1.

$$\%R = \left(\frac{Cf - Cp}{Cf}\right) \times 100 \tag{1}$$

Where C_f and C_p are TOC concentrations in feed solution and permeate, respectively.

The adsorbed mass of TOC is equal to initial TOC mass in the feed tank minus the sum of TOC mass in the feed tank and cumulative samples of permeate at the end of experiment period, as determined by following mass balance equation (Eq. 2).

$$A. \, Mads = CfiVfi - \left[CffVff + \sum CpVp\right] \tag{2}$$

Where A is the effective membrane area. M_{ads} is the adsorbed mass of TOC per surface area of membrane. C_{fi}, C_{ff} and Cp are the initial concentration of TOC, final concentration of TOC in the feed tank and concentration of TOC in permeate respectively. V_{fi}, V_{ff} and V_p are the initial volume of feed solution in the tank, final volume of solution in the tank and volume of permeate, respectively.

Results and discussion
Membrane characteristics

Figures 2a and b illustrate the cross section and surface FESEM image of the membrane respectively. A through hole membrane is clearly shown in Fig. 2a. This image also depicts the direction of cavities and grown CNTs in the holes. The anodizing process and operation conditions (namely voltage, electrolyte concentration and time) effectively influence the properties of the channels [31].

The straight form of the channels indicates the well ordering of CNTs in this fabricated membrane. Fig. 2b shows the pores of the synthesized membrane (dark points). From analysis of FESEM image using the ImageJ analyzer [32] a highly uniform distribution of pores arrangement and nearly uniform pore size (≈20 nm) was created which is in the range of a UF membrane. As seen in Fig. 2b, the surface of the membrane looks uneven. Appling low voltages for AAO synthesis usually results in lower pore sizes and higher pore densities. This property has been shown in other similar works [22, 30]. Therefore, in low voltages due to higher spaces between the pores, the surface is rougher. Cleaning and polishing the surface of aluminum foil can relatively decrease the total surface roughness, but during the anodizing process and growing the CNTs near the top surface, the roughness increases. Although such a roughness may be defined as an advantage for some applications [31] due to higher surface area as an important factor in adsorption process, this can not be of interest for water purification. In particular, high roughness affects the membrane performance in water purification that results in fouling problem.

As depicted in Fig. 3, two strong peaks appeared in the Raman spectrum. The first peak is at around 1351 cm^{-1} (disordered D line band) and the next peak is about 1602 cm^{-1} (ordered G band). From the shift of the Raman spectrum peaks (D and G) the ratio of their intensities (IG/ID) is about one which shows the good structure of CNTs formed in the porous AAO and free

from amorphous carbon which is consistent with other researchers [22, 33]. Compare to CNTs grown by other methods, commonly growing on metal cores as catalyst, growing of CNTs in AAO template produce higher purity CNTs with less amporph carbons. Accordingly the need of further purification is not necessary for AAO-CNTs. Nevertheless, this technique produces nearly uniform CNTs in the membrane structure.

Contact angles of several droplets on the surface of the membrane were measured and the average was reported (Table 1). Results showed that AAO-CNT membrane has a hydrophobic behavior as the result of high contact angle. Fig. 4 shows the droplet of pure water on the surface of the membrane and the contact to the surface. Generally, grown VA-CNTs on different substrates may somewhat change the hydrophilicity behavior and contact angle of the VA-CNT membrane due to surface chemical structure and method of synthesis [34]. Since CNTs intrinsically have hydrophobic properties, it has been revealed that membranes fabricated by unmodified or non-functionalized VA-CNTs have close contact angles and are primarily hydrophobe (Table 1). Functionalization of VA-CNT membrane can change the membrane properties and introduce function groups on the CNTs [30].

Permeation analysis

Pure water flux, (J_0) was calculated using the following equation (Eq. 3):

$$J_0 = V/At \tag{3}$$

where V is the total volume of permeated pure water, A is the effective membrane area, and t is the operation time.

The pure water permeability for the VA-CNT membrane was 3600 ± 100 $L/m^2.h$ at TMP of 1 bar. This value is several times than conventional UF membranes for water process. A high practical flux of a commercial UF membrane was reported to be 800 $L/m^2.h$-bar flux [35]. Furthermore, due to frictionless structure of VA- CNT channels, it has been reported that fluid flow

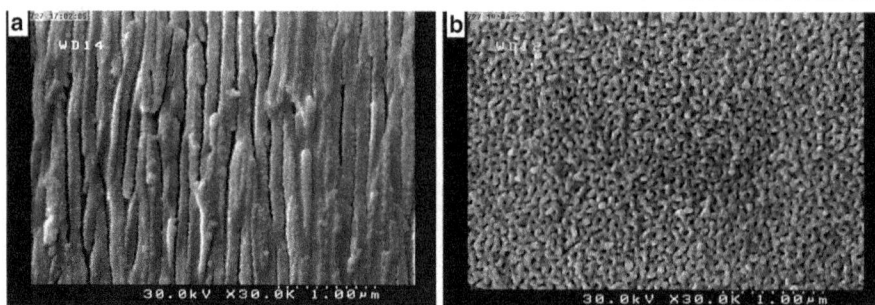

Fig. 2 FESEM image of AAO-CNT membrane. **a** cross section of the membrane and **b** surface area image of the membrane

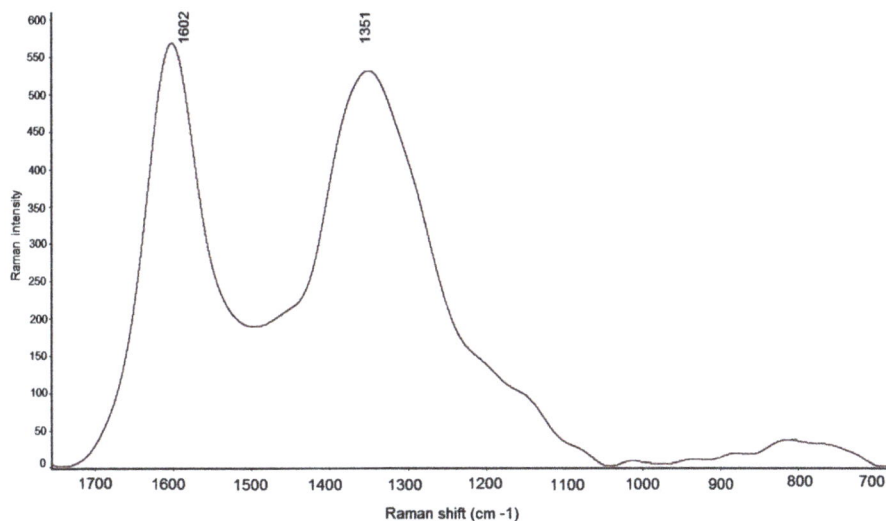

Fig. 3 Raman spectrum for AAO-CNT membrane

can be 4–5 times faster than conventional flows [36]. This dramatic velocity can facilitate the application of CNTs channel for other fields.

However, the flux of VA-CNT is significantly higher than that of similar polymeric membranes. In spite of lower pore density, lower pore diameter and higher thickness of aligned CNT membranes, a higher flux (3–4 orders of magnitude) than polymeric UF membrane have been reported [37, 38]. Besides pore size different parameters such as membrane thickness, method of preparation, size distribution can influence the flux of VA-CNT membranes that should be considered for comparing membrane permeabilities. Higher pore diameter (~20 nm) and lower thickness of the membrane result in higher flux in this study. In spite of low pore number of the synthesized membrane a high flux was observed due to lower thickness and also a guarantee of vertically aligned CNTs standing. As shown in Fig. 2a,

aligned structure of CNTs in AAO template can facilitate transport of water through the membrane. The number of pores per unit of membrane surface can also affect the membrane flux significantly in combined with other parameters (e.g. membrane structure, thickness and pore number) [21].

TOC Removal mechanism

In cross flow filtration of HA, flux declined after a short period of filtration (Fig. 5). The flux declined to near 0.6 of initial flux (J_0) after 5 min of filtration (40 % of flux decline). The flux decrease continues to more than 90 % of J_0 after 60 min of filtration. Interaction of hydrophobic surface of membrane with HA results in plugging of the pores and consequently rapid flux decrease [38].

Table 1 VA-CNT membrane characteristics compared to other related works

	This work	Youngbin Baek. et al [37]	Seung-Min Park. et al [38]
Materials	CNT-AAO	CNT-epoxy	CNT-epoxy
Pore density (pore/cm^2) *10^{10}	1.3	6.8	6.8
Average pore size (nm)	20 ± 3	4.8 ± 0.9	4.87 ± 0.87
Pore volume (cm^3/g)	0.425	Nd[a]	Nd
BET surface area (m^2/g)	220.4	Nd	Nd
Contact angle (degree)	85 ± 8	74.6 ± 2.8	92.1
Porosity	0.18 ± 0.4	Nd	Nd
Thickness (μ)	~120 ± 20	~200	200 ± 50
Flux(L/m^2.h) at 1 bar	3600 ± 100	1100 ± 130	1000 ± 100

[a]Not defined

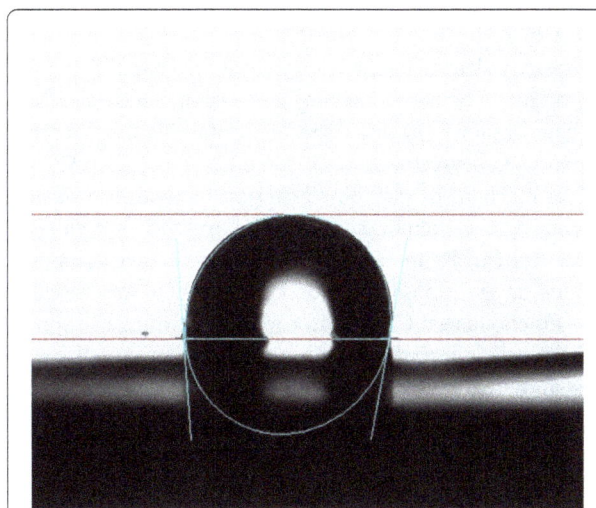

Fig. 4 Contact angle image for AAO-CNT membrane surface

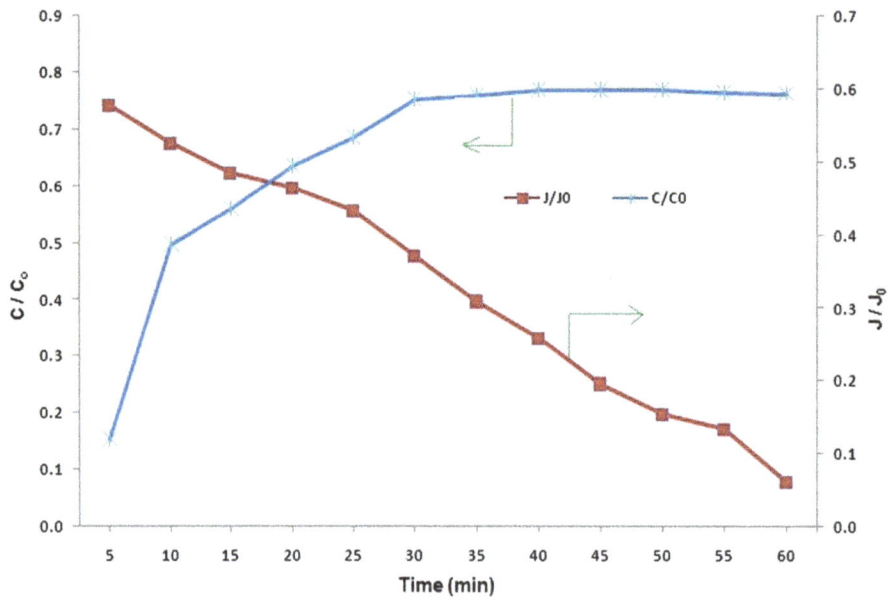

Fig. 5 Trend of TOC concentrations in permeate and flux in cross flow operation for initial TOC of 5 ± 0.3 mg/L, pH of 7.0 ± 0.2 at TMP of 1 bar

Principally, adsorption of NOM by CNT is a rapid process and the adsorption capacity is quickly exhausted. From this study, using mass balance equation (Eq. 1), 61 % of the loaded TOC was adsorbed on the VA-AAO-CNT membrane and only 11 % was repulsed. It was also calculated that 30 ± 0.5 g TOC was adsorbed per square meter of the membrane at a constant TMP of 1 bar for 60 min of filtration. Although adsorption may be an effective rejection mechanism during the initial steps of AAO-CNT membrane operation, but due to quick flux decline during the process, it is not an efficient mechanism in long-term filtration. However it may be preferred for some biochemical and biomedical applications for retention of toxin from plasma, some trace contaminants such as endocrine disrupting compounds (EDC) bisphenol A or heavy metals [39–41].

Anyway, determination of removal mechanism is very important. Therefore, making a decision about the best mechanism for removing a specific contaminant or whether depends on several factors. In some cases, the superior removal mechanism may be adsorption or repulsion of pollutants by the membrane, so the adsorption or repulsion capacities can be increased by inducing certain functional groups or by imposing certain conditions. It was revealed that functionalized CNTs with carboxylic groups can somewhat decrease the adsorption on the membrane [42]. From this study materials similar with HA substances can effectively be adsorbed on the CNT membrane types.

Fouling mechanism analysis

Blocking laws as one of the most popular models were applied to characterize the fouling mechanisms in this study. These empirical models were presented by Hermia [43]:

$$\frac{d^2 t}{dv^2} = k \left(\frac{dt}{dv} \right)^{\mathrm{n}} \tag{4}$$

Where t is time (s), V is volume (L) k is blocking law filtration coefficient (units depends on n) and n is blocking law filtration exponent (dimensionless) that expressing the fouling regime. Based on the n values different modes of fouling namely pore sealing ($n = 2$), internal pore constriction ($n = 1.5$), pore sealing with super position ($n = 1$) and cake filtration ($n = 0$) can be expressed. To determine the mechanism responsible for fouling, the experiment was conducted under dead end condition at constant pressure of 1 bar. At certain intervals, permeate volumes were recorded. Data was tabulated in a spreadsheet for analysis.

Using the Eq. 4 by plotting log $d^2 t/dv^2$) versus (dt/dv) the slope of the line is constant value (n) that represent the fouling mechanism [44].

In dead end filtration of HA, about 50 % of rejection occurs at first 10 min of operation (Fig. 6). As previously noted, high affinity of CNTs to HA results in rapid adsorption and fast depletion of adsorption capacity. In other word, due to increase in accumulation of HA in the solution, HA penetrate into the membrane, attach to adsorption sites and pass through the membrane. This trend is nearly similar to whatever occurs in the tangential operation, although in cross flow mode the rejection is higher as the operation time elapsing due to the cross

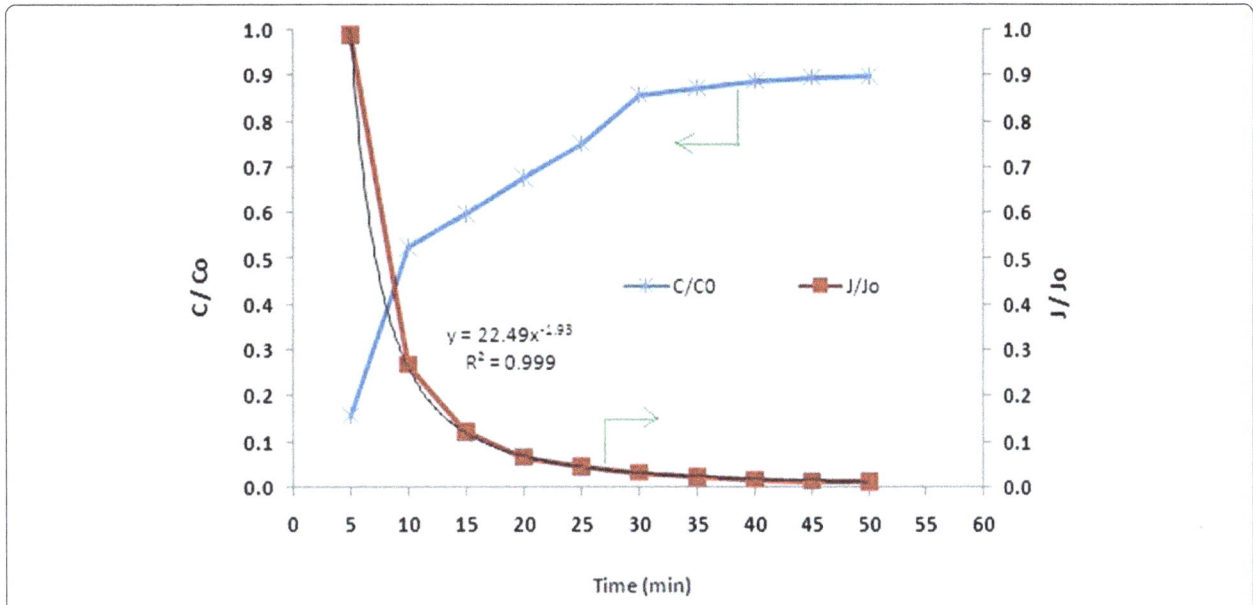

Fig. 6 Flux decline and TOC rejection trend in dead end filtration of HA at TMP of 1 bar and initial TOC = 5 ± 0.3 5 mg/L and pH of 7 ± 0.2

flow velocity, in which, some of deposited HA is swept away from the membrane surface.

In this operation mode, the flux declines rapidly (Fig. 6). About 80 % of decline occurs at first 10 min of filtration and after 20 min, it reduce to 90 % that differs from what ever seen in cross flow mode. Increasing of HA in bulk solution due to its accumulation in dead end filtration, result in rapid blocking of the pores and rapid flux decline. Generally, a 90 % of the flux decline occurred at 57 min of filtration in cross flow operation, while this percentage of flux reduction happened after about 15 min of filtration for dead end mode.

From Fig. 7 based on the equation obtained by plotting the log (d^2t/dv^2) versus log (dt/dv) the line slope (n) is 1.55. Accordingly, fouling mechanism of the membrane can be expressed by pore constriction or standard blocking filtration law that is defined as the reduction of the cross- sectional area of the membrane pores due to adsorption of HA.

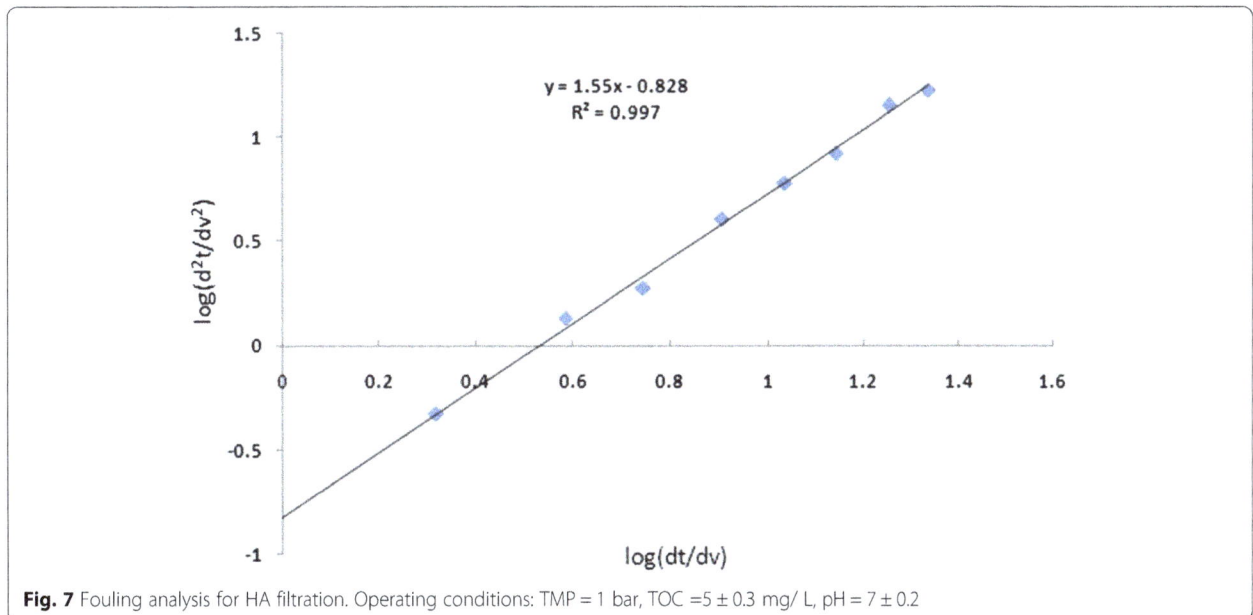

Fig. 7 Fouling analysis for HA filtration. Operating conditions: TMP = 1 bar, TOC =5 ± 0.3 mg/ L, pH = 7 ± 0.2

Sufficient small size of the pollutant that penetrate into the membrane and the adsorption affinity of the pollutant to membrane material have been considered as the main factors for pore constriction phenomenon [44]. Primarily rapid adsorption of HA into the membrane pores decrease the pore size and subsequently results in rapid flux decline. Such an adsorption through the membrane pores and channels (Fig. 2a) makes the cleaning difficult and lowers the membrane reversibility to some extent, otherwise, an effective cleaning technique should be developed.

The conditions, mainly membrane characteristics, can reveal different fouling mechanisms [45]. In fouling analysis, a combination of mechanisms may also occur in a filtration process. In some researches, intermediate blocking followed by cake filtration reported under experimental conditions [46]. In general, attempts to reduce the existing fouling by different methods and changing the removal mechanism from adsorption to electrostatic repulsion are of favor to overcome the problem. For existing membrane inducing the negative charges, can be a solution for reduction of fouling.

Conclusion

Despite of high contact angle and high hydrophobic property of the fabricated CNT-AAO membrane, pure water flux is considerably higher than that of common types of polymeric membranes. Due to high affinity, the membrane rapidly absorbs HA and consequently rapid flux decline occurs because of internal pore constriction as dominant fouling mechanism. It is important to be noted that a challenge related to present fabricated membrane is its frangibility. It is a critical obstacle for larger surface area application and high driven pressures for municipal and industrial applications. However, its application for some laboratory purposes may be beneficial to absorb some materials with the same characteristics as HA. Further studies can focus on methods of the AAO-CNT membranes fabrication with higher tolerability and flexibility for other pollutants removal from water.

Competing interests
The authors declare that they have no competing interests.

Authors' contributions
AJ has participated in all stages of the study (design of the study, conducting the experiments, analyzing the data and preparation of the manuscript). AHM participated in the design of the study, revision of the manuscript and intellectual helping in different stages of the study. SN participated in design of the study, final deeply revision of the manuscript and intellectual helping thorough the study. AMR participated in synthesis process and preparation of the membrane and interpretation of membrane characteristics. RN carried out statistical and technical analysis of data, participated in design of the study. RR participated in data collection and carried out technical analysis and manuscript preparation. All authors read and approved the final manuscript.

Acknowledgments
This research was part of a PhD dissertation of the first author and has been financially supported by a grant (NO, 22715-46-02-92) from Center for Water Quality Research (CWQR), Institute for Environmental Research, Tehran University of Medical Sciences, Tehran, Iran. Therefore, the authors greatly acknowledge the CWQR for the support of this research. The authors also thank the Laboratory staff of department of Environmental Health Engineering.

Author details
[1]Department of Environmental Health Engineering, School of Public Health, Tehran University of Medical Sciences, Tehran, Iran. [2]Center for Water Quality Research (CWQR), Institute for Environmental Research (IER), Tehran University of Medical Sciences, Tehran, Iran. [3]Center for Solid Waste Research (CSWR), Institute for Environmental Research (IER), Tehran University of Medical Sciences, Tehran, Iran. [4]Nanotechnology Research Institute of Petroleum Industry (RIPI), West Blvd. Azadi Sport Complex, Tehran, Iran. [5]Kurdistan Environmental Health Research Center, Kurdistan University of Medical Sciences, Sanandaj, Iran.

References
1. Matilainen A, Vepsäläinen M, Sillanpää M. Natural organic matter removal by coagulation during drinking water treatment: A review. Adv Colloid Interface Sci. 2010;159:189–97.
2. Karnik BS, Davies SH, Baumann MJ, Masten SJ. The effects of combined ozonation and filtration on disinfection by-product formation. Water Res. 2005;39:2839–50.
3. Zazouli M, Nasseri S, Mahvi A, Mesdaghinia A, Younecian M, Gholami M. Determination of hydrophobic and hydrophilic fractions of natural organic matter in raw water of Jalalieh and Tehranspars water treatment plants (Tehran). J Appl Sci. 2007;7:2651–5.
4. Kim H-C, Yu M-J. Charecerization of natural organic matter in conventional water treatment process for selection of treatment processes focused on DBPs control. Water Res. 2005;39:4779–89.
5. Crittenden JC, Trussel RR, W HD, J HK, GT: Warer treatment: Principles and design. 2nd edition: John Wiley & sons Inc., New Jersey; 2005
6. Chen C, Zhang X, Zhu L, Liu J, He W, Han H. Disinfection by-products and their precursors in a water treatment plant in North China: seasonal changes and fraction analysis. Sci Total Environ. 2008;397:140–7.
7. Anu Matilainen ETG, Tanja L, Leif H, Amit B, Mika S. An overview of the methods used in the characterisation of natural organic matter (NOM) in relation to drinking water treatment. Chemosphere. 2011;83:1431–42.
8. Zazouli M, Ulbricht M, Nasseri S, Susanto H. Effect of hydrophilic and hydrophobic organic matter on amoxicillin and cephalexin residuals rejection from water by nanofiltration. Iranian J Environ Health Sci Eng. 2010;7:15–24.
9. Moussavi S, Ehrampoush M, Mahvi A, Ahmadian M, Rahimi S. Adsorption of Humic Acid from Aqueous Solution on Single-Walled Carbon Nanotubes. Asian J Chem. 2013;25.
10. Moussavi S, Ehrampoush M, Mahvi A, Rahimi S, Ahmadian M. Efficiency of Multi-Walled Carbon Nanotubes in Adsorbing Humic Acid from Aqueous Solutions. Asian J Chem. 2014;26:821–6.
11. Bazrafshan E, Biglari H, Mahvi AH. Humic acid removal from aqueous environments by electrocoagulation process using iron electrodes. J Chem. 2012;9:2453–61.
12. Mahvi A, Maleki A, Rezaee R, Safari M. Reduction of humic substances in water by application of ultrasound waves and ultraviolet irradiation. Iranian J Environ Health Sci Eng. 2009;6:233–40.
13. Zazouli M, Nasseri S, Mahvi A, Gholami M, Mesdaghinia A, Younesian M. Retention of humic acid from water by nanofiltration membrane and influence of solution chemistry on membrane performance. Iranian J Environ Health Sci Eng. 2008;5:11–8.
14. Rezaee R, Maleki A, Jafari A, Mazloomi S, Zandsalimi Y, Mahvi AH. Application of response surface methodology for optimization of natural organic matter degradation by UV/H2O2 advanced oxidation process. J Environ Health Sci Eng. 2014;12:67.
15. Alipour V, Nasseri S, Nodehi RN, Mahvi AH, Rashidi A. Preparation and application of oyster shell supported zero valent nano scale iron for

removal of natural organic matter from aqueous solutions. J Environ Health Sci Eng. 2014;12:146.

16. Naghizadeh A, Nasseri S, Mahvi AH, Nabizadeh R, Kalantary RR, Rashidi A. Continuous adsorption of natural organic matters in a column packed with carbon nanotubes. J Environ Health Sci Eng. 2013;11:14.

17. Peter-Varbanets M, Zurbrügg C, Swartz C, Pronk W. Decentralized systems for potable water and the potential of membrane technology. Water Res. 2009;43:245–65.

18. Chin CJM, Shih LC, Tsai HJ, Liu TK. Adsorption of oxylene and p-xylene from water by SWCNTs. Carbon. 2007;45:1254–60.

19. Xuemei Rena CC, Masaaki N, Xiangke W. Carbon nanotubes as adsorbents in environmental pollution management: A review. Chem Eng J. 2011;170:395–410.

20. Upadhyayula VKK, Deng S, Mitchell MC, Smith GB. Application of carbon nanotube technology for removal of contaminants in drinking water: A review. Sci Total Environ. 2009;408:1–13.

21. Ahn CH, Baek Y, Lee C, Kim SO, Kim S, Lee S,Kim S-H, Bae SS, Park J, Yoon J: Carbon nanotube-based membranes: Fabrication and application to desalination. J ind Eng Chem. 2012;18:1551–9.

22. Gilani N, Daryan JT, Rashidi A, Omidkhah MR. Separation of methane–nitrogen mixtures using synthesis vertically aligned carbon nanotube membranes. Appl Surf Sci. 2012;258:4819–25.

23. Hou PX, Liu C, Shi C, Cheng HM. Carbon nanotubes prepared by anodic aluminum oxide template method. Chin Sci Bull. 2012;1–18.

24. Celik E, Park H, Choi H. Carbon nanotube blended polyethersulfone membranes for fouling control in water treatment. Water Res. 2011;45:274–82.

25. Kim E-S, Hwang G, Gamal El-Din M, Liu Y. Development of nanosilver and multi-walled carbon nanotubes thin-film nanocomposite membrane for enhanced water treatment. J Membra Sci. 2012;394–395:37–48.

26. Majeed S, Fierro D, Buhr K, Wind J, Du B, Boschetti-de-Fierro A, Abetz V: Multi-walled carbon nanotubes (MWCNTs) mixed polyacrylonitrile (PAN) ultrafiltration membranes. J Membra Sci. 2012;403–404:101–9.

27. Qiu S, Wu L, Pan X, Zhang L, Chen H, Gao C. Preparation and properties of functionalized carbon nanotube/PSF blend ultrafiltration membranes. J Membra Sci. 2009;342:165–72.

28. Rahimpour A, Jahanshahi M, Khalili S, Mollahosseini A, Zirepour A, Rajaeian B. Novel functionalized carbon nanotubes for improving the surface properties and performance of polyethersulfone (PES) membrane. Desalination. 2012;286:99–107.

29. Vatanpour V, Madaeni SS, Moradian R, Zinadini S, Astinchap B. Fabrication and characterization of novel antifouling nanofiltration membrane prepared from oxidized multiwalled carbon nanotube/polyethersulfone nanocomposite. J Membra Sci. 2011;375:284–94.

30. Gilani N, Towfighi J, Rashidi A, Mohammadi T, Omidkhah MR, Sadeghian A. Investigation of H2S separation from H2S/CH4 mixtures using functionalized and non-functionalized vertically aligned carbon nanotube membranes. Appl Surf Sci. 2013;270:115–23.

31. Poinern GEJ, Ali N, Fawcett D. Progress in nano-engineered anodic aluminum oxide membrane development. Materials. 2011;4:487–526.

32. Abràmoff MD, Magalhães PJ, Ram SJ. Image processing with ImageJ. Biophotonics Int. 2004;11:36–43.

33. Chen X, Cendrowski K, Srenscek-Nazzal J, Rümmeli M, Kalenczuk RJ, Chen H, Chu PK, Borowiak-Palen E: Fabrication method of parallel mesoporous carbon nanotubes. Colloids Surf A Physicochem Eng Asp. 2011;377:150–5.

34. Rana K, Kucukayan-Dogu G, Bengu E. Growth of vertically aligned carbon nanotubes over self-ordered nano-porous alumina films and their surface properties. Appl Surf Sci. 2012;258:7112–7.

35. Salahi A, Abbasi M, Mohammadi T. Permeate flux decline during UF of oily wastewater: Experimental and modeling. Desalination. 2010;251:153–60.

36. Majumder M, Chopra N, Andrews R, Hinds BJ. Nanoscale hydrodynamics: enhanced flow in carbon nanotubes. Nature. 2005;438:44–4.

37. Baek Y, Kim C, Seo DK, Kim T, Lee JS, Kim YH, Ahn KH, Bae SS, Lee SC, Lim J, Lee K, Yoon J: High performance and antifouling vertically aligned carbon nanotube membrane for water purification. J Membra Sci. 2014;460:171–7.

38. Park S-M, Jung J, Lee S, Baek Y, Yoon J, Seo DK,Kim YH: Fouling and rejection behavior of carbon nanotube membranes. Desalination. 2014;343:180–6.

39. Avramescu M-E, Sager W, Borneman Z, Wessling M. Adsorptive membranes for bilirubin removal. J Chromatogr B. 2004;803:215–23.

40. Jamshidi Gohari R, Lau W, Matsuura T, Halakoo E, Ismail A. Adsorptive removal of Pb (II) from aqueous solution by novel PES/HMO ultrafiltration mixed matrix membrane. Sep Purif Technol. 2013;120:59–68.

41. Liu L, Zheng G, Yang F. Adsorptive removal and oxidation of organic pollutants from water using a novel membrane. Chem Eng J. 2010;156:553–6.

42. Naghizadeh A, Nasseri S, Rashidi A, Kalantary RR, Nabizadeh R, Mahvi A. Adsorption kinetics and thermodynamics of hydrophobic natural organic matter (NOM) removal from aqueous solution by multi-wall carbon nanotubes. Water Sci Technol: Water Supply. 2013;13:273–85.

43. Hermia J. Constant pressure blocking filtration law application to powder-law non-Newtonian fluid. Trans Inst Chem Eng. 1982;60:183–7.

44. Crittenden JC, Trussell RR, Hand DW, Howe KJ, Tchobanoglous G: MWH's Water Treatment: Principles and Design. 3rd edition, John Wiley & sons Inc, New Jersey; 2012.

45. Susanto H, Ulbricht M. High-performance thin-layer hydrogel composite membranes for ultrafiltration of natural organic matter. Water Res. 2008;42:2827–35.

46. Heo J, Kim H, Her N, Lee S, Park Y-G, Yoon Y. Natural organic matter removal in single-walled carbon nanotubes–ultrafiltration membrane systems. Desalination. 2012;298:75–84.

8

Application of concrete surfaces as novel substrate for immobilization of TiO$_2$ nano powder in photocatalytic treatment of phenolic water

Mohammad Delnavaz[1,2], Bita Ayati[1*], Hossein Ganjidoust[1] and Sohrab Sanjabi[3]

Abstract

Background: In this study, concrete application as a substrate for TiO$_2$ nano powder immobilization in heterogeneous photocatalytic process was evaluated. TiO$_2$ immobilization on the pervious concrete surface was done by different procedures containing slurry method (SM), cement mixed method (CMM) and different concrete sealer formulations. Irradiation of TiO$_2$ was prepared by UV-A and UV-C lamps. Phenolic wastewater was selected as a pollutant and efficiency of the process was determined in various operation conditions including influent phenol concentration, pH, TiO$_2$ concentration, immobilization method and UV lamp intensity.

Findings: The removal efficiency of photocatalytic process in 4 h irradiation time and phenol concentration ranges of 25–500 mg/L was more than 80 %. Intermediates were identified by GC/Mass and spectrophotometric analysis.

Conclusions: According to the results, photocatalytic reactions followed the pseudo-first-order kinetics and can effectively treat phenol under optimal conditions.

Keywords: Concrete, Immobilization, Intermediates, Water cleaning technology

Background

The ability of Advanced Oxidation Processes (AOP$_s$) in treating a wide range of hazardous wastes has brought this technology to the forefront of research over the last decade. Among AOP$_s$, application of heterogeneous photo-catalysis by using semiconductors has been proved to be real interest as an efficient tool for degrading both aquatic and atmospheric organic contaminants [1]. Semiconductors are photo-reactive metal oxides for contaminants eradication that refer to photo-catalysts [2]. Titanium dioxide (TiO$_2$) is an established photocatalyst utilized in the photo-oxidation process. When TiO$_2$ is exposed to the appropriate wavelength of ultra-violet light (UV-A), electrons in the low-energy valence band will absorb the photon's energy and move into the high-energy conduction band. The result of this electron excitation is a hole, or positive charge, in the valence band (h$^+$)

and an electron in the conduction band (e$^-$) [3]. Reaction yield and photocatalytic activity will be increased when the diameter of TiO$_2$ particles becomes smaller especially below 100A° [1].

Photocatalytic reactors for water and wastewater treatment can be classified to slurry and photocatalytic ones. In the slurry reactors, the catalyst particles are freely dispersed in the fluid phase (water) and consequently the photo-catalyst is fully integrated in the liquid mobile phase [4]. Whereas the immobilized catalyst reactor design features a catalyst anchored to a fixed support and dispersed on the stationary phase. Slurry systems require the separation of the fine sub micron particles TiO$_2$ from the treated milk-like water suspension. Separation steps cause to complicate the treatment process and decrease the economical viability of the slurry reactor approach [5]. Therefore, these difficulties have led many researchers to study reactors with thin immobilized films of catalyst bonded to a solid substrate such as activated carbon [6], fiber optic cables [7] fiberglass [8], glass beads [9], quartz sand [10], silica gel [11] and stainless steel [12]. Dip coating from suspension, spray coating, sputtering, sol–gel related

* Correspondence: ayati_bi@modares.ac.ir
[1]Civil and Environmental Engineering Faculty, Tarbiat Modares University, Tehran, Iran
Full list of author information is available at the end of the article

methods, and electrophoretic deposition [4, 13] are techniques developed for immobilizing TiO_2 catalysts on these substrates. Although these techniques and substrates had a suitable performance for treating different types of wastewater in laboratory scale experiments, but their application in wastewater treatment plants and pilot scale studies are questionable.

Most applications of concrete modified by TiO_2 were done for air pollutant removal and self-cleaning surfaces [14]. Although many studies in the field of air pollution have been done by concrete-TiO_2 photocatalyst process [15, 16], but the number of wastewater treatment researches is very low. Application of concrete as TiO_2 substrate has unique characteristics such as porosity, natural abundance, absence of toxicity, and low price. This construction material is used in all of the water and wastewater treatment plants (WWTP) all over the world. Therefore, the photocatalytic process by TiO_2 photocatalyst that immobilized on concrete surfaces can be used in large scale WWTP.

Phenols and its compounds are widely found in paint, leather and textile, disinfectants, medicine, oil refinery and lubricant production wastewater industries in Iran [17–19]. Phenol is rapidly absorbed through the skin and can cause skin and eye burns upon contact. Comas, convulsions, cyanosis and death can be resulted from its overexposure [20, 21]. Therefore Phenol-containing wastewater may not be conducted in open water without treatment because of its toxicity.

Several investigations using different physical, chemical and biological systems and their combination for phenol elimination from wastewater have been reported [22–27]. Biological processes are preferred to other conventional methods due to their ability to effectively destroy the pollutants in an environmentally benign and cost effective way [19]. Sensitive of process to organic shocking load and need to high control for microbial acclimation for hard degradable compound such as phenol are limiting factor in biological treatment. Photocatalytic process didn't have these limitations and can be used separately or in joint with biological process. Different researches were done related study to photocatytic process in wastewater treatment. Chiou et al. (2008) have studied degradation of phenol and m-nitrophenol using a photocatalytic process in aqueous solution by commercial TiO_2 powders (Degussa P-25) under UV irradiation [28]. The optimal solution pH of single phenol and m-NP was at around 7.4 and 8.9, respectively. Immobilization of TiO_2 on perlit granules for photocatalytic degradation of phenol was done by Hosseini et al. (2009) [13]. The Results showed uniform coating on perlit and good photocatalytic activity for the catalysts. Pumic stone was applied as TiO_2 substrate for degradation of dyes and dye industry pollutants [29]. Real wastewaters collected before biological treatment treated in photocatalytic reactor and color disappeared after 4-h.

The main objective of this research was to study the feasibility of using concrete as substrate for TiO_2 nano powder immobilization to treat phenolic wastewater. For this reason, four methods were applied for immobilization of nano TiO_2 on concrete. Kinetic of the reactions, long term use of the process and intermediates formed during the photo degradation were other objectives of this research.

Material and methods
Chemicals
Cement in the concrete was an ordinary Portland type V that is usually applied in WWTP construction in Iran. Selection of concrete mix proportion was done according to standard practice for selecting proportions of normal concrete (ACI 211.1 :1996) [30]. Light expanded clay aggregate (LECA) was used as light coarse aggregate to lead specific gravity of concrete to 1200 kg/m^3. LECA application increased surface porosity of concrete and promoted the specific surface area as an important immobilization parameter. Concrete surface was fabricated in wooden moulds with an internal dimension of $500 \times 250 \times 50$ mm. TiO_2 Degussa P25, anatase/rutile ca. 70/30, with an average particle size of 20 nm and a BET surface area 55 ± 15 $m^2.g^{-1}$ was used as photocatalyst. UV radiation was provided by different powers of UV-A Philips medium pressure lamps. The spectral irradiance of UV-A lamp ranges from 300 to 400 nm. The primary wavelength distribution of the UV-A lamp was 365 nm and the light intensity was among 4.42-8.9 $mW.cm^{-2}$. The incident UV-A light intensity was measured by a UV power meter (UVA-365- Lutron Taiwan). Also UV illumination was provided by UV-C lamps (20 Watts, maximal light intensity at 256 nm).

Phenol (purity above 99 %) as a contaminant, NaOH, HCl for pH adjustment, NH_4OH, K_2HPO_4, KH_2PO_4, $K_3Fe(CN)_6$ and 4-aminoantiprine as phenol concentration reagents and other chemicals all offered by Merck Co., as an analytical reagent grade. Wastewater produced by adding deionized water and phenol in various concentrations (25–500 mg/L).

Immobilization methods
Immobilization of TiO_2 nano powder on concrete surfaces was carried out by four different procedures. Adhesion properties of cement and concrete sealers were applied to fix photocatalyst on pervious concrete surfaces.

Slurry method (SM)
The details of this technique are given as follows:

- 20 g/L slurry TiO_2 was prepared using 25 % (v/v) methanol in deionized water. Methanol would help with the attachment of TiO_2 to the surface.
- The solution was stirred vigorously for 10 min at 20 °C.
- The slurry was sonicated for 5 min to separate the flocculated TiO_2 and to obtain more uniform slurry.
- One half of slurry sprayed on fresh concrete to use adhesive properties of cement.
- The coated surface was placed in oven at 100 °C for 2 h to remove the moisture content.
- The rest of the slurry sprayed on hardened concrete and annealed at 450 °C for 2 h to remove any organics from the surface.
- The support was let to dry and was washed with pure water to eliminate the excess of the catalyst.

Cement mixed method (CMM)

In this method, 20 g/L TiO_2 was mixed by cement grout (40 g cement + 20 mL water) and distributed on hardened concrete base by an ordinary brush. After 8 h the concrete surface was placed in oven at 100 °C for 2 h to be dried. The support was washed with pure water to eliminate the excess of the catalyst.

Epoxy sealer method (ESM) & Waterproof sealer method (WSM)

In these procedures, the effect of epoxy concrete sealer (Nitofix- Fars Iran Company) and waterproof concrete sealer (Nitotile- Fars Iran Company) for adhesion of TiO_2 was examined. The details of this technique are as follows:

- 100 mL of the selected concrete sealer mixed with 1000 mL of pure water and stirred vigorously for 10 min at 20 °C.
- Prepared emulsion was distributed on the hardened concrete surface with a trowel and let to dry for 20 min.
- After that time, 20 gr/L of TiO_2 poured on concrete surface for adhesion on sealer.
- The support was let to dry and was washed with pure water to eliminate the excess of the catalyst.

Photocatalytic reactor setup

The immobilized concrete surfaces were placed in the pilot scale photocatalytic reactor (Fig. 1). A dosing pump was used to feed the phenol-laden wastewater into photocatalytic zone and distributed by a pipe with 1 cm mesh in its length. Reactor hydraulics parameters were controlled by water depth limited to 4 mm. Three concrete surfaces and circulation mode was used to prepare efficient contact time between photocatalyst and wastewater. Dissolved oxygen (DO) and pH was monitored in all the experimental period time by Crison-Oxi45 and Metrohm690, respectively. UV radiation was provided by UV-A and UV-C Philips medium pressure lamps. Samples were taken periodically form the sample ports for phenol concentration analysis. Prior to measurement, the liquid samples were centrifuged by Sigma101 at about 4000 rpm for 10 min to remove detached TiO_2 particles. The phenol concentrations were measured by colorimetric 4-aminoantipyrine procedure using a Perkin Elmer-Lambda EZ 150 UV/vis spectrophotometer as described in standard methods (2005). The structure and

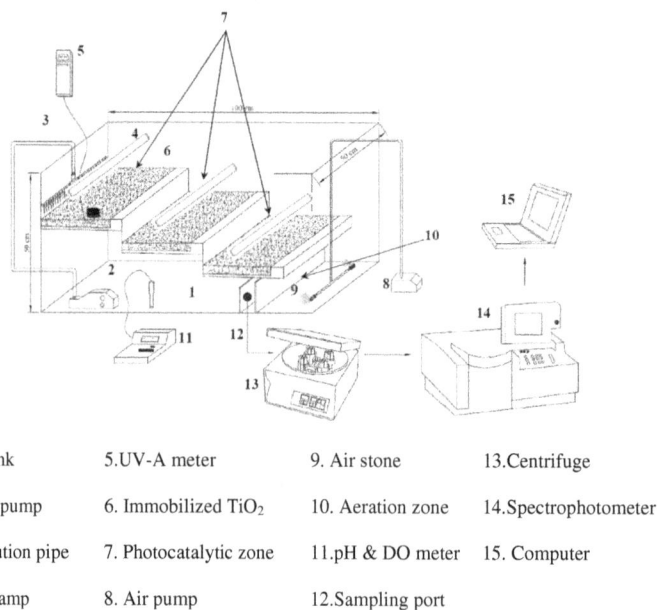

1.Feed tank	5.UV-A meter	9. Air stone	13.Centrifuge
2.Dosing pump	6. Immobilized TiO_2	10. Aeration zone	14.Spectrophotometer
3.Distribution pipe	7. Photocatalytic zone	11.pH & DO meter	15. Computer
4.UV-A lamp	8. Air pump	12.Sampling port	

Fig. 1 Schematic design of photo-catalyst reactor

morphology of prepared catalysts and concrete surface were determined using Philips XL30 scanning electron microscope (SEM) followed by AU-coated by sputtering method using a coater sputter (Bal-Tec, Switzerland). All other parameters were determined according to the standard methods (2005). The intermediates determined by GC/Mass (column: chrompack CP-Sil 8 CB 50 m × 250 μm × 0.12 μm). The gas vector used was helium and the detector was a FID. The program of temperature was: detector temperature = 240 °C, injector temperature = 230 °C, oven initial and final temperatures = 105 °C & 190 °C, oven rise 10 °C min^{-1}, initial and final time = 2 & 15 min). All other parameters were determined according to the Standard Methods [31].

Results and discussion
Characteristics of immobilized concrete surfaces
In all immobilization procedures, SEM and energy dispersive X-ray microanalysis (EDX) were done to confirm the presence of TiO$_2$ on concrete surfaces (Fig. 2). Images of non-immobilized surface proved high surface porosity of concrete as a good support for TiO$_2$. SEM analysis showed a uniform appearance of TiO$_2$ catalyst in ESM, WSM and SM but dispersed coating in CMM.

Mixing TiO$_2$ by cement as an adhesive agent caused the minimum level of active surface catalyst in CMM (Fig. 2c). In other methods, catalyst was poured on concrete surface that covered by adhesive agent and consequently uniform TiO$_2$ cover was prepared. EDX analysis

of TiO$_2$ coatings showed no significant levels of noticeable impurities in all immobilizations cases.

Effect of influent phenol concentration
Effect of different initial concentrations (C$_0$ = 25–500 ppm) in removal efficiency is shown in Fig. 3.

At first, in all experiments the removal efficiency was measured in pH = 7 without UV lamps for 60 min to determine the pollutants adsorption to concrete and phenol volatility. Removal efficiency in these conditions was less than 5 % and stripping of phenol by aeration was negligible because of its very low Henry's constant [32]. Photocatalytic process was then provided by turning on UV lamps and removal efficiency was measured in different retention times. In SM method after 4 h, 95 and 69 % of phenol was degraded in 25 and 500 ppm initial concentration, respectively. Equal degradation was achieved in ESM when C$_0$ was 25 ppm but the removal efficiency in WSM was 75 % at the same conditions. The removal efficiency of the CMM was lower than that of other methods. So that after 4 h, 28 % phenol with initial concentration of 500 ppm was degraded. At the same conditions removal efficiency for ESM and WSM were 56 and 43 %, respectively. The results of this study are comparable with similar researches in recent years. For example Chiou et al. (2008) in slurry photo-reactor treated 60 % of 0.52 mM (~49 mg/L) phenol after 3 h and 400 W UV lamp. In other research, Hosseini et al. (2007) reached to 88.3 % phenol degradation after 4 h in initial concentration = 1 mM (~94 mg/L) and UV lamp intensity = 120 W [13].

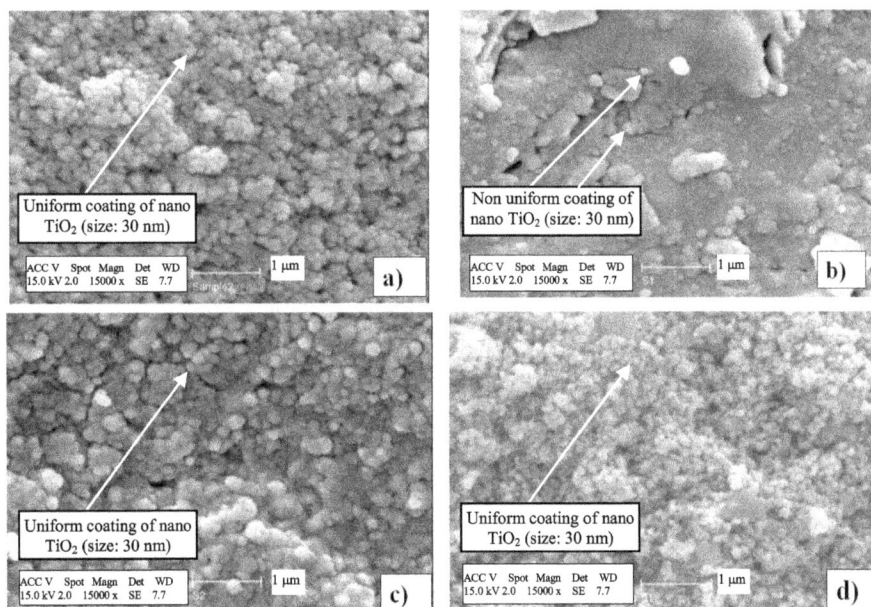

Fig. 2 SEM picture and EDX spectrum of TiO$_2$ film coated on concrete surface: **a** SM, **b** CMM, **c** ESM, **d** WSM

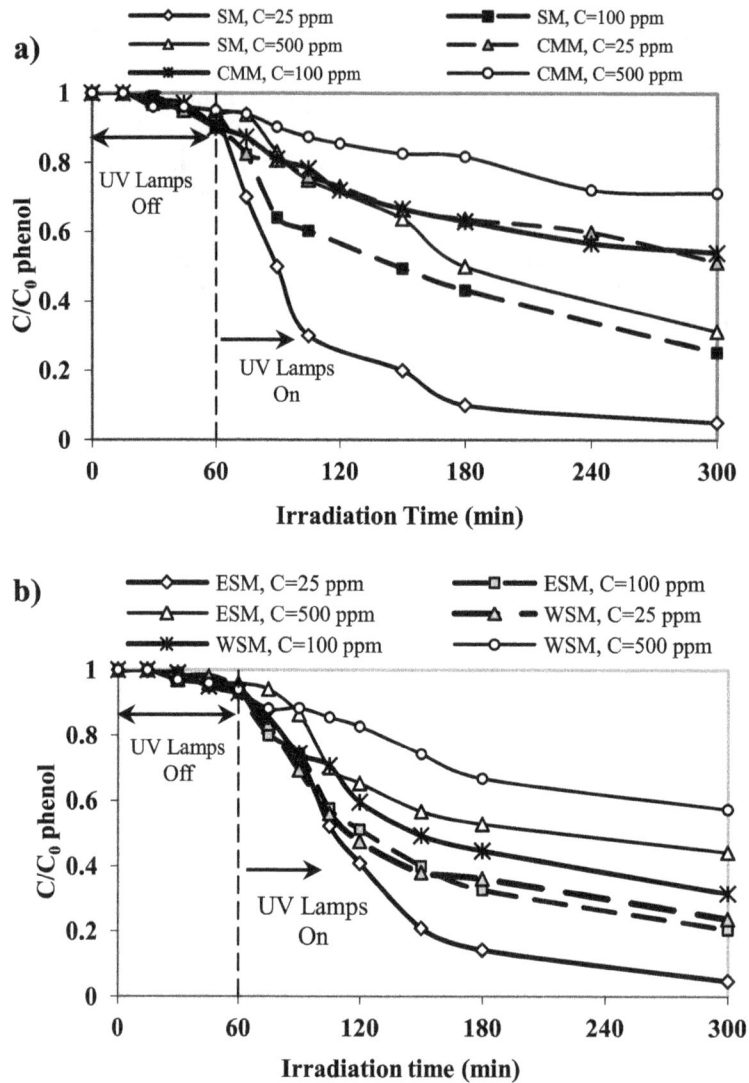

Fig. 3 Effect of phenol initial concentration in removal efficiency, **a** SM & CMM, **b** WSM & ESM

Effect of pH solution

Electrostatic interaction between semiconductor surface, solvent molecules, substrate and charged radicals formed during photocatalytic oxidation is strongly dependent on the pH of the solution [28]. Solution pH dominates photo-degradation process due to the strong pH dependence of many related properties such as the semiconductor's surface charge state, flat band potential, and the solution dissociation. In alkaline pH, phenol was converted to phenoxide group (remove of H^+ and creation of negative charge on hydroxyl group) that was more reactive than phenol in a solution [33]. In the other hand, the ionization state of the surface of the photocatalyst can also be protonated and deprotonated under acidic and alkaline conditions. The pH_{ZPC} of Degussa P-25 TiO_2 used here is 6.5,

and phenol is 9.89, respectively [28]. While under acidic conditions, the positive charge of the TiO_2 surface increases as the pH decreases ($TiOH_2^+$); above pH 6.5 the negative charge at the surface of the TiO_2 increases with increasing pH (TiO^-). The optimum pH can be obtained between pH_{ZPC} of photocatalyst and contaminant. Effect of pH in the range of 4–12 on phenol removal efficiency in ESM was evaluated (Fig. 4).

The highest efficiency was observed at pH of 9–12. The difference between phenol removal efficiency at pH of 12 and 4 was determined 39 % in ESM (52 and 86 % removal efficiency) when initial concentration was 100 mg/L and UV-A lamp intensity was equal to 5.33 $mW.cm^{-2}$. The rate of removal efficiency in different pH was similar in different immobilization methods and determined between pH_{ZPC} of phenol

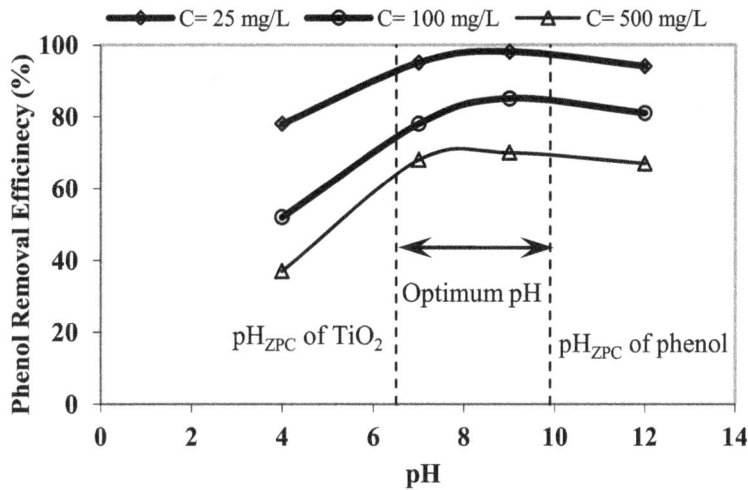

Fig. 4 Effect of pH in phenol removal efficiency in ESM

and TiO₂. Several researchers have observed different results about the effect of the pH on the TiO₂ photocatalytic decomposition of phenol. This discrepancy on the optimum pH may be a function of the various operating conditions considered.

Effect of immobilization methods

The photocatalytic activities for all combinations of coating methods are summarized in Fig. 5. Influent phenol concentration was 50 ppm, pH = 7, UV intensity = 5.33 mW.cm^{-2} and UV lamp distance to concrete surface = 10 cm in these experiments. Results showed ESM had the highest photocatalytic activities so that the removal efficiency of this method was more than 90 % after 4 h. The photocatalytic efficiency of the coating methods was in the following order: ESM > SM > WSM > CMM. The removal efficiency of CMM method after 4 h was 42 % that was the minimum efficiency between other immobilization techniques. The most important reason that proved by SEM-EDX analysis was reduction levels of active surface of nano particles that mixed with cement. While in other methods, TiO₂ nano particles attached to a concrete top surface, in the CMM nano TiO₂ particles mixed with cement as cohesive agent and a lot of percents of photoctalyst disappeared. Comparison between ESM and WSM showed that ESM had better performance due to hydrophobic properties of waterproof sealers that decreased connection between contaminants and immobilized pohotocatalyst.

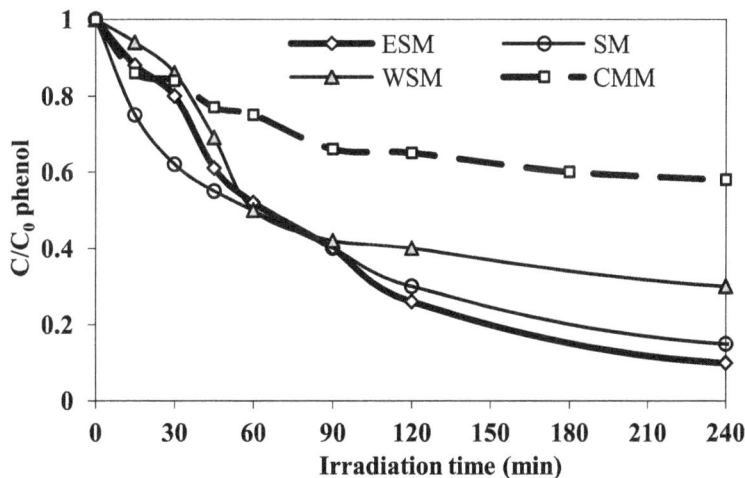

Fig. 5 Comparison between phenol removal efficiency by different immobilization methods

Effect of UV lamp Intensity

The effect of UV lamp intensity on the phenol photo degradation at constant initial concentration (100 ppm) and pH = 7 is presented in Fig. 6. The results showed that phenol removal efficiency increased when the UV lamps intensity promoted. In other words, photocatalytic efficiency in all immobilization methods improved about 50 % when UV-A lamp intensity increased from 4.42 to 8.9 mW.cm^{-2}. This is reasonable because the stronger the irradiating UV, the more the UV penetrating. A few differences between phenol degradation since 8.9 and 8.1 UV intensity was confirmed that further increase in UV intensity couldn't increase the amount of phenol destroyed.

Application of UV-C lamp promoted phenol removal efficiency more than 18 % compared to UV-A lamps at a constant irradiation time. Capability of UV-C lamp in degradation of 50 % of phenol without TiO$_2$ photocatalyst showed proper spectral irradiance of UV-A lamp that degraded only 7 % of phenol when photocatalyst was deleted. On the other hand, UV-A lamps had appreciated feature in the photocatalytic process compare to UV-C lamps.

Effect of long-term use

One of the most important parameters in photocatalytic reactors with immobilized photocatalyst is a reduction in removal efficiency because of TiO$_2$ particles detachment and catalyst surface fouling by formation of by-products during the degradation process. This limitation inhibited the application of immobilized procedure as a long term process for wastewater treatment. The influence of long term use in the degradation efficiency has been examined at a constant initial concentration (100 ppm), pH = 7 and UV lamp intensity of 5.33 mW.cm^{-2}. Proper connection between TiO$_2$ nano particles and concrete surface led to 2 and 3.5 % reduction in

phenol removal efficiency for ESM and WSM, respectively. In SM that TiO$_2$ poured on concrete surfaces and cement was used as a cohesive agent instead of concrete sealers, the reduction in removal efficiency after 45 times iterations was more than 21 %. Mixing cement by TiO$_2$ caused 4 % removal efficiency reduction after 45 times in CMM. In other research, the elimination of some TiO$_2$ from the pumice stone surface showed significant decrease of photocatalytic efficiency in long term use of process when SM was used [29].

Kinetics of the reactions

Langmuir-Hinshelwood kinetic has been used to characterize the destruction of many contaminants in different structure of catalyst [34]. The final form of this model can be expressed as Eq. 1:

$$r = -\frac{dC}{dt} = \frac{K_r K_s C_0}{1 + K_s C_0} \tag{1}$$

Where (r) is the rate of the photocatalytic degradation, (C$_0$) is the organic concentration, (K$_r$) the reaction rate constant, (K$_s$) the apparent adsorption constant and (t) is the time of reaction. The term K$_s$C$_0$ is often negligible when the concentration is low, and the reaction rate can be expressed as pseudo-first-order model as shown in Eq. 2:

$$-\frac{dC}{dt} = K_r K_s C_0 = K_{app} C_0 \tag{2}$$

Where K$_{app}$ is called apparent first order reaction rate. Integration of the equation yields to Eq. 3:

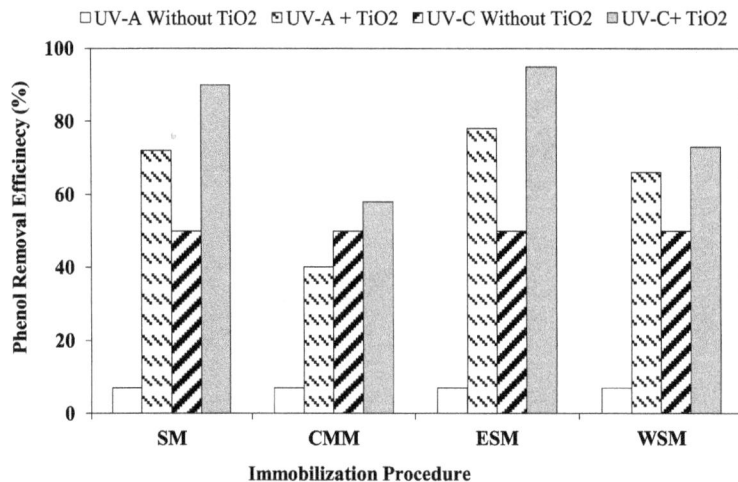

Fig. 6 Effect of UV lamp intensity in phenol removal efficiency

$$\ln\left(\frac{C}{C_0}\right) = -K_{app}t \qquad (3)$$

A plot of $-\ln(C/C_0)$ against (t) and slope of linear regression analysis is equal to the value of K_{app}. The calculated result indicated that photocatalytic degradation of phenol at the reactions conditions follows a pseudo-first-order kinetics that compared with other researches [13, 35].

Regression coefficients (R^2) and K_{app} for the photo degradation of phenol at different immobilization method, pH values and initial concentration are shown in Table 1.

The results showed that K_{app} increased in all immobilization methods and different initial concentration when pH value changes from 4 to 12. Comparison among different immobilization methods approved that ESM had the highest removal efficiency compared to the other methods that had the most K_{app}.

Phenol photo degradation intermediates

The intermediates of photo degradation of 100 mg/L phenol at pH = 9 and intensity = 8.1 mW.cm^{-2} were obtained by GC/Mass analysis. The results showed phenol hydroxylated via hydroxyl radical attack to different extent forming a variety of intermediates from the very beginning of the reaction. Firstly, aromatic ring (hydroquinone, catechol and benzoquinone) produced and after 60 min, intermediates convert to linear compounds such as oxalic acids and formic acids. Lastly, phenol is completely mineralized to CO_2 and H_2O. This can be explained given the fact that adsorbed phenol molecules on TiO_2 are subject to variable interaction with hydroxyl radicals.

Changes of the UV spectrum with respect to destruction of aromatic ring and phenol and variation of the absorbance at 268 nm in different irradiation time are shown in Fig. 7. It can be seen that when the irradiation time increased, maximum peak of the absorption spectrum decreased and finally a spectrum without noticeable peak was observed. At the end, a residual absorbance around 200 nm corresponding to the organic acid formation as final degradation products could be seen.

Conclusions

In this study, the performances of various coating methods alternatives on concrete surfaces for treating phenolic wastewater were compared. Based on the findings, the following conclusions can be drawn:

1. SEM determination has showed well uniformity of coating processes and EDX spectrum approved nano TiO_2 on concrete surface.
2. The photocatalytic tests of phenol degradation showed a good photocatalytic activity for TiO_2.
3. The ESM coating method was found to be the best technique, with the highest photocatalytic activity

Table 1 Regression coefficients (R^2) and K_{app} for phenol photo degradation

C_0 (ppm)	pH		SM	CMM	ESM	WSM
25	4	R^2	0.94	0.9	0.96	0.95
		K_{app}	0.0044	0.002	0.012	0.003
	7	R^2	0.99	0.81	0.97	0.93
		K_{app}	0.0054	0.0029	0.0135	0.01
	9	R^2	0.97	0.88	0.95	0.93
		K_{app}	0.0061	0.0035	0.0161	0.0055
	12	R^2	0.97	0.87	0.99	0.96
		K_{app}	0.007	0.0041	0.0162	0.006
100	4	R^2	0.94	0.9	0.99	0.98
		K_{app}	0.0044	0.0025	0.0058	0.0032
	7	R^2	0.94	0.91	0.91	0.9
		K_{app}	0.0057	0.0025	0.0073	0.0052
	9	R^2	0.97	0.89	0.92	0.95
		K_{app}	0.0061	0.0033	0.01	0.0058
	12	R^2	0.97	0.88	0.97	0.95
		K_{app}	0.007	0.005	0.012	0.0065
500	4	R^2	0.86	0.92	0.98	0.99
		K_{app}	0.0023	0.001	0.0042	0.002
	7	R^2	0.99	0.95	0.98	0.95
		K_{app}	0.0047	0.0013	0.006	0.0023
	9	R^2	0.93	0.92	0.99	0.98
		K_{app}	0.0057	0.0023	0.0075	0.004
	12	R^2	0.96	0.88	0.98	0.93
		K_{app}	0.0058	0.004	0.0089	0.0052

and adhesion. The removal efficiency of this method was more than 90 % after 4 h.

4. Effect of pH showed that the removal efficiency of phenol increased as pH value increased from 4 to 9 while phenol was converted to phenoxide group.
5. Photocatalytic efficiency in all immobilization methods improved about 50 % when UV lamp intensity increased 2 times.
6. Process kinetics by Langmuir-Hinshelwood model was approved pseudo-first-order reaction for phenol photo degradation.
7. Reduction of phenol removal efficiency was done as a reverse method for evaluation of nano TiO_2 detachment from concrete surfaces.
8. Results showed that application of concrete sealers had better performance in comparison with SM and CMM that cement was used as a cohesive agent. In WSM and ESM, reduction of removal efficiency was less than 2 and 3.5 % after 45 times iteration of process, respectively.

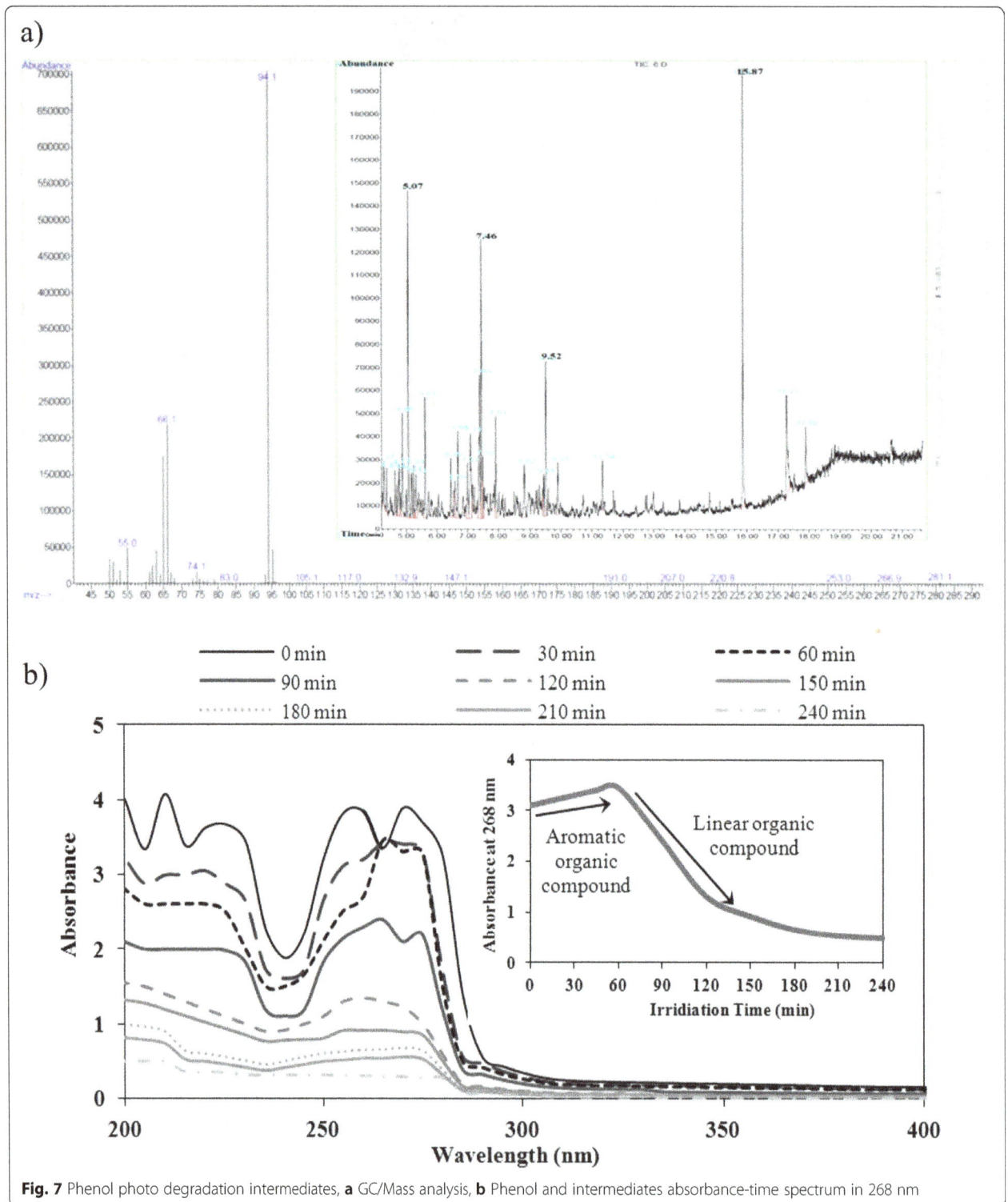

Fig. 7 Phenol photo degradation intermediates, **a** GC/Mass analysis, **b** Phenol and intermediates absorbance-time spectrum in 268 nm

9. Consequently, concrete due to its special characteristics and large consumption in WWTP as a construction material seems to be an ideal support for the immobilization of TiO_2 photo-catalysts. Also concrete sealers showed good capability for attachment of TiO_2 nano particles to concrete surfaces.

Competing interests
The authors declare that they have no competing interests.

Authors' contributions
All authors participated in conception and design, generation of data, analysis of data, interpretation of data and revision of the manuscript. All authors read and approved the final manuscript.

Acknowledgements

The authors would like to acknowledge Iranian Nano Technology Initiative Council and Vice-Chancellor for Research Affairs of Tarbiat Modares University for their partial financial support in this research. The valuable supports of Environmental Engineering Laboratory of Civil and Environmental Engineering Faculty of Tarbiat Modares University are also appreciated.

Author details

[1]Civil and Environmental Engineering Faculty, Tarbiat Modares University, Tehran, Iran. [2]Civil Engineering Department, Faculty of Engineering, Kharazmi University, Tehran, Iran. [3]Material Engineering Department, Nano Materials Division, Tarbiat Modares University, Tehran, Iran.

References

1. Gaya UI, Abdullah AH. Heterogeneous photocatalytic degradation of organic contaminants over titanium dioxide: a review of fundamentals. Progress and problems, J Photoch Photobio C. 2008;9:1–12.
2. Li Y, Li S, Li Y, Guo Y, Wang J. Visible-light driven photocatalyst (Er^3+:YAlO$_3$/Pt–NaTaO$_3$) for hydrogen production from water splitting. Int J Hydrogen Energ. 2014;39:17608–16.
3. Rizzo L, Koch J, Belgiorno V, Anderson MA. Removal of methylene blue in a photocatalytic reactor using polymethylmethacrylate supported TiO$_2$ nanofilm. Desalination. 2007;211:1–9.
4. De Lasa H, Errano B, Salaices M. Photocatalytic reaction engineering. USA: Springer Science Pub; 2005.
5. Bickley RI, Slater MJ, Wang WJ. Engineering development of a photocatalytic reactor for wastewater treatment. Process Saf Environ. 2005;83:205–16.
6. Shi J, Zheng J, Wu P, Ji X. Immobilization of TiO$_2$ films on activated carbon fiber and their photocatalytic degradation properties for dye compounds with different molecular size. Catal Commun. 2008;9:1846–50.
7. Lim LLP, Lynch RJ, In SI. Comparison of simple and economical photocatalyst immobilisation procedures. Appl Catal A-Gen. 2009;365:214–21.
8. Horikoshi S, Watanabe N, Onishi H, Hidaka H, Serpone N. Photodecomposition of a nonylphenol polyethoxylate surfactant in a cylindrical photoreactor with TiO$_2$ immobilized fiberglass cloth. Appl Catal B-Environ. 2002;37:117–29.
9. Sakthivel S, Shankar MV, Palanichamy M, Arabindoo B, Murugesan V. Photocatalytic decomposition of leather dye: comparative study of TiO$_2$ supported on alumina and glass beads. J Photoch Photobio A. 2002;148:153–9.
10. Choy CC, Wazne M, Meng X. Application of an empirical transport model to simulate retention of nanocrystalline titanium dioxide in sand columns. Chemosphere. 2008;7:1794–01.
11. Zainudin NF, Abdullah AZ, Mohamed AR. Characteristics of supported nano-TiO$_2$/ZSM-5/silica gel (SNTZS): photocatalytic degradation of phenol. J Hazard Mater. 2010;174:299–06.
12. Chen Y, Dionysiou DD. TiO$_2$ photocatalytic films on stainless steel: the role of Degussa P-25 in modified sol–gel methods. Appl Catal B-Environ. 2006;62:255–64.
13. Hosseini SN, Borghei SM, Vossoughi M, Taghavinia N. Immobilization of TiO$_2$ on perlite granules for photocatalytic degradation of phenol. Appl Catal B-Environ. 2007;74:53–62.
14. Chen J, Poon CS. Photocatalytic activity of titanium dioxide modified concrete materials – influence of utilizing recycled glass cullets as aggregates. J Environ Manage. 2009;90:3436–42.
15. Poon CS, Cheung E. NO removal efficiency of photocatalytic paving blocks prepared with recycled materials. Constr Build Mater. 2007;21:1746–53.
16. Husken G, Hunger M, Brouwers HJH. Experimental study of photocatalytic concrete products for air purification. Build Environ. 2009;44:2463–74.
17. Mortazavi SB, Sabzali A, Rezaee A. Sequence-Fenton Reaction for decreasing phenol formation during benzene chemical conversion in aqueous solutions. Iranian J Environ Health Sci Eng. 2005;2:62–71.
18. Kidak R, Ince NH. Ultrasonic destruction of phenol and substituted phenols: a review of current research. Ultrason Sonochem. 2006;13:195–9.
19. Moussavi G, Mahmoudi M, Barikbin B. Biological removal of phenol from strong wastewaters using a novel MSBR. Water Res. 2009;43:1095–302.
20. Busca G, Berardinelli S, Resini C, Arrighi L. Technologies for the removal of phenol from fluid streams: a short review of recent developments. J Hazard Mater. 2008;160:265–88.
21. Gholizadeh A, Kermani M, Gholami M, Farzadkia M. Kinetic and isotherm studies of adsorption and biosorption processes in the removal of phenolic compounds from aqueous solutions: comparative study. J Environ Health Sci Eng. 2013;11:29.
22. Asadgol Z, Forootanfar H, Rezaei S, Mahvi AH, Faramarzi MA. Removal of phenol and bisphenol-a catalyzed by laccase in aqueous solution. J Environ Health Sci Eng. 2014;12:93.
23. Hemmati Borji S, Nasseri S, Mahvi A, Nabizadeh R, Javadi A. Investigation of photocatalytic degradation of phenol by Fe(III)-doped TiO$_2$ and TiO$_2$ nanoparticles. J Environ Health Sci Eng. 2014;12:101.
24. Mahvi AH, Maleki A, Alimohamadi M, Ghasri A. Photo-oxidation of phenol in aqueous solution: toxicity of intermediates. Korean J Chem Eng. 2007;24:79–82.
25. Mahvi AH, Maleki A. Photosonochemical degradation of phenol in water. Deaslin Water Treat. 2010;20:197–02.
26. Maleki A, Mahvi AH, Mesdaghinia A, Naddafi K. Degradation and toxicity reduction of phenol by ultrasound wavesSoc. Bull Chem Ethiop. 2007;21:33–8.
27. Shahamat YD, Farzadkia M, Nasseri S, Mahvi A, Gholami M, Esrafili A. Magnetic heterogeneous catalytic ozonation: a new removal method for phenol in industrial wastewater. J Environ Health Sci Eng. 2014;12:50.
28. Chiou CH, Wu CY, Juang RS. Photocatalytic degradation of phenol and m-nitrophenol using irradiated TiO$_2$ in aqueous solutions. Sep Purif Technol. 2008;62:559–64.
29. Venkata Subba Rao K. Immobilization of TiO$_2$ on pumice stone for the photo-catalytic degradation of dyes and dye industry pollutants. Appl Catal B-Environ. 2003;46:77–85.
30. ACI 211. Manual of concrete practice. Farmington Hill: ACI; 1996.
31. APHA, AWWA, WEF. Standard methods for the examination of water and wastewater. 21st ed. Washington: American Public Health Association; 2005.
32. Feigenbrugel V, Le Calve S, Mirabel P, Louis F. Henry's law constant measurements for phenol, o-, m-, and p-cresol as a function of temperature. Atmos Environ. 2004;38:5577–88.
33. Morrison RT, Boyd RN. Organic chemistry. 6th ed. USA: McGraw-Hill; 2000.
34. Zheng S, Cai Y, O'Shea KE. TiO$_2$ photocatalytic degradation of phenylarsonic acid. J Photoch Photobio A. 2010;210:61–8.
35. Royaee SJ, Sohrabi M. Application of photo-impinging streams reactor in degradation of phenol in aqueous phase. Desalination. 2010;253:57–61.

9

Potential of polyaniline modified clay nanocomposite as a selective decontamination adsorbent for Pb(II) ions from contaminated waters; kinetics and thermodynamic study

Somayeh Piri[1], Zahra Alikhani Zanjani[2], Farideh Piri[1], Abbasali Zamani[2*], Mohamadreza Yaftian[1] and Mehdi Davari[3]

Abstract

Background: Nowadays significant attention is to nanocomposite compounds in water cleaning. In this article the synthesis and characterization of conductive polyaniline/clay (PANI/clay) as a hybrid nanocomposite with extended chain conformation and its application for water purification are presented.

Methods: Clay samples were obtained from the central plain of Abhar region, Abhar, Zanjan Province, Iran. Clay was dried and sieved before used as adsorbent. The conductive polyaniline was inflicted into the layers of clay to fabricate a hybrid material. The structural properties of the fabricated nanocomposite are studied by X-ray diffraction (XRD), Fourier transform infrared spectroscopy (FT-IR) and scanning electron microscope (SEM). The elimination process of Pb(II) and Cd(II) ions from synthetics aqueous phase on the surface of PANI/clay as adsorbent were evaluated in batch experiments. Flame atomic absorption instrument spectrophotometer was used for determination of the studied ions concentration. Consequence change of the pH and initial metal amount in aqueous solution, the procedure time and the used adsorbent dose as the effective parameters on the removal efficiency was investigated.

Results: Surface characterization was exhibited that the clay layers were flaked in the hybrid nanocomposite. The results show that what happen when a nanocomposite polyaniline chain is inserted between the clay layers. The adsorption of ions confirmed a pH dependency procedure and a maximum removal value was seen at pH 5.0. The adsorption isotherm and the kinetics of the adsorption processes were described by Temkin model and pseudo-second-order equation. Time of procedure, pH and initial ion amount have a severe effect on adsorption efficiency of PANI/clay.

Conclusions: By using suggested synthesise method, nano-composite as the adsorbent simply will be prepared. The prepared PANI/clay showed excellent adsorption capability for decontamination of Pb ions from contaminated water. Both of suggested synthesise and removal methods are affordable techniques.

Keywords: Polyaniline, Clay, Nanocomposite, Nanolayers, Natural adsorbent, Water treatment, Heavy metals

* Correspondence: zamani@znu.ac.ir
[2]Department of Environmental Science, Faculty of Science, University of Zanjan, 45371-38791 Zanjan, Iran
Full list of author information is available at the end of the article

Background

Due to unique characteristics such as their interesting electrical and electrochemical properties, conducting polymers was used by many research groups worldwide. Among conducting polymers, polyaniline (PANI) has attracted considerable industrial interest and has been used in sensors fabrication [1, 2], electronic devices [3], batteries [4, 5], and as anti-corrosive additive inorganic coatings [6–8]. This wide range of applications motivates researchers to the development of PANI with improved characteristics. The process ability and some other properties of PANI could be enhanced by the synthesis of blending and composites compounds [9].

Polymers with two-dimensional nanomaterial's structure, in particularly anisotropic platelet-like layered compounds such as layered silicates [10–12] have received more attention in recent years. Layered silicates platelets are exploited by a variety of methods and techniques [13–17]. Surface charge of these layers is permanent negative due to it is relocated by exchangeable inorganic cations same as Na^+ and Ca^{2+}. Silicate layers trend to hoard and form bundles. Therefore dispersing is an important need of individual Nano-platelets compounds within the polymer. The monomer molecules trend to penetration into the space between aggregate clay layers. Different levels of dispersion can be cratered based on the dispersion method used to fabricate the Nano-layer's structure. The two ends of levels of dispersion are intercalated nanocomposite and exfoliated nanocomposite [18]. As an outcome, by controlling the amount of polymerized polymer in the clay layers at a low level, fully intercalated nanocomposite may be obtained. Clay nanocomposites can be used as a model for investigation on behavior of polymer confined in a two-dimensional space. Layered silicates/polymer nancomposites have been used for the sanitization of the wastewater due to their wide range of sources [19], readily available and much cheaper than adsorbents else.

Numerous methods such as solvent extraction [20], osmosis [21], chemical precipitation [22] and adsorption are famous and available methods for decontamination of heavy metals from wastewaters. Among these methods, adsorption [23] is preferable to have access to the goal. Among various effectual adsorbents such as activated carbon [24] and silica [25], clay is a suitable candidate for adsorption applications [26, 27]. This is due to the unique properties of clay [28–30]. Clay layered structures and ability to imprison water in the interlayer space raise the heavy metal adsorption and ion exchange. Therefore improvement of clay adsorption capacities by using different techniques is a favorable subject for researchers [28–30]. Same as clay other adsorbent was used in water refinement such as mainly polysaccharides such as chitosan [31], pistachio-nut shell ash [32],

salvadora persica stem ash [33] and starch [34]. Low surface area and difficult separation from the water phase are disadvantages for natural polymers that decrease their use in field wastewater treatment applications.

Notable adsorption performance, low cost, wide availability and the presence of various functional groups on conducting polymeric composite materials are main cause that it has gained a distinctive attention [35]. Moreover, materials such as polyaniline have been used as profitable adsorbent for treatment aqueous solution of heavy metals ions. The different structural shape, special mechanism and environmental stability of PANI are mainly its reason [36–38].

Lead as a hazardous heavy metal is highly toxic to different types of living species on earth. Consuming contaminated waters with lead is a cause various types of serious diseases [39]. The suggested limit of lead ions is 10 μg L^{-1} in drinking water [40–42]. If 5 μg L^{-1} lead dissolved in drinking-water, the total intake of it can be calculated to range from 3.8 to 10 μg day^{-1} for an infant and an adult, respectively [41]. This is reported that by increasing the concentration of lead from the limits set by world health organization (WHO) and United States environmental protection agency (USEPA) (10 μg L^{-1}), it impact the surrounding environment adversely and it can help to the outbreak of several diseases such as anemia, kidney damage and disorder in the nervous system [22].

In this report an easy, environmentally friend fabricating and economical method for synthesize nanocomposite from polyaniline and clay, via chemical grafting of PANI onto clay as a useful mineral adsorbent that is to find in nature abundantly, is demonstrated. The surface structure and morphology of the synthesized PANI/clay nanocomposite were studied by X-ray diffraction (XRD), Fourier transform infrared spectroscopy (FT-IR) and scanning electron microscope (SEM) techniques. Subsequently, the nanocomposites potency as decontamination agents was assessed in the removal of Pb(II) and Cd(II) ions in contaminated waters. The research of adsorption isotherms and kinetics of procedure were also done to understanding the adsorption behavior between the synthesized PANI/clay and the adsorbate ions. The whole of study was done in the summer of 2015 in the Environmental Science Research and Taghipour Dr. Laboratories, University of Zanjan, Zanjan-Iran.

Method

Materials and chemicals

Clay samples were obtained from the central Plain of Abhar region, Abhar, Zanjan Province, Iran. All used chemicals in this research with synthesis and analytical grade reagents were purchased from Merck or Fluka and were utilized in their initial form. Primal solutions of lead and cadmium ions with concentration 1000 mg L^{-1} were provided by dissolving a proper amount of corresponding nitrate salts in deionized water. Working solutions were obtained by appropriate dilution of the primal solutions with deionized water. For pH adjustments of solutions nitric acid and sodium hydroxide solutions were applied.

Preparing of clay

Clay was first pretreated by the following procedure; at first, dried clay was sieved to 150 mm particle size then 30 g prepared clay was added into 300 mL concentrated sulfuric acid solution and the slurry mixtures was stirred for 1 week. Then, after separation of initial modified clay by filtration and it was washed thoroughly with distilled water until a time when the pH value of water filtrated was been 7.0. The crude product was dried before synthesis of the PANI/clay nanocomposite.

Synthesis of the PANI/clay nanocomposite

The PANI/clay was synthesized via in situ chemical oxidative polymerization technique. In this manner that 2 g of acid modified clay was disorganized in water/ethanol mixture in a conical flask by sonication for 30 min. This procedure was done at room temperature. Then, 0.5 mL of aniline monomer was added and mixture again was sonicated for 20 min else for better diffusion of the monomer into the clay sheets. The monomer was polymerized by adding 0.42 g of ammonium persulphate and mixture stirring for 90 min else. The black mass was obtained as resulting nanocomposite and it was separated by filtration and washed with ethanol and distilled water repeatedly. The produced nanocomposite was air-dried at room temperature.

Instruments

Field emission scanning electron micrographs (FESEM) was used for microscopy characterization morphological analyses in this propose the nanocomposite films were took by Mira 3-XMU system. Power X-ray diffraction patterns were performed for the PANI/clay nanocomposite on a Bruker D advance XRD meter between angle $2\theta = 5\text{-}60°$ at 40 kV. Fourier transform infrared spectroscopy was carried out on a Bruker Vector 22 spectrophotometer. A flame atomic absorption spectrophotometer Varian 220A was used in quantitative analysis of metal ions concentration. A digital pH meter, Metrohm 780, was performed for pH adjustments.

Adsorption measurements

Batch experiments in laboratory scale were selected to realize the effect of pH, time contacting and adsorbent dosage on behavior between PANI/clay adsorbent and studied ions. The solutions pH values were adjusted in the range 2–7 with HNO$_3$/NaOH solutions (0.1 mol L^{-1}). For

investigation the adsorption behavior of the synthesized PANI/clay nanocomposite, 100 mg of it was put into 40 mL of 20 mg L^{-1} lead and cadmium ion solutions. This concentration was selected because that nerve conduction velocity is being appeared with increasing lead concentration from 20 mg L^{-1}. Also lead ores comprise 20 mg Kg^{-1} of the earth's crust [43]. The mixture of adsorbent and ions solution was mixed by using a magnetically stirring at laboratory temperature. After separation of the two mixed phases, the residual metal ion concentration in the aqueous phase was measured by FAAS. The amount of removed metal ions by per unit mass of used PANI/clay nanocomposites is computed by using eq. 1:

$$q_e = \frac{(C_0 - C_e)V}{m} \tag{1}$$

In this equation q_e note adsorbent adsorption capacity in the equilibrium time, C_0 and C_e is the studied metal ion concentration (mg L^{-1}) in zero and equilibrium time, respectively, m is the mass of the adsorbent (g), and V is the used volume solutions (L).

Results and discussion

It is very clear where nitrogen atoms exist in amine compounds due to the presence of electron in SP^3 orbital of nitrogen can makes coordinate bond with positive charge of analytes. Figure 1 as a result mechanism introduce removal procedure of Pb(II) that may be explained with ion exchange between proton of amines in polyaniline or hydroxyl groups of clay nanocomposite with Pb(II) ions in water.

Characterize analyses of PANI/clay nanocomposites

SEM technique was used to morphological analyses and characterization of the size and shape of the resulted PANI/clay nanocomposites (Fig. 2a). It shows the clay sheets as gray narrow plates have a very nice distribution in the synthesized nanocomposite (Fig. 2b). Also it confirmed that after modification of clay by PANI, the flaky clay structure coated by PANI and many individual platelets are seen in SEM images. This result reveals that PANI readily entered the layers of the clay and expanded and pushes it apart resulting in intercalated layered silicate PANI particles. Clay sheet thickness was near 40–50 nm.

The XRD pattern of the clay and PANI/clay nanocomposite in the 2θ range of about 5–60° are presented in Fig. 3. In the PANI/clay nanocomposite XRD image, the main peaks are similar to the clay particles, which confirmed that the crystalline structure of clay is nicely protected after the coating step under polymerization process. Due to the relatively thin layer and amorphous crystallinity of the PANI prepared

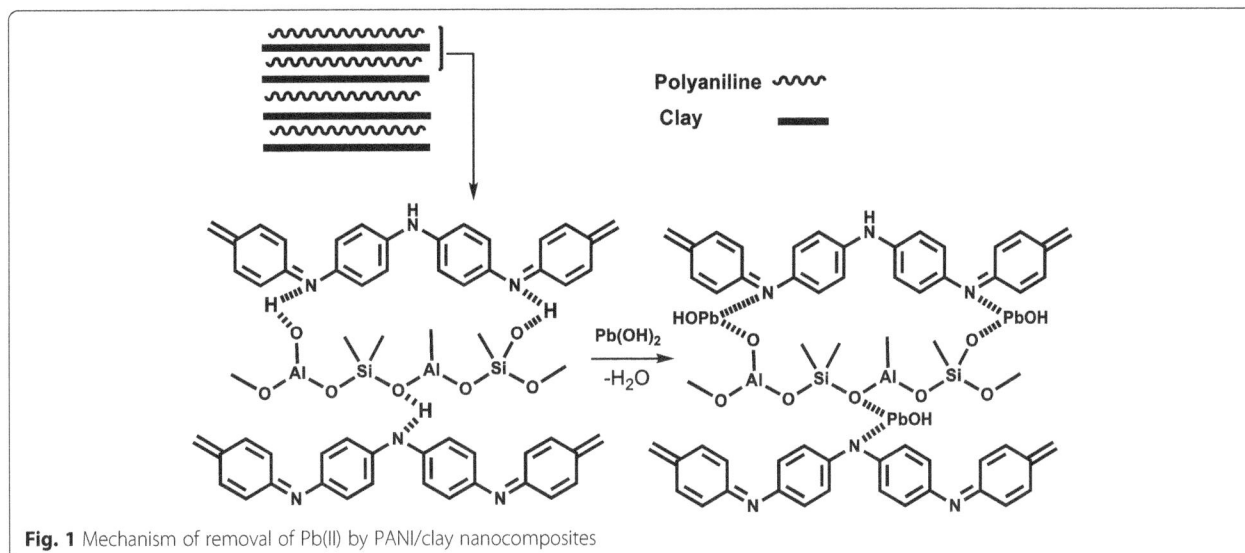

Fig. 1 Mechanism of removal of Pb(II) by PANI/clay nanocomposites

under this polymerization method, no obvious diffraction peak for the PANI is detected. XRD Result for the clay represented in Table 1.

Figure 4 exhibits the Fourier transform infrared spectroscopy spectra of the KBr pellet PANI/clay (a) and clay (b) specimens in the wavenumber of 2000–400 cm^{-1}. In the PANI/clay specimens, the absorption bands at 1567 and 1489 cm^{-1} are appointed to the stretching vibration of Quinone and benzene rings of PANI compound, respectively. For both spectra of clay (b) and PANI/clay (a), the absorption peak at 1641 cm^{-1} is dependent to the H−O−H vibration in water bending, and the Si-O and Al-O asymmetrical stretching vibration was appeared at 1079 cm^{-1} and 1167 cm^{-1}. Also, the appeared peaks at 798 cm^{-1} and 695 cm^{-1} were assigned to the Si-O and Al-O symmetrical stretching and bending vibrations, respectively. In the PANI/clay nanocomposite FT-IR spectra (a) the peaks correspond to Si-O and Al-O at 798 cm^{-1} are slightly shifted to

higher asymmetric frequency which may be due to interactions like Vander Waal's forces and hydrogen bonding.

Application PANI/clay nanocomposite as adsorbent for removal of heavy metal ions

In order to assessment of adsorption capacity, the obtained synthesized PANI/clay were applied as an adsorbents for decontamination of Pb(II) and Cd(II) ions from polluted aqueous solutions. The effective parameters on the adsorption process such as pH of aqueous solution, contact time and sorbent dose is studied. Then the synthesized PANI/clay was used for treatment of real water samples that it is polluted with lead ions.

Influence of working solution pH

Due to effect on solubility of adsorbate, concentration of the studied ions on the adsorbent functional groups and degree of ionization and deformation of the adsorbate and adsorbent during reaction, the pH effect of

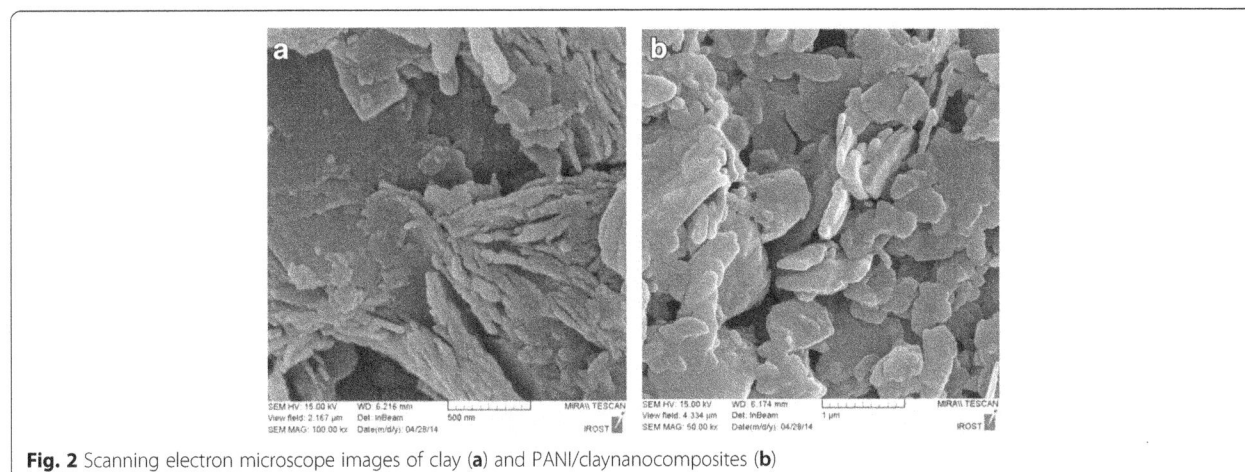

Fig. 2 Scanning electron microscope images of clay (**a**) and PANI/claynanocomposites (**b**)

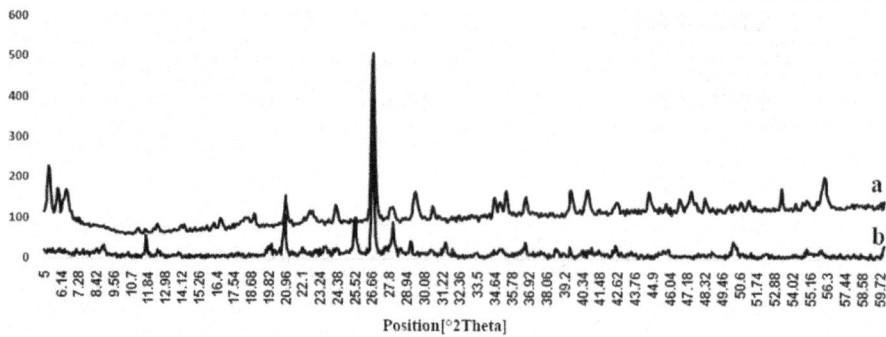

Fig. 3 X-ray diffraction patterns of the clay **a** and PANI/clay **b** nanocomposites

a solution content adsorbate is an important study in adsorption studies [30]. Therefore, in the first step of this study the role of pH in the maximum removal of studied ions was examined over a pH range of 2.0–7.0. Figure 5 shows that the total removal percent of lead ions by PANI/clay increases with an increase in pH from 2.0 to 3.0 and the maximum removal around an initial pH = 4.0 can be seen. By direction initial pH of solution to basic range excited Pb-N bond formation between the reactive groups on the PANI/clay and Pb(II). Also in basic solutions, by releasing the protons from the imine groups, more activated binding sites are available for Pb(II) ions. As shown in Fig. 5, the high removal percent was seen at pH = 5–6. By increasing pH from 2 to 6, the Pb(II) adsorbed raised from 4.2 to 7.4 mg g^{-1}.

At aqueous solution with pH < 6, the majority presented lead specie is Pb(II) form and the decontamination of Pb(II) is mainly done by sorption reaction. Therefore, the low removal Pb(II) ions at acidic solutions can be illustrated to the competition between H$^+$ and Pb^{2+} ions on the activated surface sites of adsorbent [44, 45].

The same behavior for pH effect have been reported by Jiang et al., [30] in using modified kaolin as adsorbent for Pb(II) ions. It is shown highest adsorption was seen at final pH > 4 and increasing pH of aqueous solution

increases amount of adsorbed ions [30]. Also comparison of both result confirmed that the capability of present studied adsorbent in lead removal is lower than modified kaolin.

Also surfactant emulsion membrane technology was used by Lende [22] for removal of Pb(II) from printed circuit board (PCB). In this study the pH of the filtered waste water was found to be around 5 and pH 4 is optimum amount in the removal of Pb(II) ions (initial concentration 150 mg L^{-1}) with 82 % extraction [22]. Therefore the quantitative removal at pH 5–6 (in the present study) is good for decrease lead ions from PCB wastewater.

Time dependency

Equilibrium time as one other important parameters show the need time for removal in adsorption procedure. This is important in design procedure for pilot or industry scale. The contact time dependency of the removal procedure of Pb(II) ions by PANI/clay adsorbent is given in Fig. 6.

Figure 6 show that the removal of Pb(II) and Cd(II) ions by the used PANI/clay nanocomposite as adsorbents are a quick process, removal percent for both studied ions increase with the contact time, where over 90 % of lead ions removal was done within the first 20 min and equilibrium time is about 25 min. The

Table 1 Representative XRD analysis of clay

Ref. code	Score	Compound name	Scale factor	Chemical formula
01-086-1628	18	Quartz low	0.607	SiO_2
00-013-0259	7	Montmorillonite-14A	0.194	$Na_{0.3}$ (Al, Mg)$_2$ Si$_4$O$_{10}$ (OH)$_2$!x H$_2$O
00-003-0015	4	Montmorillonite (bentonite)	0.097	(Na, Ca)$_{0.3}$ (Al, Mg)$_2$ Si$_4$O$_{10}$ (OH)$_2$!x H$_2$O
00-029-0989	4	Merlinoite	0.087	$K_5Ca_2(Al_9Si_{23}O_{64})$!24 H$_2$O
01-076-0885	2	Biotite 2 M1	54.925	$KMg_2Al_2Si_3O_{11}(OH)$
01-079-1343	0	Dolomite	0.092	$CaMg(CO_3)_2$
01-086-2339	1	Calcite	0.415	$Ca(CO_3)$
00-006-0046	2	Gypsum	0.046	$CaSO_4$!2 H$_2$O

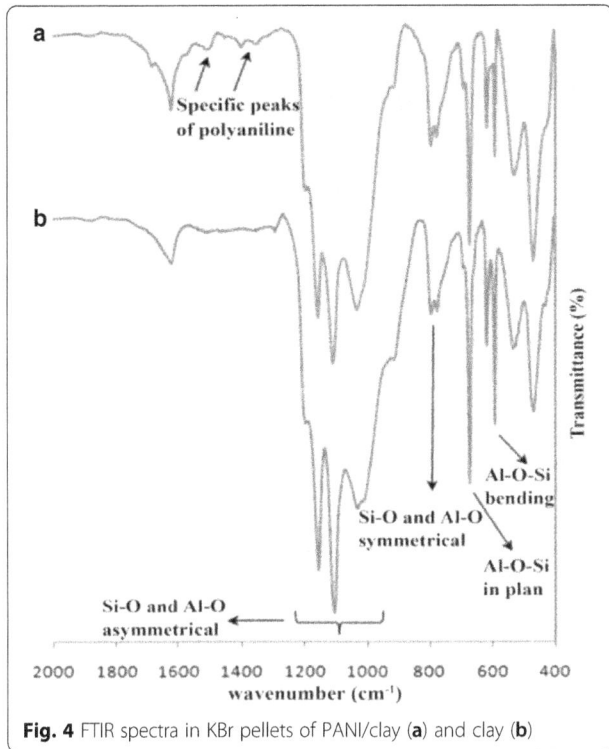

Fig. 4 FTIR spectra in KBr pellets of PANI/clay (**a**) and clay (**b**)

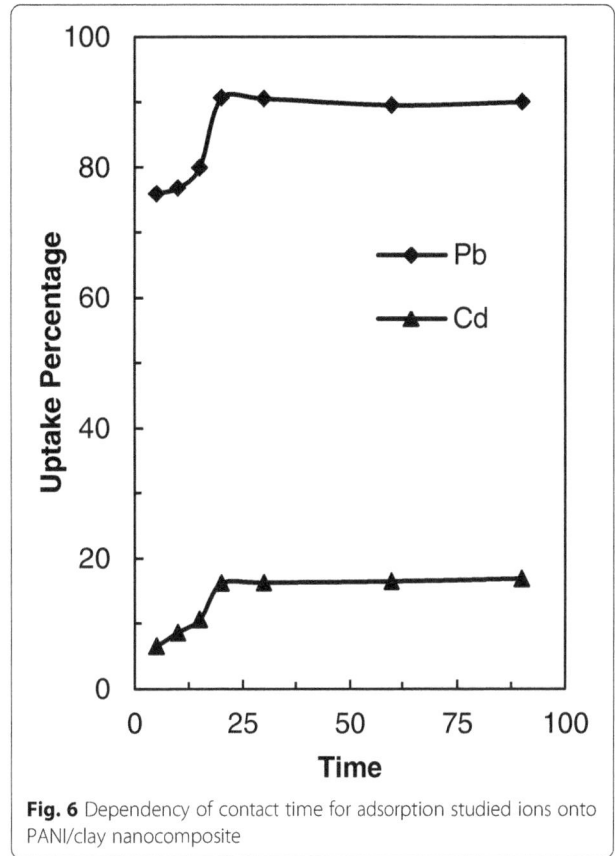

Fig. 6 Dependency of contact time for adsorption studied ions onto PANI/clay nanocomposite

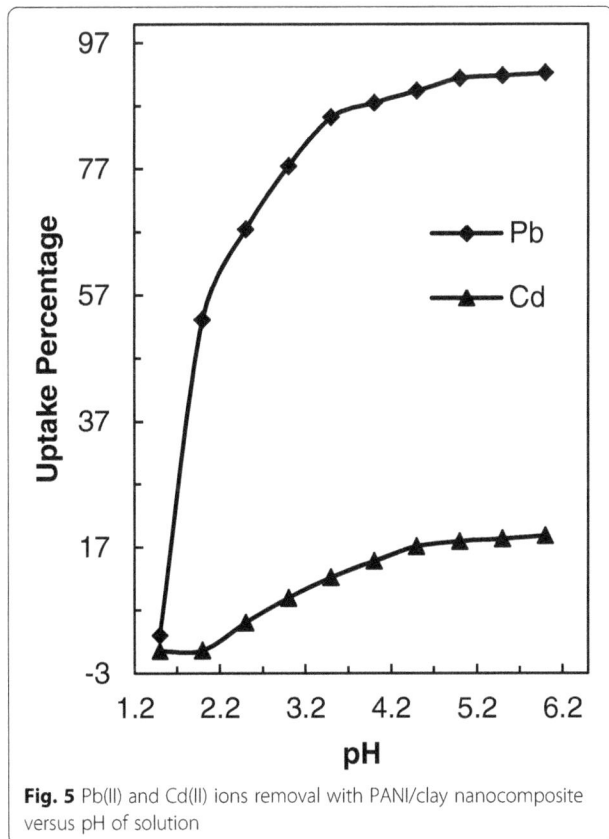

Fig. 5 Pb(II) and Cd(II) ions removal with PANI/clay nanocomposite versus pH of solution

reason of quick removal of Pb(II) ions at the initial times may be due to excess active sites on the uncovered surface of adsorbents. With increasing contact time and decreasing the active adsorption sites on PANI/clay nanocomposite as well as initial studied ion concentration, the adsorption became firstly slow and then fixed and steady curve can be seen. As another results in study of time dependency the maximal removal of Pb(II) is noticeable than removal of Cd(II). To complete the adsorption study versus contact time, the pseudo-first order and pseudo-second order kinetic models was used as the usefulness and famous models for study and determine of the kinetic parameters for adsorption procedure. By quickly covering of the active sites on the PANI/clay nanocomposite by Pb(II) ions the removal percent is dependent on the transported rate of the ions that penetrate from the bulk liquid phase to activated adsorption sites [30].

Effect of dosage adsorbents

The removal of Pb(II) and Cd(II) ions versus the adsorbent dosage (0.01–0.15 g) at aqueous solution with initial lead concentrations of 100 mg L^{-1} and pH 5 is demonstrated at Fig. 7. This Figure shows that the removal of Pb(II) ions per gram of used PANI/clay nanocomposite sharply raises with increasing adsorbent amount from

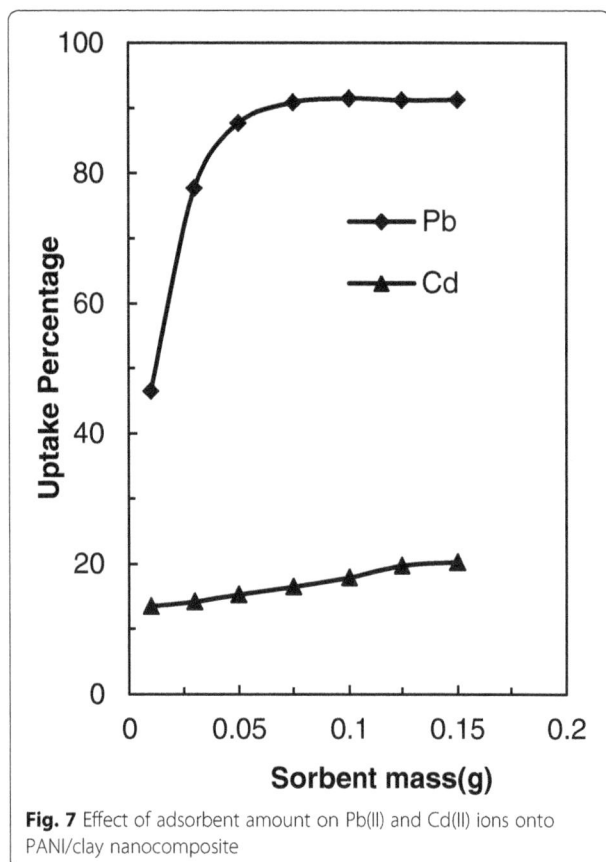

Fig. 7 Effect of adsorbent amount on Pb(II) and Cd(II) ions onto PANI/clay nanocomposite

Where q refer to the amount of analyte in mg g^{-1} and subscripts e and t show equilibrium and at any time, respectively and K_1 (min^{-1}) and K_2 (g mg^{-1} min^{-1}) in this equations denote the equilibrium rate constant corresponded to pseudo-first order and pseudo-second order adsorption, respectively.

A linear plot of log(q_e- q_t) versus t for this model was employed and the achieved R^2 values for Pb(II) and Cd(II) ions are 0.977, 0.985, respectively.

The pseudo-second order adsorption can be obtained from the plot of t/q_t against t. The application of the model show that Linear plots of t/q_t versus t are obtained with R^2 values of 0.989, for the Pb(II) and 0.999 for Cd(II) ions. This confirms that interaction between the studied ions and PANI/clay nanocomposite follow from the pseudo second-order mechanism (Fig. 8). It can be said that the rate limiting step is chemical interaction involving valence forces through sharing of electrons.

The result of studied mechanism indicates that removal of lead ions is subsequent to chemical reaction rather than physical-sorption. Also the quickly procedure in Pb(II) adsorption onto adsorbent show a chemical sorption which was done due to the strong electrostatic

0.01 to 0.05 g. At higher amount of PANI/nanocomposite by increasing the surface area and active sites higher removal of the lead ions was done. Further in removal of Cd(II), increasing adsorbent amount did not any significant effect. Above of 0.06 g of adsorbed equilibrium status can be seen between solid and solution phase. This optimum condition was kept constant for the study of all other parameters.

Adsorption kinetics

Kinetic studies of adsorption procedure show data about the mechanism, which is useful for planning the practical process. In this study, the kinetics of adsorption of Pb(II), Cu(II) and Zn(II) on PANI/clay nanocomposite were studied by using the frequently used models, the Lagergren pseudo-first order model expressed in equation 2 and the pseudo-second order model expressed in equation 3 [38].

$$ln(q_e - q_t) = lnq_e - K_1t \qquad (2)$$

$$\frac{t}{q_t} = \frac{1}{k_2 q_e^2} + \frac{t}{q_e} \qquad (3)$$

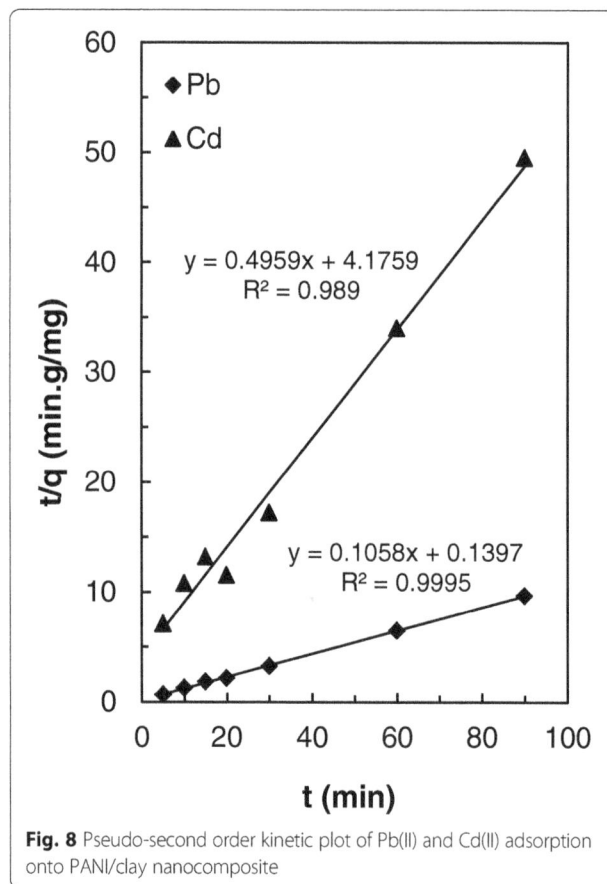

Fig. 8 Pseudo-second order kinetic plot of Pb(II) and Cd(II) adsorption onto PANI/clay nanocomposite

Table 2 The obtained parameters in study of procedure isotherms

Ions	Isotherms parameters								
	langmuir			Freundlich			Temkin		
	$b(L\ mg^{-1})$	$q_{max}(mg\ g^{-1})$	R^2	n	$K_f(mg\ g^{-1})$	R^2	$A_T\ (L\ g^{-1})$	$b_T\ (kJ\ mol^{-1})$	R^2
Pb(II)	0.04	70.42	0.74	1.01	4.16	0.87	0.87	0.15	0.98
Cd(II)	0.05	0.12	0.96	0.04	0.00	0.78	0.05	0.02	0.53

interaction between the negative charge on the PANI/clay nanocomposite surface and Pb(II) ions [30].

Adsorption isotherms

The Langmuir, Freundlich and Temkin isotherm models was used for assessment of data of adsorption isotherm. These models describe the dependence between the adsorption amount of studied ions on the adsorbent surface and the equilibrium concentration of ions in the liquid phase.

The Langmuir isotherm and Freundlich equation used for monolayer and multilayer adsorption onto a surface, respectively. In the Langmuir isotherm identical active sites have finite number and Freundlich equation show heterogeneous surfaces [44]. Temkin isotherm model is a useful tool to estimate the adsorption heat due to correlation of adsorption heat of all molecules and temperature [46].

The linear equations of the Langmuir, Freundlich and Temkin isotherms can be expressed in the equations 4, 5 and 6, respectively:

$$\frac{C_e}{q_e} = \frac{1}{b\ q_{max}} + \frac{C_e}{q_{max}} \tag{4}$$

$$\log q_e = \frac{1}{n}\log C_e + \ \log K_f \tag{5}$$

$$q_e = \frac{RT}{b_T}\ln A_T C_e \tag{6}$$

Where C show the equilibrium concentration (mg L^{-1}), q_{max} (mg g^{-1}) denote the maximum adsorption capacity, b (L mg^{-1}) relates the energy of adsorption, K_f indicates relative adsorption capacity ($mg^{1-(1/n)}\ L^{1/n}g^{-1}$) and n is an empirical parameter related to the intensity of adsorption. A_T is Temkin isotherm equilibrium binding constant (L g^{-1}) and b_T is Temkin isotherm constant respectively.

The effective parameters of the studied isotherm models obtained from regression analysis of the experimental data and they are summarized in Table 2. According these reported result, the Temkin isotherm justified experimental data better than the Langmuir and

Freundlich isotherm in the studied lead concentration range. In Temkin isotherm model, B parameter (equations 7) shows heat of sorption (J mol^{-1}).

$$B = \frac{RT}{b_T} \tag{7}$$

The values $A_T = 0.87$ L g^{-1}, $R_2 = 0.98$ and B = 16.52 J mol $^{-1}$ were estimated From the Temkin plot. The heat of sorption indicates a physical adsorption process.

Effect of initial metal ion concentration

Pb(II) ion concentration was set in the ranges of 10, 30, 40, 50, 100, 200 and 500 mg L^{-1} to determine of maximum quantity removal. The rising initial Pb(II) concentration caused an increasing in the Pb(II) removal by using PANI/clay nanocomposite (the results not shown). With increasing initial lead ion concentration, the amount of metal ion adsorbed raised due to increasing driving force of the adsorber towards the active sites on both the modified and unmodified adsorbents [30]. Due to the saturation of binding sites, at higher concentrations, more Pb(II) as the adsorbers was returned in to solution. Also when initial Pb(II) concentration in aqueous solution was 200 mg L^{-1}, the empirical maximum adsorption capacity calculated that it was 9.6 mg g^{-1}.

Desorption studies

Due to metal ion recycling and recovery of adsorbant, desorption study is important stage in adsorption process. As the quantitative desorption of the adsorbed lead ions on the PANI/clay nanocomposite by distilled water was not successful, thus, hydrochloric, nitric and sulfuric acids were used to this end. HCl and HNO_3 presents higher desorption capacity towards lead ions.

Table 3 Adsorption of Pb(II) and Cd(II) from the real water samples

Ions	Removal (%)		
	Drinking water	River water	Sea water
Pb(II)	90.7	91.2	91.7
Cd(II)	16.2	15.9	16.4

Table 4 PANI/clay nanocomposite for lead removal against various reported adsorbents

Adsorbent	Maximum Adsorption Capacity (mg g^{-1})^a	Adsorption isotherm	Adsorption Kinetic model	References
Polyaniline/clay	70.4	Temkin	pseudo-second order	Present study
Unmodified kaolinite clay	4.7	Langmuir	pseudo-second order	[30]
Modified kaolinite clay	32.2	Langmuir	pseudo-second order	[30]
polyaniline on multiwalled carbon nano-tubes	22.2	Langmuir	-	[37]
Peganum harmala seeds	90.0	Freundlich	pseudo-second order	[44]

^{a}Calculated from Langmuir isotherm

More than 80 % of all the adsorbed studied ions were left adsorbent surface under the using 5 mL of HCl and HNO$_3$ as stripping solutions (0.1 M).

Application of procedure for real samples

In order to investigate the matrix effect on suggested procedure, the addition method, with an addition of lead and Cadmium ions to drinking water, river water and sea water as real samples was used. The real matrixes commonly decrease of adsorption efficiency due to present high amount of interfering agent. Obtained results of study matrix effect are presented in Table 3, it proved that the presence of interfering ions and other reagents commonly found in real water have negligible influence on removal of Pb(II) ion by using PANI/clay nanocomposite.

Conclusions

Clay was found as a suitable substrate or support for coating of polyaniline. The results of nanocomposite characterization confirmed that the clay sheets were become layered in the prepared nanocomposite. The sorption capacity by modified sorbent was strongly dependent on contact time, pH, and initial ion concentration. The metal uptake was found to increase with pH. It was also found that the sorption of Pb(II) by polyaniline/clay appeared to follow the Temkin isotherm. Adsorption kinetics followed the pseudo-second-order model with very good correlation coefficients for Adsorption. In other hand Temkin is good isotherm model for studied process. The pseudo-second-order as kinetic model and Temkin as isotherm model confirm a companionship physical and chemical adsorption in studied removal process. In this concern, complete removal of contaminating lead ions was achieved in the real samples under investigations. It might be concluded that polyaniline modified clay nanocomposite are promising adsorption system in future water and wastewater treatment in order to remove lead ion. The present study highlights is the introducing new method in synthesize clay nanocomposite which have low price and first application of polyaniline modified clay nanocomposite as a sorbent

for water treatment of lead ions. Comparison of the adsorption efficiency of polyaniline/clay and other materials is presented in Table 4.

Abbreviations

FT-IR: Fourier transforms infrared spectroscopy; PANI: Polyaniline; PANI/clay: Polyaniline/Clay; SEM: Scanning electron microscope; USEPA: United States Environmental Protection Agency; WHO: World Health Organization; XRD: X-ray diffraction

Acknowledgements

The authors would like to thank students and members of Environmental Science Research and Taghipour Dr. Laboratories, University of Zanjan, Zanjan-Iran, for their contributions to this research.

Funding

All steps of this research were supported by a grant from the vice chancellor for research and technology of university of Zanjan in 2014.

Authors' contributions

SP, ZAZ, FP, AZ, MY, and MD as authors in this manuscript carried out the modification and characterization of sorbent, participated in the sequence alignment and drafted the manuscript. SP, ZAZ conceived of the study and carried out the laboratory experiments; FP, AZ and MY participated in the design of the study and performed the results analysis; MD participated in surface characterizations study. All authors participated in the design of the study and performed the statistical analysis and writing the manuscript. Also all authors read and approved the final manuscript.

Competing interests

The authors declare that they have no competing interests.

Author details

[1]Department of Chemistry, Faculty of Science, University of Zanjan, 45371-38791 Zanjan, Iran. [2]Department of Environmental Science, Faculty of Science, University of Zanjan, 45371-38791 Zanjan, Iran. [3]Iranian Research Organization for Science and Technology, Tehran, Iran.

References

1.   Patil UV, Ramgir NS, Karmakar N, Bhogale A, Debnath AK, Aswal DK, Gupta SK, Kothari DC. Room temperature ammonia sensor based on copper nanoparticle intercalated polyanilinenanocomposite thin films. Appl Surf Sci. 2015;339:69–74.
2.   Song E, Choi J-W. Self-calibration of a polyaniline nanowire-based chemiresistive pH sensor. Microelectronic Eng. 2014;116:26–32.
3.   Sydorov D, Duboriz I, Pud A. Poly(3-methylthiophene)–polyaniline couple

spectroelectrochemistry revisited for the complementary red–green–blue electrochro- mic device. Electrochim Acta. 2013;106:114–20.

4. Xia Y, Zhu D, Si S, Li D, Wu S. Nickel foam-supported polyaniline cathode prepared with electrophoresis for improvement of rechargeable Zn battery performance. J Power Sources. 2015;283:125–31.

5. Liu H, Zhang F, Li W, Zhang X, Lee C-S, Wang W, et al. Porous tremella-like MoS2/polyaniline hybrid composite with enhanced performance for lithium-ion battery anodes. Electrochim Acta. 2015;167:132–8.

6. Kohl M, Kalendová A, Stejskal J. The effect of polyaniline phosphate on mechanical and corrosive properties of protective organic coatings containing high amounts of zinc metal particles. Prog Org Coat. 2014;77:512–7.

7. Kalendová A, Veselý D, Stejskal J. Organic coatings containing polyaniline and inorganic pigments as corrosion inhibitors. Prog Org Coat. 2008;62:105–16.

8. Akbarinezhad E, Ebrahimi M, Sharif F, Ghanbarzadeh A. Evaluating protection performance of zinc rich epoxy paints modified with polyaniline and polyaniline-clay nanocomposite. Prog Org Coat. 2014;77:1299–308.

9. Fan Z, Wang Z, Sun N, Wang J, Wang S. Performance improvement of polysulfone ultrafiltration membrane by blending with polyaniline nanofibers. J Membr Sci. 2008;320:363–71.

10. Yeh J-M, Chang K-C. Polymer/layered silicate nanocomposite anticorrosive coatings. J Ind Eng Chem. 2008;14:275–91.

11. Alexandre M, Dubois P. Polymer-layered silicate nanocomposites: preparation, properties and uses of a new class of materials. Mater Sci Eng. 2000;28:1–63.

12. Shokuhfar A, Zare-Shahabadi A, Atai A-A, Ebrahimi-Nejad S, Termeh M. Predictive modeling of creep in polymer/layered silicate nanocomposites. PolymTest. 2012;31:345–54.

13. Sinha Ray S, Okamoto M. Polymer/layered silicate nanocomposites: a review from preparation to processing. Prog Polym Sci. 2003;28:1539–641.

14. Kiliaris P, Papaspyrides CD. Polymer/layered silicate (clay) nanocomposites: An overview of flame retardancy. Prog Polym Sci. 2010;35:902–58.

15. Pavlidou S, Papaspyrides CD. A review on polymer–layered silicate nanocomposites. Prog Polym Sci. 2008;33:1119–98.

16. Zhao Z, Tang T, Qin Y, Huang B. Relationship between the continually expanded interlayer distance of layered silicates and excess intercalation of cationic surfactants. Langmuir. 2003;19:9260–5.

17. Lin J-J, Chu C-C, Chiang M-L, Tsai W-C. First isolation of individual silicate platelets from clay exfoliation and their unique self-assembly into fibrous arrays. J Phys Chem B. 2006;110:18115–20.

18. Ismail NM, Ismail AF, Mustafa A, Matsuura T, Soga T, Nagata K, et al. Qualitative and quantitative analysis of intercalated and exfoliated silicate layers in asymmetric polyethersulfone/cloisite15A® mixed matrix membrane for CO2/CH4 separation. Chem Eng J. 2015;268:371–83.

19. Unuabonah EI, Taubert A. Clay–polymer nanocomposites (CPNs): Adsorbents of the future for water treatment. Appl Clay Sci. 2014;99:83–92.

20. Baba AA, Adekola FA. Solvent extraction of Pb(II) and Zn(II) from a Nigerian galena ore leach liquor by tributylphosphate and bis(2,4,4-trimethylpentyl) phosphinic acid. Journal of King Saud University – Science. 2013;25:297–305.

21. Gamal Khedr M. Radioactive contamination of groundwater, special aspects and advantages of removal by reverse osmosis and nanofiltration. Desalination. 2013;321:47–54.

22. Lende AB. Improvement in removal of Pb(II) using surfactant emulsion membrane from PCB wastewater by addition of NaCl Original Research Article. J Water Proc Eng. 2016;11:55–9.

23. Alizadeh B, Ghorbani M, Salehi MA. Application of polyrhodanine modified multi-walled carbon nanotubes for high efficiency removal of Pb(II) from aqueous solution. J Mol Liq. 2016;220:142–9.

24. Fu R, Liu Y, Lou Z, Wang Z, Ali Baig S, Xu X. Adsorptive removal of Pb(II) by magnetic activated carbon incorporated with amino groups from aqueous solutions. J Taiwan Inst Chem Eng. 2016;62:247–58.

25. Bao S, Tang L, Li K, Ning P, Peng J, Guo H, et al. Highly selective removal of Zn(II) ion from hot-dip galvanizing pickling waste with amino-functionalized Fe3O4@SiO2 magnetic nano-adsorbent. J Colloid Interface Sci. 2016;462:235–42.

26. Glatstein DA, Francisca FM. Influence of pH and ionic strength on Cd.Cu and Pb removal from water by adsorption in Na-bentonite. Appl Clay Sci. 2015;118:61–7.

27. Drweesh SA, Fathy NA, Wahba MA, Hanna AA, Akarish AIM, Elzahany EAM, et al. Equilibrium, kinetic and thermodynamic studies of Pb(II) adsorption

from aqueous solutions on HCl-treated Egyptian kaolin. J Environ Chem Eng. 2016;4:1674–84.

28. Olu-Owolabi BI, Alabi AH, Unuabonah EI, Diagboya PN, Böhm L, Düring R-A. Calcined biomass-modified bentonite clay for removal of aqueous metal ions. J Environ Chem Eng. 2016;4:1376–82.

29. Yang F, Sun S, Chen X, Chang Y, Zha F, Lei Z. Mg–Al layered double hydroxides modified clay adsorbents for efficient removal of Pb(II), Cu2+ and Ni2+ from water. Appl Clay Sci. 2016;123:134–40.

30. M-q J, Wang Q-p, X-y J, Z-l C. Removal of Pb(II) from aqueous solution using modified and unmodified kaolinite clay. J Hazard Mater. 2009;170:332–9.

31. Zhang L, Zeng Y, Cheng Z. Removal of heavy metal ions using chitosan and modified chitosan: A review. J Mol Liq. 2016;214:175–91.

32. Bazrafshan E, Mostafapour FK, Mahvi AH. Phenol removal from aqueous solutions using pistachio-nut shell ash as a low cost adsorbent. Fresen Environ Bull. 2012;21:2962–8.

33. Kord Mostafapour F, Bazrafshan E, Farzadkia M, Amini S. Arsenic removal from aqueous solutions by salvadora persica stem ash. J Chem. 2013;2013:1–8.

34. Bao S, Tang L, Li K, Ning P, Peng J, Guo H, Zhu T, Liu Y. Synthesis of linear low-density polyethylene-g-poly (acrylic acid)-co-starch/organo-montmorillonite hydrogel composite as an adsorbent for removal of Pb(II) from aqueous solutions. J Environ Sci. 2016;27:9–20.

35. Kotal M, Bhowmick AK. Polymer nanocomposites from modified clays: Recent advances and challenges. Prog Polym Sci. 2015;51:127–87.

36. Chávez-Guajardo AE, Medina-Llamas JC, Maqueira L, Andrade CAS, Alves KGB, et al. Efficient removal of Cr (VI) and Cu (II) ions from aqueous media by use of polypyrrole/maghemite and polyaniline/maghemite magnetic nanocomposites. Chem Eng J Sciences. 2015;281:826–36.

37. Shao D, Chen C, Wang X. Application of polyaniline and multiwalled carbon nanotube magnetic composites for removal of Pb(II). Chem Eng J. 2012; 185–186:144–50.

38. Li R, Liu L, Yang F. Preparation of polyaniline/reduced graphene oxide nanocomposite and its application in adsorption of aqueous Hg(II). Chem Eng J. 2013;229:460–8.

39. Han Y, Zhang L, Yang Y, Yuan H, Zhao J, Gu J, et al. Pb uptake and toxicity to Iris halophila tested on Pb mine tailing material. Environ Pollut. 2016;214: 510–6.

40. Chen R-H, Li F-P, Zhang H-P, Jiang Y, Mao L-C, Wu L-L, et al. Comparative analysis of water quality and toxicity assessment methods for urban highway runoff. Sci Total Environ. 2016;553:519–23.

41. WHO. Lead in Drinking-water: Background document for development of WHO Guidelines for Drinking-water Quality. Geneva: World Health Organization; 2011.

42. Zamani AA, Yaftian MR, Parizanganeh AH. Statistical assessment of heavy metal pollution sources of groundwater around a lead and zinc plant in Iran. Iran J Environ Health Sci Eng. 2012;9:29–38.

43. WHO. Childhood lead poisoning. Geneva: World Health Organization; 2010.

44. Zamani AA, Shokri R, Yaftian MR, Parizanganeh AH. Adsorption of lead, zinc and cadmium ions from contaminated water onto Peganum harmala seeds as biosorbent. Int J Environ Sci Technol. 2013;10:93–102.

45. Xu D, Tan X, Chen C, Wang X. Removal of Pb(II) from aqueous solution by oxidized multiwalled carbon nanotubes. J Hazard Mater. 2008;154:407–16.

46. Dada AO, Olalekan AP, Olatunya AM, Dada O. Langmuir, Freundlich, Temkin and Dubinin–Radushkevich isotherms studies of equilibrium sorption of Zn2+ unto phosphoric acid modified Rice Husk. IOSR J Appl Chem. 2012;3:38–45.

10

Decolorization of synthetic textile wastewater using electrochemical cell divided by cellulosic separator

Ali Asghar Najafpoor[1], Mojtaba Davoudi[2] and Elham Rahmanpour Salmani[3*]

Abstract

Background: Annually, large quantities of dyes are produced and consumed in different industries. The discharge of highly colored textile effluents to the aquatic environments causes serious health problems in living organisms. This paper investigates the performance of each of the electro-oxidation and electro-reduction pathways in the removal of reactive red 120 (RR120) from synthetic textile effluents using a novel electrochemical reactor.

Methods: In the current study, a two-compartment reactor divided by cellulosic separator was applied in batch mode using graphite anodes and stainless steel cathodes. Central Composite Design was used to design the experiments and find the optimal conditions. The operational parameters were initial dye concentration (100–500 mg L^{-1}), sodium chloride concentration (2500–12,500 mg L^{-1}), electrolysis time (7.5–37.5 min), and current intensity (0.06–0.3 A).

Results: The results showed that electro-oxidation was much more efficient than electro-reduction in the removal of RR120. According to the developed models, current intensity was the most effective factor on the electro-oxidation of RR120 as well as in power consumption (Coefficients of 12.06 and 0.73, respectively). With regard to the dye removal through electro-reduction, electrolysis time (coefficient of 8.05) was the most influential factor. Under optimal conditions (RR120 = 200 mg.L^{-1}, NaCl = 7914.29 mg.L^{-1}, current intensity = 0.12 A, and reaction time = 30 min), the dye was removed as 99.44 and 32.38% via electro-oxidation and electro-reduction mechanisms, respectively, with consuming only 1.21 kwhm^{-3} of electrical energy.

Conclusions: According to the results, electro-oxidation using graphite anodes in a cell divided by cellulosic separator is very efficient, compared to electro-reduction, in the removal of RR120 from aqueous solutions.

Keywords: Cellulosic separator, Electro-oxidation, Electro-reduction, Graphite anodes, Reactive red 120

Background

Wastewater generation in huge volumes is one of the consequences of uncontrolled demand for textile articles, which causes extreme water consumption by textile industries [1]. Different wet-processing operations in the manufacturing process of textile industry result in the production of effluent which contains various pollutants including dyes, surfactants, detergents, and suspended solids [2]. Azo dyes as the largest group of organic dyes [3] constitute 20–40% of the dyes used in the textile industry [4] and are the most frequent chemical class of

dyes applied to industrial scale [5]. The general chemical formula of azo compounds has been shown in the form of R-N = N-R functional group. In the structure of these compounds, the double bond between nitrogen atoms indicates the azo chromophores, while R is the aromatic ring [1] containing groups such as sulfonate and hydroxyl [3]. The relatively low degree of dye fixation to fabrics especially for the reactive dyes results in the release of unfixed dyes into the effluent [6, 7]. It has been stated that textile industries produce a strongly colored wastewater [8]. It has also been declared that even the presence of inconsiderable dye concentrations in the effluent can reduce the penetration of light into the receiving water bodies. This leads to devastating effects on the aquatic biota [9] such as photosynthetic activity of

* Correspondence: Rahmanpoure1991@gmail.com
[3]Student Research Committee, School of Health, Mashhad University of Medical Sciences, Mashhad, Iran
Full list of author information is available at the end of the article

aquatic plants [8]. The probable persistence and the long-term bioaccumulation of synthetic organic dyes severely damage the health of ecosystems and living organisms [10].

A wide range of technologies for the removal of dyes from contaminated effluents can be found in literature [11]. Conventional treatment methods, i.e., physical, chemical, and biological processes, are still highly used. The physical methods mainly are practical for separating the solid pollutants, since there must be a difference between the pollutant and its media regarding the physical property. It is noticeable that chemical treatment occurs just under conditions that electrostatic property of both pollutant and coagulant is compatible [12]. Undesirable efficiency, high cost, and secondary pollutants are major shortcomings of physicochemical processes [9]. In spite of the fact that synthetic dyes have properties such as stability against light, temperature, and biodegradability [4], which makes decolorization difficult and incomplete [6], it was stated by Kariyajjanavar et al. that azo dyes are non-resistant to biological treatment methods under anaerobic conditions. However, applying this method is not suggested for dye removal as the products resulted from breakdown of azo dyes can be more toxic than the dye molecules [4]. The adverse environmental and health effects of dyes and their degradation products have pushed scholars' efforts towards developing powerful and effective treatment technologies [13]. According to the literature review, numerous advanced methods including adsorption, biosorption, reverse osmosis, ion-exchange [6], membrane separation, electro kinetic coagulation, irradiation, ozonation [8], sonication, enzymatic treatments, engineered wetland systems [9], and advanced oxidation processes (AOPs) such as TiO_2 photo-catalysis and electrochemical methods [10] have been utilized by researchers for the efficient treatment of textile wastewater. The electrochemical advanced oxidation processes (EAOPs) have received special interest for water and wastewater remediation [13]. Among them, electrocoagulation (EC) [14], electro-oxidation (EO) [15], and electro-Fenton (EF) [16] have been frequently studied. EAOPs have some significant advantages such as simple equipment [4], easy implementation [17], close control of the favored reactions through applying optimum electrical current, on-site treatment in less space [18], and high efficiency for the degradation of persistent pollutants, while the cost of electricity used can be a drawback [19]. The presence of iron ions in the EC and EF processes leads to the sludge generation that imposes the cost of further treatment [20]. EO is the most widely used mechanism of EAOPs [21] and anodic oxidation (AO) is the most typical kind of EO [22]. The explanation of complicated electrochemical reactions that occur during the EO treatment process and determining the definite removal mechanism of many of the contaminants do not seem an easy task [6]. EO of pollutants can occur through AO directly or indirectly, and also by the participation of chlorine-based oxidants when chloride solutions are treated [18]. In direct AO, pollutant molecules are oxidized at anode via electron transfer from the organic matter to the electrode, while in the indirect AO, the chemical reactions with electro-generated species such as hydroxyl radicals resulted from water discharge at the anode leads to the pollutant degradation [23]. It is known that direct AO leads to poor decontamination, while the effectiveness of indirect AO is dramatically dependent on the used anode. In the so-called "active" anodes which have low oxidation power, the chemisorbed "active oxygen" (MO_{x+1}) is the yield of water oxidation, while the physisorbed hydroxyl radical is the product of water discharge at the high oxidation power anodes also named "non-active". The Pt, IrO_2, and RuO_2 are some examples of the former anodes in the formation of selective oxidation products (Eq. 1), while boron-doped diamond (BDD), PbO_2, and SnO_2 are typical kinds of the latter anodes causing complete combustion of the organic compounds (R) (Eq. 2) [24]:

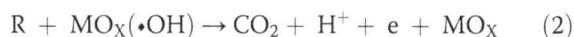

$$R + MO_{X+1} \rightarrow RO + MO_X \tag{1}$$

$$R + MO_X(\bullet OH) \rightarrow CO_2 + H^+ + e + MO_X \tag{2}$$

Although non-active anodes have been preferred in most of the EO studies for their ability in quick and total mineralization, it was shown by Méndez-Martínez et al. [10] that the use of active anodes can be equally interesting as they provide the chance for thorough elucidation of the general degradation mechanisms.

As previously stated, in chloride medium, the oxidation of organics can also occur by chlorine-based oxidants. The presence of NaCl in the reaction mixture leads to the formation of chlorohydroxyl radicals $(ClOH^\bullet)$ on the anode surface (Eq. 3), which oxidize the organic matter as given in Eq. (4):

$$H_2O + M + Cl^- \rightarrow M[ClOH^\bullet] + H^+ + 2e^- \tag{3}$$

$$R + M[ClOH^\bullet] \rightarrow M + RO + H^+ + Cl^- \tag{4}$$

Furthermore, primary oxidants such as oxygen (Eq. 5), chlorine (Eq. 6), hydrogen peroxide (Eq. 7), and hypochlorite (Eq. 8) can result from the reactions between water and radicals near the anode:

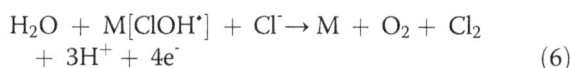

$$H_2O + [MOH^\bullet] \rightarrow M + O_2 + 3H^+ + 3e^- \tag{5}$$

$$H_2O + M[ClOH^\bullet] + Cl^- \rightarrow M + O_2 + Cl_2 + 3H^+ + 4e^- \tag{6}$$

$$H_2O + [MOH^\bullet] \rightarrow M + H_2O_2 + H^+ + e^- \tag{7}$$

$$H_2O + Cl^- \rightarrow HOCl \ (HOCl \leftrightarrow ClO^- + H^+)$$
$$+ \ H^+ + 2e^- \qquad\qquad (8)$$

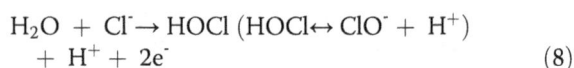

Then, the reaction between free chlorine and oxygen results in the formation of secondary oxidants such as chlorine dioxide and ozone according to the reactions presented by Eq. (9) and Eq. (10) [7]:

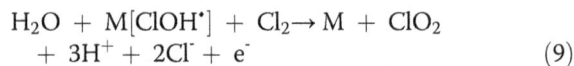

$$H_2O + M[ClOH^\bullet] + Cl_2 \rightarrow M + ClO_2$$
$$+ \ 3H^+ + 2Cl^- + e^- \qquad (9)$$

$$O_2 + M[OH^\bullet] \rightarrow M + O_3 + H^+ + e^- \qquad (10)$$

Simultaneously with the oxidation reactions at the anodic chamber, the reduction reactions occur in the cathodic compartment. Based on the scientific evidence, the electrochemical reduction, or in other words, the electro-reduction (ER) method has been applied for the removal of dyes and many of the other contaminants from both synthetic and real effluents [20]. However, a limited number of studies have tried ER compared to AO mainly because of the performance dissatisfaction [24]. ER was considered by Bechtold et al. as a proper method for the treatment of strongly colored effluents containing reactive dyes. They remarked that partial reduction of dye (Eq. 11) produces hydrazine, while its total reduction generates the amino compounds (Eq. 12) [25]:

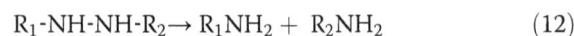

$$R_1\text{-}N = N\text{-}R_2 \rightarrow R_1\text{-}NH\text{-}NH\text{-}R_2 \qquad (11)$$

$$R_1\text{-}NH\text{-}NH\text{-}R_2 \rightarrow R_1NH_2 + R_2NH_2 \qquad (12)$$

Electrolytic hydrogenation is the most common cathodic reaction resulted from water electrolysis at the cathode surface [26] (Eq. 13 [2]):

$$2H_2O + 2e^- \rightarrow H_2 + 2OH^- \qquad (13)$$

Roessler et al. [27] have shown that the formed hydrogen can then react with the dye adsorbed at the cathode surface.

Radha et al. [2] achieved color removal efficiency equal to 96% via EO method in a batch system using stainless steel and graphite as cathode and anode, respectively, under 60 min reaction time at 0.6 A current intensity. Wang et al. [18] investigated the simultaneous removal of color and COD from real textile wastewater at the presence of Pt/Ti anode and graphite cathode in a divided reactor. Hypochlorite and hydrogen peroxide were determined as the main factors responsible for treatment process. The overall removal efficiency was proportional to the applied current and the color removal efficiency in the anodic chamber was much higher than cathodic one. Del Rio et al. [28] studied the effect of oxidation, reduction and oxido-reduction processes in divided and undivided electrolytic cells for the removal of Reactive Orange 4 dye. In the separated cell, a cationic membrane

was used and the electrodes of Ti/SnO2–Pt–Sb and stainless steel were employed as anode and cathode, respectively. The maximum dye removal efficiency was obtained in the undivided cell via oxido-reduction reactions. Maljaei et al. [1] explored the removal of Reactive Yellow 3 dye through indirect EO method using graphite as both anode and cathode in a batch reactor. They reported higher efficiency of dye removal with increasing current intensity and decreasing electrolysis time. Kariyajjanavar et al. [4] introduced graphite as a relatively cheap anode, which provides satisfactory results for the electrochemical dye degradation.

The current study aimed to evaluate the performance of EO and ER using graphite anodes and stainless steel cathodes in the removal of RR120 dye from synthetic textile wastewater. We used a novel two-compartment reactor divided by cellulosic separator, while most of the electrochemical decontamination studies had been conducted in undivided cells. It is known that high cost of separation and non-ecofriendly properties of common membranes have restricted the use of divided cells. But interesting advantages of divided cells including the increased rate of electrochemical oxidation and reduction reactions, decreased generation of toxic by-products, and increased life-time of anode caused by acidic pH of the anodic compartment [26] encouraged the researchers of the current study to apply a low-cost material to divide the reactor to enjoy the benefits of divided cells. Therefore, cellulosic separator that was previously employed by Davoudi et al. [20] for synthetic tannery wastewater treatment was used in this project for synthetic textile wastewater treatment.

Methods

Chemicals

Double Distilled Water (DDW) was used to prepare stock solutions and required dilutions. Graphite electrode was purchased from Noavaran Shimi Company, Iran. Ultrapure grade of NaCl, sulfuric acid (H_2SO_4) and sodium hydroxide (NaOH) was obtained from Merck, Germany. To prepare 1000 ppm standard solution, 1 g of RR120 was weighed by a digital balance (Sartorius bp 110 s) and dissolved in 1000 mL of DDW. RR120 (red HE3B) was provided by Shadilon Textile Group Co., Iran. The molecular structure of the dye is demonstrated in Fig. 1 [29] and its physicochemical properties are shown in Table 1.

Experimental design

Experiments were designed based on CCD, a well-known design of Response Surface Methodology (RSM) in the Design Expert 7.0 (trial version). RSM is an efficient tool to optimize experimental conditions while minimizes the number of experiments [30]. Accordingly,

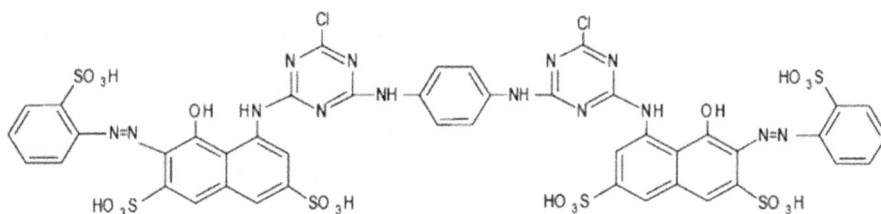

Fig. 1 The molecular structure of RR120 dye

30 experiments were determined based on 2^4 factorial designs with 6 central and 8 axial points. This number was obtained by defining the actual values of independent variables in the initial $(-\alpha)$ and end $(+\alpha)$ points by user while the three other levels of each parameter were suggested by CCD. Table 2 represents the real and coded values of operational factors. The expected responses in the current study were dye removal efficiency in each of the reactor chambers and consumed energy. In RSM, the experimental data corresponding to each dependent variable were fitted to a polynomial model to find the most influential factors and their various effects including linear, interaction, and quadratic effects. ANOVA was used to validate the adequacy of the models. To assess the quality of fit in the developed regression models, the coefficient of determination (R^2) and adjusted R^2 were applied. The Fisher distribution test (F-test) and adequate precision ratio were used to determine the statistical significance of the models and its associated terms [31].

Experimental set-up and procedure

Figure 2 gives the schematic of the electrochemical cell consisted of a rectangular reactor coupled with the power supply and multimeter. The hold-up container was a Plexiglass vessel in which two electrodes were placed close to the both sides of the separator. The separator was made of cellulose fibers (cotton) located in the middle of the reactor to divide it into two distinct equal parts. A rod of graphite was used as the anode electrode while a sheet of stainless steel was applied as the cathode electrode. An adjustable laboratory DC power supply was used to provide the electrical energy needed to operate the system. The test solutions were prepared from the stock solution and completely mixed using a magnetic stirrer before experiments. Treatment was carried out in a batch system with a net working volume of 200 mL. In each experiment, sampling was done from both parts of the reactor at the time determined by CCD, and the pH of the samples was neutralized by dilute solutions of H_2SO_4 and NaOH. The acidic pH of the anolyte contents was due to water hydrolysis at the anode which produced H^+ ions and the alkaline pH of the catholyte was due to the H_2 and/or O_2 reduction reactions that consume H^+ and generate OH^- at the cathode zone [32]. To measure the remained concentration of dye, the Milton Roy Company Spectronic 20 Spectrophotometer (UV–VIS) was applied at 530 nm.

Analytical methods

To determine the dye concentration, at first a calibration curve ($R^2 = 0.9999$) was drawn based on the absorbance levels corresponding to samples of known concentration. Then, the following expression (Eq. 14) was used to calculate the final dye concentration:

$$((48.416.A) - 0.1718)df \tag{14}$$

Where A is the absorbance of the solution and df shows the dilution factor. Then, the removal efficiency was calculated according to Eq. (15):

$$\frac{C_{in} - C_f}{C_{in}} \times 100 \tag{15}$$

Table 1 The physicochemical properties of RR120

Chemical formula	$C_{44}H_{24}Cl_2N_{14}Na_6O_{20}S_6$
CAS Registry Number	61951-82-4
Natural state	Powdered
Chemical structure	Diazo
Solubility in 20 °C water (g/l)	100
Molecular weight (g/mol)	1469.98
Charge	Negative
pH	6–9
Density (kg/m^3)	450–500
λ_{max}(nm)	530

Table 2 Independent variables in coded and real levels

Independent variables	Study levels				
	$-\alpha$	-1	0	$+1$	$+\alpha$
X_1: RR120 Conc. (mg. L^{-1})	100	200	300	400	500
X_2: NaCl Conc. (mg. L^{-1})	2500	5000	7500	10,000	12,500
X_3: Current intensity (A)	0.06	0.12	0.18	0.24	0.3
X_4: Electrolysis time (min)	7.5	15	22.5	30	37.5

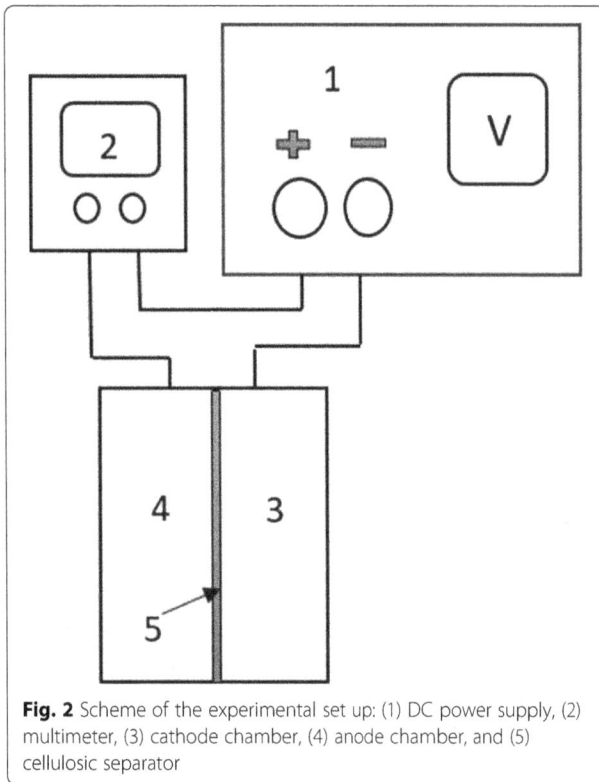

Fig. 2 Scheme of the experimental set up: (1) DC power supply, (2) multimeter, (3) cathode chamber, (4) anode chamber, and (5) cellulosic separator

Where the initial and final color concentrations are shown by C_{in} and C_f, respectively. The average potential of the cell (V) during the electrolysis was recorded to calculate the energy consumption according to Eq. (16):

$$EC = \left(\frac{VAT}{1000V_S}\right) \tag{16}$$

Where EC shows the energy consumption per volume of treated wastewater (Kwhm^{-3}), T is the time of electrolysis (h), A and V_s are the current intensity, and sample volume (m^3), respectively. Finally, the obtained data were analyzed in Design Expert 7 program.

Results and Discussion

In the current study, a set of electrochemical batch experiments were performed in a two-compartment reactor divided by cellulosic separator to study the effectiveness of EO and ER using graphite and stainless steel electrodes, respectively, in the removal of RR120 from synthetic textile effluents. Cellulose was applied for cell separation in the present research because of its great benefits such as its natural abundance which subsequently results in the lower cost of operation. Its physical characteristics such as porosity and permeability cause satisfactory separation between the contents of anolyte and catholyte chambers. Cellulose ability for retaining a layer

of water in its structure allows establishing electric flux between the contents of separated compartments [26].

In this work, the effect of the operational parameters including initial dye concentration, electrolyte concentration, current intensity, and electrolysis time on the RR120 removal efficiency was studied within a specified range. Accordingly, the effect of initial RR120 concentration on its removal efficiency was studied in the range between 100 mg. L^{-1} and 500 mg. L^{-1}. Cardoso et al. [33] investigated the removal of RR120 from aqueous effluents through adsorption process within the range of 50–1200 mg. L^{-1}. Tehrani-Bagha and Amini [34] studied the effectiveness of UV-Enhanced Ozonation for the treatment of simulated dyebath effluents containing 200 mg. L^{-1} and 800 mg. L^{-1} of RR120. Another factor affecting the RR120 removal rate was NaCl concentration that was studied from 2500 to 12,500 mg. L^{-1}. The concentration of this electrolyte in the real textile wastewater has been reported from 5000 to 12,000 mg. L^{-1} [29]. The impact of current intensity on the treatment efficiency in this study was assessed in the range between 0.06 A and 0.3 A that was close to the range of 0.1–0.35 A in a study conducted by Ghalib [35] for electrochemical removal of direct blue dye from textile wastewater. It was decided to apply electrical current at such low intensities because it is known that graphite electrodes have small values of overvoltage for oxygen evolution, indicating their effective performance for pollutant oxidation only at very low current intensities [1]. In the present work, the electrolysis time was studied at five points from 7.5 to 37.5 min. In the Zaviska et al. [36] study for atrazine removal using EO process, the effect of treatment time was assessed at two levels: 10 min and 40 min. The time of electrolysis varied from 5–20 min in Kariyajjanavar et al. [4] study for electrochemical degradation of reactive azo dyes from aqueous solutions using graphite electrodes.

Regression models and ANOVA

In this study, a total of 30 runs were performed according to the CCD suggestions to assess the relationship between each response and four independent variables. For this purpose, a mathematical equation was developed for every response in RSM to study the behavior of the system as a function of RR120 concentration (x_1), NaCl concentration (x_2), current intensity (x_3), and electrolysis time (x_4). After removing model terms which were not statistically significant because of the Prob > F > 0.05, each equation was achieved as a sum of a constant value, and main, interaction, and quadratic effects in the model. The modified models are shown in the following:

$Y_1(RR120 \ removal \ in \ anolyte \ chamber)\%$
$= 99.32 - 2.98x_1 + 2.22x_2 + 12.06x_3$
$\quad + 9.94x_4 + 4.34x_1x_3$
$\quad + 2.60x_1x_4 - 2.80x_2x_3 - 9.05x_3x_4 - 6.74x_3^2 - 5.62x_4^2$

$$(17)$$

$Y_2(RR120 \ removal \ in \ catholyte \ chamber)\%$
$= 22.34 - 6.72x_1 + 4.72x_3 + 8.05x_4 \qquad (18)$

$Y_3(Energy \ consumption)kwhm^{-3}$
$= 1.65 + 0.73x_3 + 0.51x_4 + 0.22x_3x_4 \qquad (19)$

After screening the models to exclude insignificant effects, the experimental data were analyzed using ANOVA to check the adequacy of the models. Based on F-test results which are given in Table 3, the quadratic model for Y_1, the linear model for Y_2, and the 2FI model for Y_3 were all highly significant. The F-values were 49.24, 20.48, and 236.49 for the functions corresponding to Y_1, Y_2, and Y_3, respectively. The chance of achieving these large values of F due to error is only 0.01%. Furthermore, the R^2 coefficient was 0.963 for Y_1, 0.703 for Y_2, and 0.964 for Y_3. With respect to R^2 value which measures the proportion of total variations in the dependent variable that can be explained by the model predicators [37], the models predictions were in good agreement with the experimental data. Since some of the variables were excluded from the regression model in the modification process, the R^2 index was calculated using the variables retained in the model and was named

the adjusted R^2. The difference between this index and the R-squared predicted by the model must be a number lower than 0.2 to ensure well data fitting by the developed model [31]. The disagreement between the adjusted and predicted R^2-values for all models was less than 0.09 (see under Table 3). The models precision was adequate because of the signal/noise ratio more than 4 in all cases. Although graphs of normal % probability and studentized residuals are not shown, regarding the fairly straight lines of these graphs, the distribution of data was normal for all responses.

Dye removal efficiency in anodic (oxidative) cell

The first part of Eq. (17) shows 99.32% removal efficiency for RR120 from the anolyte content when all terms in the second part of the equation are fixed at their central values. The magnitude of the coefficient devoted to each term and the corresponding positive or negative sign determines the variations that may occur in the RR120 removal rate when the levels of the experiment factors in the of the equation change. Equation 17 indicates that the positive coefficient (+12.06) related to the current intensity factor had the highest value among different coefficients; thus, this factor created the most meaningful effect on the response. The next rank was allocated to the effect of contact time with the coefficient of +9.94. According to Table 4 which represents the experimental results as a function of various levels of independent parameters, there was a direct relationship between the two aforementioned variables and the study

Table 3 Statistical indices obtained from the ANOVA for regression models

Source	Sum of squares	Degrees of freedom	Mean square	F value	P > F
In the anolyte compartment [a]					
Model	10010.49	10	1001.05	49.24	<0.0001
Residual	386.23	19	20.33	Na	Na
Lack of fit	385.87	14	27.56	381.01	<0.0001
Pure Error	0.36	5	0.072	Na	Na
In the catholyte compartment [b]					
Model	3173.34	3	1057.78	20.48	<0.0001
Residual	1342.74	26	51.64	Na	Na
Lack of fit	1162.51	21	55.36	1.54	0.3362
Pure Error	180.23	5	36.05	Na	Na
Energy consumption [c]					
Model	19.81	3	6.60	236.49	<0.0001
Residual	0.73	26	0.028	Na	Na
Lack of fit	0.53	21	0.025	0.64	0.7861
Pure Error	0.2	5	0.039	Na	Na

Na Not applicable
[a] $R^2 = 0.963$, $R_{adj}^2 = 0.943$, $R_{pred}^2 = 0.879$, adequate precision = 23.549
[b] $R^2 = 0.703$, $R_{adj}^2 = 0.668$, $R_{pred}^2 = 0.579$, adequate precision = 14.855
[c] $R^2 = 0.964$, $R_{adj}^2 = 0.961$, $R_{pred}^2 = 0.952$, adequate precision = 47.801

Table 4 Experimental conditions determined by CCD and the observed results

Run No.	Independent variables				Dependent variables		
	RR120 Conc. (mg. L^{-1})	NaCl Conc. (mg. L^{-1})	Current intensity (A)	Electrolysis time (min)	Decolorization efficiency via EO (%)	Decolorization efficiency via ER (%)	Energy consumption (kwhm^{-3})
1	300	7500	0.06	22.5	43.4	11.8	0.365
2	400	10,000	0.24	15	99	13.26	1.44
3	400	5000	0.12	15	37.2	9.29	0.525
4	300	7500	0.18	22.5	98.5	27	1.333
5	300	7500	0.18	22.5	99.2	31.2	1.50
6	400	10,000	0.12	15	52	2.7	0.577
7	200	5000	0.24	15	99.2	22.76	1.71
8	300	12,500	0.18	22.5	99.2	20.65	1.636
9	200	10,000	0.24	15	99.1	22.8	1.65
10	500	7500	0.18	22.5	98.4	17.26	1.603
11	200	10,000	0.24	30	98.58	48.56	2.64
12	400	5000	0.24	30	98.45	16.94	3.12
13	100	7500	0.18	22.5	98.72	52.6	1.67
14	400	10,000	0.24	30	99.34	34.5	3.24
15	300	7500	0.18	22.5	99.17	22.39	1.62
16	200	5000	0.12	15	64.65	10	0.675
17	300	7500	0.18	7.5	52	14.49	0.534
18	200	5000	0.24	30	98.2	38.85	3.39
19	300	7500	0.18	22.5	99.19	18	1.636
20	300	7500	0.3	22.5	98.88	30	3.26
21	300	7500	0.18	22.5	99	22.11	1.755
22	200	10,000	0.12	15	93.3	16.14	0.637
23	300	7500	0.18	22.5	99	14.49	1.906
24	400	10,000	0.12	30	98.48	16.31	1.17
25	400	5000	0.12	30	82.54	20.87	1.08
26	200	10,000	0.12	30	99.34	22.32	1.275
27	300	2500	0.18	22.5	97.73	16.29	1.84
28	400	5000	0.24	15	97.59	2.18	1.725
29	200	5000	0.12	30	97.37	25.19	1.275
30	300	7500	0.18	37.5	99.32	48.84	2.56

response, which can also be concluded from the perturbation plot of Fig. 3. With regard to the literature review, current intensity has been the most important factor affecting the performance of EO process in lab scale [2, 6]. Curve C in the Fig. 3 shows a steep increase in the dye removal rate from the level −1 to the central level of current intensity, while this rate increases gradually up to the level +1 and then stops. This behavior has been indicated by the significant quadratic effect devoted to the applied current in Eq. 17. This negative second order effect can be explained by considering the nature of the graphite as it has low values of overpotential for O_2 evolution. It is known that in higher current intensities, the parasite nonoxidizing reaction of O_2 evolution is a dominant mechanism which causes a significant reduction in current efficiency. Thus, applying low current intensities can be effective for oxidation of pollutants on this anode [1]. In addition to the discussed first and second order effects of the current intensity, this factor also showed the most important combined effect on the response in the interaction with the electrolysis time parameter. 3D surface plot of Fig. 4 shows this interaction. As can be seen, instead of applying high levels of both factors to reach a favorable removal efficiency, electrolysis can be performed at lower intensities and higher reaction times or vice versa or at moderate levels of both parameters to yield the same removal percentage. The negative sign of the respective quadratic effect confirms this concept.

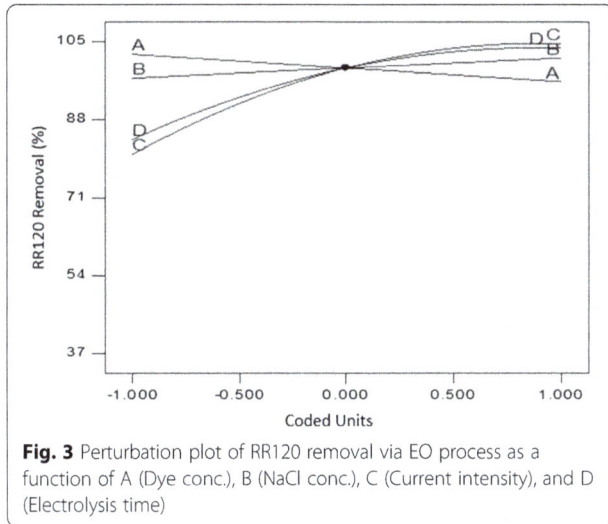

Fig. 3 Perturbation plot of RR120 removal via EO process as a function of A (Dye conc.), B (NaCl conc.), C (Current intensity), and D (Electrolysis time)

Equation 17 also shows the sensitivity of decolorization efficiency to the effect of RR120 concentration. It can be observed in Fig. 3 that when the dye concentration increased from 200 to 400 mg. L^{-1}, the removal percentage reduced to 6%. This result is in agreement with results already reported by Körbahti et al. who studied electrochemical decolorization of textile dyes [6]. As can be seen from curve B in the perturbation plot of Fig. 3, the addition of NaCl as electrolyte into the working solution resulted in a better removal performance which can be attributed to the increased cell conductivity and generation of powerful oxidizing agents such as Cl_2 and HOCl. The former will increase the ionic transfer and current intensity at a given operating voltage which provides more chance for the latter. The acidic pH of the anodic cell is in favor of electrochemical degradation of azo dyes since it causes chloride reduction to chlorine gas and further to

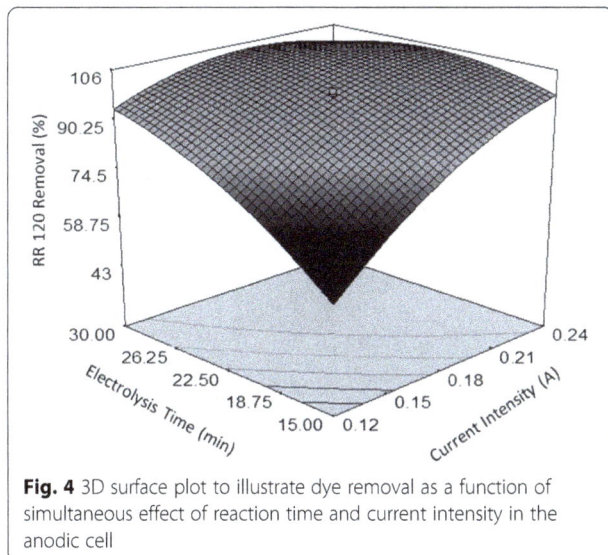

hypochlorous acid which is known for its high potential of oxidation [4]. The anodic degradation of the RR120 can be explained by the action of the chemisorbed hydroxyl radicals and also the active chlorine species. The reactions involved in the formation of chlorine based oxidants are completely expressed in the introduction section, but regarding the role of chemically adsorbed hydroxyl radical or M('OH) it must be considered that the strong interaction of the electrode surface with the 'OH does not allow its direct reaction with organics and instead, a superoxide (MO) is formed according to Eq. (20). MO further acts as a mediator in the oxidation of organics by reaction presented in Eq. (21):

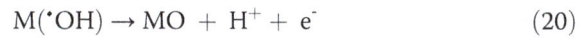

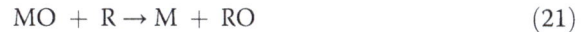

$$M(\cdot OH) \rightarrow MO + H^+ + e^- \tag{20}$$

$$MO + R \rightarrow M + RO \tag{21}$$

As previously discussed in the introduction section, achieving to complete mineralization via electrochemical combustion is unexpected when the used anode is an active one. Instead, the electrochemical conversion of organics into reaction intermediates is happened [24]. This was proved through the Fourier transform infrared spectroscopy (FTIR) analysis (data not shown). The band at 3447.95 cm^{-1} corresponds to the N-H stretching vibration [10], while the peak appeared at 1717.62 cm^{-1} belongs to the carbonyl region and can be ascribed to -C = O stretching vibration [38]. The appearance of peak at 1594.80 cm^{-1} which shows the -N-H bending mode suggests the formation of the amino group by the cleavage of azo bond [39].

Dye removal efficiency in cathodic (reductive) cell

With respect to Eq. (18), 22.34% of removal efficiency was observed for RR120 via reductive pathway, which was independent of any factor and interaction of factors. The model indicates the direct relationship of RR120 removal with the main effects of applied current (+4.72) and time (+8.05). On the contrary, the removal performance was negatively associated with RR120 initial concentration with respect to the coefficient of −6.72. This effect can also be seen in Table 4, where the maximum removal efficiency of 52% was observed for the solution containing 100 mg. L^{-1} of RR120 as the lowest examined concentration. The line A in the perturbation plot of Fig. 5 suggests that when the concentration of RR120 was doubled in the solution, 14% reduction occurred in its removal efficiency. The middle line in the Fig. 5 corresponds to the effect of current intensity which caused 10.6% higher removal efficiency when its level increased from 0.12 A to 0.24 A. The effect of electrolysis time was more pronounced when it increased from 15 min to 30 min, leading to 16% higher removal performance. Regardless of the parameters and the effect of different

Fig. 4 3D surface plot to illustrate dye removal as a function of simultaneous effect of reaction time and current intensity in the anodic cell

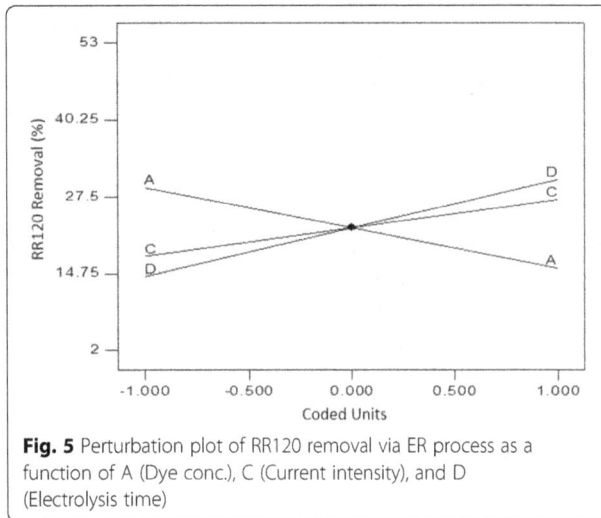

Fig. 5 Perturbation plot of RR120 removal via ER process as a function of A (Dye conc.), C (Current intensity), and D (Electrolysis time)

levels of them on the response, the total performance of reductive cell used in the present research for RR120 removal was low. This is in agreement with the study conducted by Carneiro et al. [40] who achieved only 37% RB4 removal via ER process.

Two possible pathways can be proposed for degradation of RR120 in the cathodic cell. The dye may be adsorbed on the surface of stainless steel and then the direct cathodic electron transfer may occur [41] according to the Eqs. (11) and (12) presented in the introduction section. For the latter mechanism, we can refer to the role of hydrogen. When stainless steel is used as cathode, the chemisorbed hydrogen is generated at the electrode surface by electrolysis of water [2] according to the Eq. (13) which can then participate in decolorization. At basic pH of the catholyte compartment, hypochlorite ions are dominant species in the bulk, resulting in the cleavage of azo bond. Oxidation of amid group can lead to the generation of carboxylic derivatives and thus weak acidic condition. The nitrogen in the azo bond is reduced by accepting hydrogen, the double bond transforms to single bond and then to amine. The weak acidic condition encourages the amine compounds to accept proton and as a result, it can be adsorbed onto the negative charged sites of the cathode [1].

Energy consumption

With respect to Eq. (19), the average amount of consumed energy for the removal of RR120 in the designed electrochemical cell was 1.65 kwhm^{-3} that was remarkably lower than consumed energy in similar studies [4, 5, 11]. Furthermore, it was proved that energy consumption is proportional to the current intensity and electrolysis time and also the combined effect of these two factors. A change in the applied current and time from level −1 to +1 resulted in more energy consumption as 1.46 kwhm^{-3} and 1.02 kwhm^{-3}, respectively. Hence, the energy consumption was

more affected by the variations occurred in the level of applied current in comparison to the reaction time which can be attributed to the increased oxygen and hydrogen evolution reaction in the anodic and cathodic cell, respectively, at higher current intensities [4]. Figure 6 indicates current intensity interaction with electrolysis time on energy consumption rate while the two other factors were constant at the central level. As obviously seen in the graph, there was a synergistic effect between the parameters and the response. According to RSM prediction, 3 kwhm^{-3} of electrical energy will be consumed if the solution is treated at the level +1 of both factors.

Optimization

In the optimization process of RR120 removal, the criteria goal was selected "in range" for all the independent variables while it was desired as "maximize" for removal efficiency and as "minimize" for power consumption. The optimum values of parameters proposed in first solution were: initial RR120 concentration (200 mg. L^{-1}), NaCl concentration (7914.29 mg. L^{-1}), current intensity (0.12 A), and reaction time (30 min), leading to 99.44 and 32.38% of RR120 removal via EO and ER mechanisms, respectively, with consuming 1.21 kwhm^{-3} of electrical energy. According to the confirmation study carried out in optimal conditions, the removal performance via EO was obtained as 96% that was very close to the predicted value. The removal efficiency in the sample drawn from the catholyte compartment was 22%. This value also located in the predicted interval (PI low: 16.49% and PI high: 48.28%). In addition, 1.3 kwhm^{-3} of electrical energy was consumed in the conducted experiment that was also was in the PI. Considering the low performance of cathodic degradation, chemical oxygen demand (COD) analysis was just conducted on the sample taken out from the anolyte compartment. The

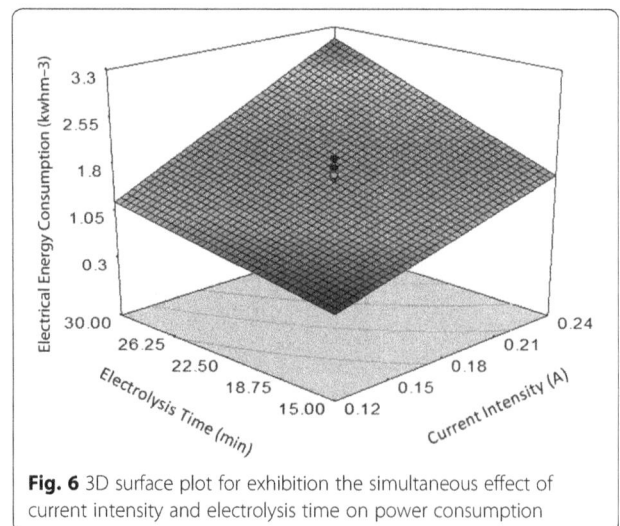

Fig. 6 3D surface plot for exhibition the simultaneous effect of current intensity and electrolysis time on power consumption

analysis revealed that only 17.56% of COD was removed during the electrolysis. However, achieving such a low efficiency in COD removal with respect to the electrolysis performed in a short time (30 min) at low current intensity (0.12 A) was not unexpected. Rajkumar and Kim achieved 73.5% COD reduction for a mixture of reactive dyes at a concentration of 200 mg. L^{-1} after 120 min electrolysis time using 2 A of current intensity [29]. Although the levels of operating parameters in the current work satisfied the goal of the study, i.e., cost-effective removal of RR120 under optimal conditions, given the importance of mineralization, it is suggested for further investigation to try longer electrolysis times and higher electrical currents to reach a remarkable COD removal.

Conclusion

This investigation assessed the performance of electro-oxidation and electro-reduction pathways by means of graphite and stainless steel electrodes in a two-compartment reactor divided by cellulosic separator in the removal of RR120 dye from synthetic textile effluent. Based on the results, some conclusions are drawn as follows:

i. Anodic oxidation using graphite electrode was successfully applied for decolorization of strongly colored effluent and gave ≥90% removal efficiency in four fifth of the experiments.

ii. The reductive pathway using stainless steel failed to achieve a satisfactory removal rate. The maximum RR120 removal rate in cathodic cell was 52% obtained for the most dilute solution.

iii. The average amount of electrical energy consumption 1.65 $kwhm^{-3}$ was much less than the corresponding values in similar studies, mainly due to the low levels of applied current and time of electrolysis.

iv. 96% dye removal efficiency was obtained in the anodic cell under the optimized operating conditions of 7914 mg L^{-1} NaCl, 0.12 A current intensity, and 30 min reaction time for a solution containing 200 mg L^{-1} of RR120 concentration.

v. The RR120 degradation was removed due to electrochemical conversion that caused formation of intermediate products. With respect to the low reduction of COD, higher levels of current intensity and electrolysis time should be tried to provide the opportunity for dye intermediates to be converted to CO_2 and H_2O.

Abbreviations

ANOVA: Analysis of variance; AO: Anodic oxidation; AOPs: Advanced oxidation processes; CCD: Central composite design; COD: Chemical oxygen demand; DDW: Double distilled water; EAOPs: Electrochemical advanced oxidation processes; EC: Electrocoagulation; EF: Electro-fenton; EO: Electro-oxidation; ER: Electro-reduction; FTIR: Fourier transform infrared spectroscopy; H_2SO_4: Sulfuric acid; NaCl: Sodium chloride; NaOH: Sodium hydroxide; PI: Prediction interval; RR120: Reactive red 120; RSM: Response surface methodology

Acknowledgements
The authors thank the staffs of environmental chemistry lab of Health school of Mashhad University of Medical Sciences.

Funding
This paper was funded by the Vice Chancellor for Research of Mashhad University of Medical Sciences under Project No. 940176.

Authors' contributions
A.A.N supervised the study. MD participated in designing the study and analyzing the data. ERS conducted the experiments and drafted the manuscript. All the authors read and approved the final manuscript.

Competing interests
The authors declare that they have no competing interests.

Author details
[1]Health Sciences Research Center, Department of Environmental Health Engineering, School of Health, Mashhad University of Medical Sciences, Mashhad, Iran. [2]Department of Environmental Health Engineering, School of Health, Torbat Heydariyeh University of Medical Sciences, Torbat Heydariyeh, Iran. [3]Student Research Committee, School of Health, Mashhad University of Medical Sciences, Mashhad, Iran.

References

1. Maljaei A, Arami M, Mahmoodi NM. Decolorization and aromatic ring degradation of colored textile wastewater using indirect electrochemical oxidation method. Desalination. 2009;249(3):1074–8.
2. Radha K, Sridevi V, Kalaivani K. Electrochemical oxidation for the treatment of textile industry wastewater. Bioresour Technol. 2009;100(2):987–90.
3. Florenza X, Solano AMS, Centellas F, Martínez-Huitle CA, Brillas E, Garcia-Segura S. Degradation of the azo dye acid red 1 by anodic oxidation and indirect electrochemical processes based on Fenton's reaction chemistry. Relationship between decolorization, mineralization and products. Electrochim Acta. 2014;142:276–88.
4. Kariyajjanavar P, Jogttappa N, Nayaka YA. Studies on degradation of reactive textile dyes solution by electrochemical method. J Hazard Mater. 2011; 190(1–3):952–61.
5. de Oliveira GR, Fernandes NS, Melo JV, da Silva DR, Urgeghe C, Martínez-Huitle CA. Electrocatalytic properties of Ti-supported Pt for decolorizing and removing dye from synthetic textile wastewaters. Chem Eng J. 2011;168(1):208–14.
6. Körbahti BK, Artut K, Geçgel C, Özer A. Electrochemical decolorization of textile dyes and removal of metal ions from textile dye and metal ion binary mixtures. Chem Eng J. 2011;173(3):677–88.
7. Chatzisymeon E, Xekoukoulotakis NP, Coz A, Kalogerakis N, Mantzavinos D. Electrochemical treatment of textile dyes and dyehouse effluents. J Hazard Mater. 2006;137(2):998–1007.
8. Khlifi R, Belbahri L, Woodward S, Ellouz M, Dhouib A, Sayadi S, et al. Decolourization and detoxification of textile industry wastewater by the laccase-mediator system. J Hazard Mater. 2010;175(1–3):802–8.
9. Basha CA, Sendhil J, Selvakumar KV, Muniswaran PKA, Lee CW. Electrochemical degradation of textile dyeing industry effluent in batch and flow reactor systems. Desalination. 2012;285:188–97.
10. Méndez-Martínez AJ, Dávila-Jiménez MM, Ornelas-Dávila O, Elizalde-González MP, Arroyo-Abad U, Sirés I, et al. Electrochemical reduction and oxidation pathways for reactive black 5 dye using nickel electrodes in divided and undivided cells. Electrochim Acta. 2012;59:140–9.
11. Tavares MG, da Silva LVA, Sales Solano AM, Tonholo J, Martínez-Huitle CA, Zanta CLPS. Electrochemical oxidation of Methyl Red using Ti/Ru0.3Ti0.7O2 and Ti/Pt anodes. Chem Eng J. 2012;204–206:141–50.
12. Mukimin A, Vistanty H, Zen N. Oxidation of textile wastewater using cylinder Ti/β-PbO2 electrode in electrocatalytic tube reactor. Chem Eng J. 2015;259:430–7.
13. Thiam A, Sirés I, Garrido JA, Rodríguez RM, Brillas E. Effect of anions on electrochemical degradation of azo dye carmoisine (acid red 14) using a

BDD anode and air-diffusion cathode. Sep Purif Technol. 2015;140:43–52.

14. Martinez-Huitle CA, Ferro S. Electrochemical oxidation of organic pollutants for the wastewater treatment: direct and indirect processes. Chem Soc Rev. 2006;35(12):1324–40.

15. Mahvi AH, Ebrahimi SJA-d, Mesdaghinia A, Gharibi H, Sowlat MH. Performance evaluation of a continuous bipolar electrocoagulation/electrooxidation–electroflotation (ECEO–EF) reactor designed for simultaneous removal of ammonia and phosphate from wastewater effluent. J Hazard Mater. 2011;192(3):1267–74.

16. Jaafarzadeh N, Ghanbari F, Ahmadi M, Omidinasab M. Efficient integrated processes for pulp and paper wastewater treatment and phytotoxicity reduction: permanganate, electro-fenton and Co 3 O 4/UV/peroxymonosulfate. Chem Eng J. 2017;308:142–50.

17. Aquino JM, Rocha-Filho RC, Ruotolo LAM, Bocchi N, Biaggio SR. Electrochemical degradation of a real textile wastewater using β-PbO2 and DSA® anodes. Chem Eng J. 2014;251(1):138–45.

18. Wang C-T, Chou W-L, Kuo Y-M, Chang F-L. Paired removal of color and COD from textile dyeing wastewater by simultaneous anodic and indirect cathodic oxidation. J Hazard Mater. 2009;169(1–3):16–22.

19. Robinson T, McMullan G, Marchant R, Nigam P. Remediation of dyes in textile effluent: a critical review on current treatment technologies with a proposed alternative. Bioresour Technol. 2001;77(3):247–55.

20. Davoudi M, Gholami M, Naseri S, Mahvi AH, Farzadkia M, Esrafili A, et al. Application of electrochemical reactor divided by cellulosic membrane for optimized simultaneous removal of phenols, chromium, and ammonia from tannery effluents. Toxicol Environ Chem. 2014;96(9):1310–32.

21. Thiam A, Sirés I, Garrido JA, Rodríguez RM, Brillas E. Decolorization and mineralization of allura red AC aqueous solutions by electrochemical advanced oxidation processes. J Hazard Mater. 2015;290:34–42.

22. Tsantaki E, Velegraki T, Katsaounis A, Mantzavinos D. Anodic oxidation of textile dyehouse effluents on boron-doped diamond electrode. J Hazard Mater. 2012;207–208:91–6.

23. Zhao H-Z, Sun Y, Xu L-N, Ni J-R. Removal of acid orange 7 in simulated wastewater using a three-dimensional electrode reactor: removal mechanisms and dye degradation pathway. Chemosphere. 2010;78(1):46–51.

24. Martínez-Huitle CA, Brillas E. Decontamination of wastewaters containing synthetic organic dyes by electrochemical methods: a general review. Appl Catal B Environ. 2009;87(3):105–45.

25. Bechtold T, Mader C, Mader J. Cathodic decolourization of textile dyebaths: tests with full scale plant. J Appl Electrochem. 2002;32(9):943–50.

26. Najafpoor AA, Davoudi M, Salmani ER. Optimization of copper removal from aqueous solutions in a continuous electrochemical cell divided by cellulosic separator. Water Sci Technol. 2017;75(5):1233–1242.

27. Roessler A, Dossenbach O, Marte W, Rys P. Electrocatalytic hydrogenation of vat dyes. Dyes Pigments. 2002;54(2):141–6.

28. del Río Al, Molina J, Bonastre J, Cases F. Influence of electrochemical reduction and oxidation processes on the decolourisation and degradation of C.I. Reactive orange 4 solutions. Chemosphere. 2009;75(10):1329–37.

29. Rajkumar D, Kim JG. Oxidation of various reactive dyes with in situ electro-generated active chlorine for textile dyeing industry wastewater treatment. J Hazard Mater. 2006;136(2):203–12.

30. Li M, Feng C, Zhang Z, Chen R, Xue Q, Gao C, et al. Optimization of process parameters for electrochemical nitrate removal using box–behnken design. Electrochim Acta. 2010;56(1):265–70.

31. Salmani ER, Ghorbanian A, Ahmadzadeh S, Dolatabadi M, Nemanifar N. Removal of reactive red 141 dye from synthetic wastewater by electrocoagulation process: investigation of operational parameters. Iran J Health Saf Environ. 2016;3(1):403–11.

32. Doan HD, Saidi M. Simultaneous removal of metal ions and linear alkylbenzene sulfonate by combined electrochemical and photocatalytic process. J Hazard Mater. 2008;158(2–3):557–67.

33. Cardoso NF, Lima EC, Royer B, Bach MV, Dotto GL, Pinto LA, et al. Comparison of spirulina platensis microalgae and commercial activated carbon as adsorbents for the removal of reactive red 120 dye from aqueous effluents. J Hazard Mater. 2012;241:146–53.

34. Tehrani-Bagha A, Amini F. Decolorization of wastewater containing Cl reactive red 120 by UV-enhanced ozonation. J Color Sci Tech. 2010;4:151–60.

35. Ghalib AM. Removal of direct blue dye in textile wastewater effluent by electrocoagualtion. J Eng. 2010;16(4):6198–205.

36. Zaviska F, Drogui P, Blais J-F, Mercier G, Lafrance P. Experimental design methodology applied to electrochemical oxidation of the herbicide atrazine using Ti/IrO 2 and Ti/SnO 2 circular anode electrodes. J Hazard Mater. 2011;185(2):1499–507.

37. Mokhtari SA, Farzadkia M, Esrafili A, Kalantari RR, Jafari AJ, Kermani M, et al. Bisphenol A removal from aqueous solutions using novel UV/persulfate/H 2 O 2/Cu system: optimization and modelling with central composite design and response surface methodology. J Environ Health Sci Eng. 2016;14(1):19.

38. Chen S, Peng HM, Webster RD. Infrared and UV–vis spectra of phenoxonium cations produced during the oxidation of phenols with structures similar to vitamin E. Electrochim Acta. 2010;55(28):8863–9.

39. Del Rio A, Molina J, Bonastre J, Cases F. Study of the electrochemical oxidation and reduction of Cl reactive orange 4 in sodium sulphate alkaline solutions. J Hazard Mater. 2009;172(1):187–95.

40. Carneiro PA, Boralle N, Stradiotto NR, Furlan M, Zanoni MVB. Decolourization of anthraquinone reactive dye by electrochemical reduction on reticulated glassy carbon electrode. J Braz Chem Soc. 2004;15(4):587–94.

41. Méndez M, Tovar G, Dávila M, Ornelas O, Elizalde M. Degradation of reactive black 5 and basic yellow 28 on metallic-polymer composites. Port Electrochim Acta. 2008;26(1):89–100.

11

Effects of ethanol on the electrochemical removal of *Bacillus subtilis* spores from water

Masuma Moghaddam Arjmand[1], Abbas Rezaee[1*], Simin Nasseri[2] and Said Eshraghi[3]

Abstract

This study aimed to characterize the effects of ethanol on the monopolar electrochemical process to remove *Bacillus subtilis* spores from drinking water. In particular, spores' destruction was tested by applying 20–100 mA current for 15–60 min to *B. subtilis* spores (10^2–10^4 CFU/mL density), with stainless steel electrodes. The experimental results showed electrochemical removal of spores in the presence of 0.4 M ethanol at 15, 45, and 60 min and 5 mA/cm^2 current density. However, the use of ethanol or the electrochemical process alone did not eliminate *B. subtilis* spores at these time points. Overall, this study suggests that adding ethanol to the electrochemical process successfully removes *B. subtilis* spores from drinking water.

Introduction

Cryptosporidium parvum is an important microbial contaminant found in drinking water and is associated with a waterborne disease in humans [1]. Recently, *Bacillus subtilis* spores were used to evaluate the inactivation of *C. parvum* during water treatment [2]. To date, several methods of water treatment have been proposed, including chlorine. Although chlorination represents an efficient method of water treatment, it presents several disadvantages such as unfavorable taste and odor and the generation of potentially toxic disinfection products. In ddition, chlorine is ineffective when used alone against resistant microorganisms such as *C. parvum* and *Giardia* spp. [3]. A number of alternatives to chlorination have been suggested, including chemical (e.g., ozone and electrochemical treatments), physical (e.g., ultraviolet irradiation), and microwave systems [4]. In recent years, increasing attention has been paid to the electrochemical process as an alternative method to chlorination in water disinfection [5]. This treatment has been proposed since the 1950s [4] and can be divided into two categories: direct electrolyzers and mixed oxidant generators [6, 7]. The electrochemical process has several advantages, including the simplicity of the equipment and the fact that no additional chemicals are required for this method, as they

can be generated during the process [8]. In the presence of iron and stainless steel electrodes, the general electrochemical mechanism for this process can be illustrated as follows:

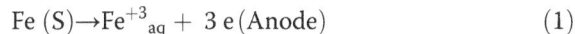

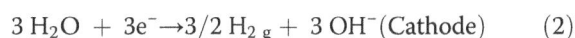

$$\text{Fe (S)} \rightarrow \text{Fe}^{+3}_{aq} + 3\,\text{e (Anode)} \qquad (1)$$

$$3\,\text{H}_2\text{O} + 3\text{e}^- \rightarrow 3/2\,\text{H}_{2\,g} + 3\,\text{OH}^- \text{(Cathode)} \qquad (2)$$

Fe^{3+} and OH^- ions, generated at the electrodes surface, react to generate Fe(OH)_3 compounds that can remove pollutants from aqueous solutions [9, 10]. Ethanol has also been used in water treatment processes, even though it has not been found to be an efficient sporicidal agent [11]. According to previous studies, 875 mg/L ethanol is needed to reduce *B. subtilis* populations over 6 log10 [12]. Ethanol is a membrane disrupter that induces rapid release of intracellular components and membrane disorganization, most likely due to the penetration of solvents into the hydrophobic region of the membrane bilayer. The aim of this study was to evaluate the efficiency of an electrochemical process in the presence of ethanol in water treatment. The effects of the operative parameters on *Bacillus B. subtilis* spore removal were also studied.

Materials and methods
Bacterial strain and culture conditions

The *B. subtilis* ATCC 6633 strain was obtained from the culture collection at the Tehran University (Iran). The strain was maintained on slant nutrient agar at 4 °C.

* Correspondence: rezaee@modares.ac.ir
[1]Department of Environmental Health Engineering, Faculty of Medical Sciences, Tarbiat Modares University, Tehran, Iran
Full list of author information is available at the end of the article

Table 1 Analysis of the water quality

Parameter	Value (mg/L)	Parameter	Value (mg/L)	Parameter	Value
Ca	59	Cl⁻	34.2	pH	7.5
Mg	20.2	SO$_4$	21.4	TDS (mg/l)	364
Na	86	F⁻	0.25	Conductivity(μs)	638
K	1.25	NH$_4$- N	0.1	Nitrate (mg/l)	12

Stocks were stored in aliquots containing 10 % glycerinated nutrient broth at −18 °C. 0.5 McFarland standards (corresponding to ~1.5×10^8 CFU/mL) spores were reactivated by incubation in 100 mL Erlenmeyer flasks containing 50 mL fresh trypticase soy broth (Merck) at 37 °C for 24 h, under aerobic conditions. Next, spore suspension was poured into sterile Erlenmeyer flasks and placed in a water-bath at 80 °C for 15 min to eliminate vegetative cells. Sporulation was confirmed by optical microscopy using the Gram staining technique, and spores were diluted into water. Total counts of bacterial spore suspensions were made using the pour plate method. Briefly, after incubation at 37 °C for 48 h, spore-forming bacteria were counted, and the results were expressed as the mean number of spores/mL. Culture media and equipment were sterilized by autoclaving at 121 °C for 15 min. The pH was adjusted at 0.1 M by adding NaOH or HCl.

Minimal inhibitory concentration (MIC)

Minimal inhibitory concentration (MIC) is defined as the lowest concentration of an antimicrobial agent that prevents the visible growth of a microorganism under certain *in vitro* condition. In this study, MICs were tested using the dilution broth method, according to the National Committee for Clinical Laboratory Standards [13]. Briefly, 1 mL of the culture was transferred into a first sterile glass tube containing 10 mL TBS medium

and 0.2 to 4 M ethanol. Next, after stirring, 1 mL was transferred from the first to the second tube and so on, to obtain a total of 12 dilutions. Tubes were incubated at optimal temperature for 24 and 48 h, prior to determine the MICs.

Experimental set up and operation

Electrochemical treatments were conducted in a single electrochemical reactor. The electrochemical reactor was equipped with two sheets of stainless steel that were used as anode and cathode electrodes. The distance between the electrodes was adjusted to 2 cm and maintained by placing plastic spacers. Experimental runs were conducted by imposing current densities ranging from 1 to 5 mA/cm^2 (Atten APS 3005S-3, China). The electrochemical process operated in batch mode and was performed using a 500 mL capacity glass beaker with 300 mL water at room temperature. Commercially available steel plates (size $15 \times 4 \times 0.1$ cm) were applied as electrodes and dipped in water to a depth of 10 cm. The effects of the operating conditions on the efficacy of the process were evaluated, including the applied current (50–150 mA), ethanol concentrations (0.2–0.4 M), spore concentrations (10^2–10^4), retention time (60 min), and current density (1–5 mA/cm^2). The quality of water is presented in Table 1.

The electrodes were connected to a DC power supply (Atten APS 3005S-3, China) with operational options for controlling the constant voltage and current density. The current density was calculated through the following equation as follows:

$$CD = I/S \qquad (3)$$

where I is the current through the solutions (A) and S is the area of the electrode (cm^2). The pH level and

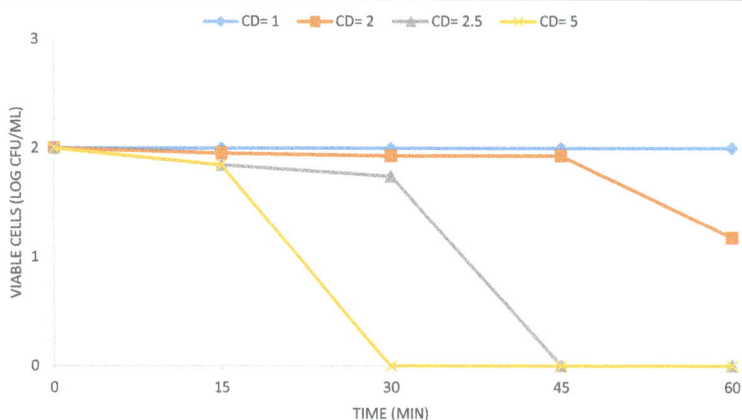

Fig. 1 Kinetic of electrochemical *B. subtilis* spore removal from water. Experimental conditions: current density CD = 1–5 mA/cm^2; spore density 10^2 CFU/mL; stainless steel as electrode anode; T = 25 °C; pH = 7.2; electrodes gap = 2 cm

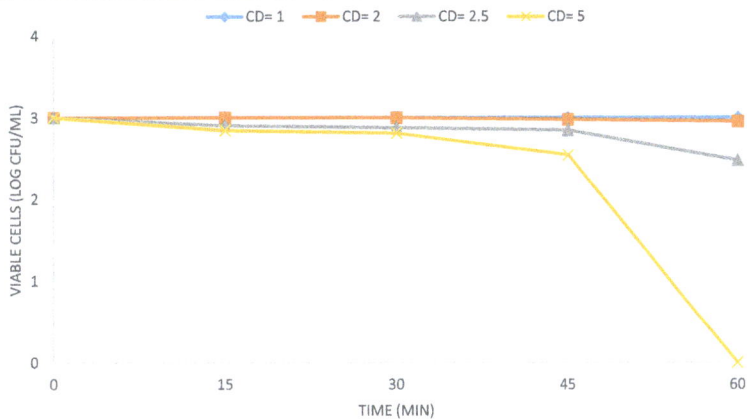

Fig. 2 Kinetic of electrochemical *B. subtilis* spore removal from water. Experimental conditions: current density CD = 1–5 mA/cm^2; spores density 10^3 CFU/mL; stainless steel as electrode anode; T = 25 °C; pH = 7.2; electrodes gap = 2 cm

conductivity of the solution were measured using a portable pH and EC meters (Eutech, Singapore). All the experiments were performed at a pH of 7.2. At the end of each experiment, the DC power source was switched off and the electrodes were removed from the water. During the experiments, samples were taken at 15 min interval and plated on TBS plates. All the experiments were repeated twice. The density of *B. subtilis* spore (10^2–10^4 m/l) removal was assessed in the electrochemical treatment using ethanol (0.2–0.4 M) at 20–100 mA current (1–5 mA/cm^2).

Results

In this study, the experiments were carried out in laboratory scale to evaluate the sporicidal effect of ethanol (0.2–0.4 M) in combination with the electrochemical treatment, using *B. subtilis* ATCC 6633 strain as a surrogate microorganism. In order to evaluate the sporicidal efficacy of ethanol, MIC determination was performed. The results showed that ethanol alone is not an efficient sporicidal agent (Figs. 1, 2 and 3).

However, in combination with the electrochemical treatment, it acquires a high antimicrobial activity (Figs. 4, 5 and 6).

These results demonstrated that the sporicidal efficiency is inversely proportional to the initial number of spore in solution. In addition, the sporicidal activity of the electrochemical treatment was directly proportional to ethanol concentrations, as shown in Figs. 4, 5 and 6. Overall, the results obtained in this study demonstrated that ethanol in combination with the electrochemical treatment improves the sporicidal efficiency of water disinfection, suggesting a synergistic effect between these two agents. The best result was obtained on 10^3 *B. subtilis* spores, using 0.4 M ethanol for 45 min at 5 mA/cm^2 current density (Fig. 5).

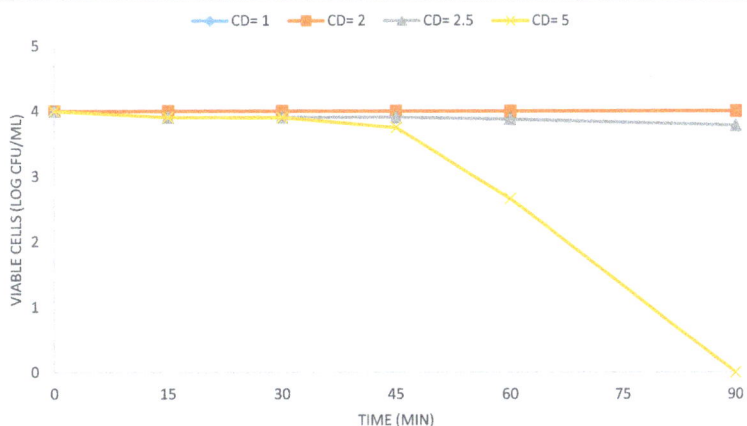

Fig. 3 Kinetic of electrochemical *B. subtilis* spore removal from water. Experimental conditions: current density CD = 1–5 mA/cm^2; spore density 10^4 CFU/mL; stainless steel as electrode anode; T = 25 ° C; pH = 7.2; electrodes gap = 2 cm

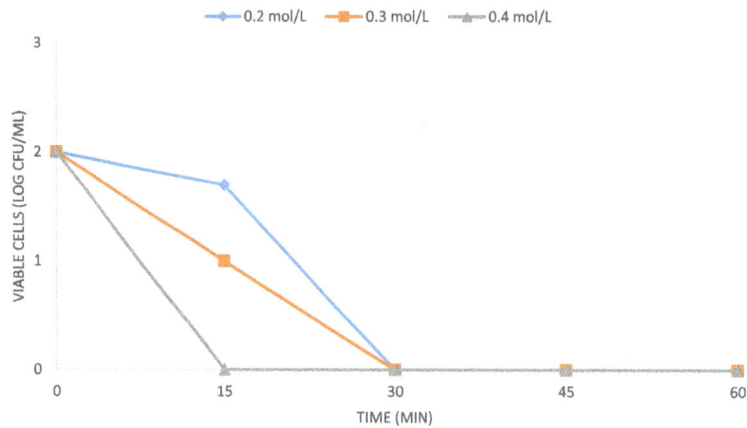

Fig. 4 Antimicrobial efficacy of ethanol and the electrochemical treatment on *B. subtilis* spore removal from water. Experimental conditions: current density CD = 5 mA/cm^2; spore density 10^2 CFU/mL; ethanol concentrations 0.2–0.4 M; stainless steel as electrode anode; T = 25 °C; pH = 7.2; electrodes gap = 2 cm

Discussion

Spores of *B. subtilis* are particularly resistant to conventional water disinfection treatments and, for this reason, they are used as surrogates for some waterborne pathogens such as *Cryptosporidium spp*, and as an indicator of hygienic quality of drinking water. Bacterial spores are more resistant to general sterilization and disinfection treatments such as heating, radiation, and the use of various chemicals than their vegetative cells [14]. Several parameters participate in spore resistance, including impermeability, low water content, high levels of pyridine-2,6-dicarboxylic acid and divalent cations, and outer membrane thickness [15]. In addition, spore DNA is protected against various types of damage [16]. In recent years, increasing attention has been paid to electrochemical oxidation as an efficient technology for

water disinfection. During this process, free chlorine is produced. This chemical damages the bacterial outer membrane, penetrates into the periplasm, destroys the inner membrane and degenerates cytoplasmic proteins. Also, the process can oxidize the microbes on the electrode surfaces [17]. Usually, the oxidants of the electrochemical treatment are reactive oxygen species generated from the oxidation of water molecules [18]. It was shown that the electrochemical treatment efficiently removes bacteria spores but not their vegetative cells [19]. The aim of this study was to evaluate the efficiency of a combination of the electrochemical process and ethanol at low concentrations for disinfection of *B. subtilis* spores. Ethanol is a general bactericidal agent and has been widely applied for disinfection of human tissues and contaminated surfaces. According to the

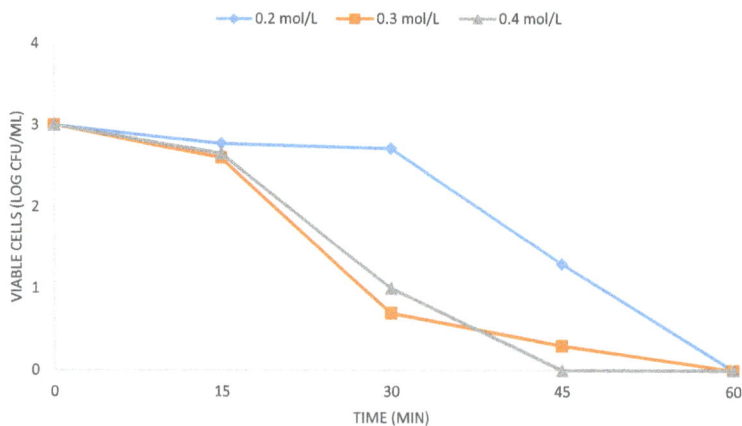

Fig. 5 Antimicrobial efficacy of ethanol and the electrochemical treatment on *B. subtilis* spore removal from water. Experimental conditions: current density CD = 5 mA/cm^2; spore density 10^3 CFU/mL; ethanol concentrations 0.2–0.4 M; stainless steel as electrode anode; T = 25 °C; pH = 7.2; electrodes gap = 2 cm

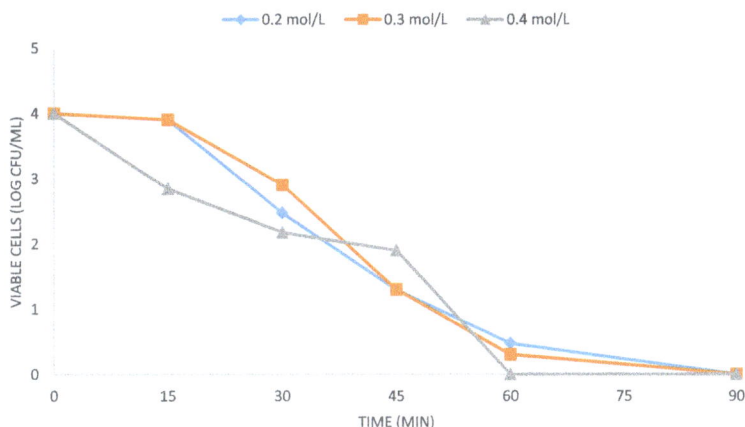

Fig. 6 Antimicrobial efficacy of ethanol and the electrochemical treatment on *B. subtilis* spore removal from water. Experimental conditions: current density CD = 5 mA/cm^2; spore density 10^4 CFU/mL; ethanol concentrations 0.2–0.4 M; stainless steel as electrode anode; T = 25 °C; pH = 7.2; electrodes gap = 2 cm

presented results, however, ethanol alone does not possess sporicidal activity [11]. It has been reported that ethanol causes microbial membrane damage and denaturation of proteins thus interfering with cell metabolism and inducing cell lysis. According to the obtained results, maximum sporicidal effects were obtained by adding 0.4 M ethanol to10^4 *B. subtilis* spores at 100 mA current. Lower ethanol concentration (0.2 M) increased the reaction time to 90 min. Ethanol alone does not possess high sporicidal efficiency. Some studies have reported that the combination of ethanol with ferric chloride and ethylenediaminetetraacetic acid can act as sporicidal agent [20]. Also, it has been shown that ethanol and anionic surfactants have sporicidal activity at low pH values. In this study, the sporicidal effects on *B. subtilis* spores was assessed by MIC. The results showed that ethanol alone has not sporicidal effect on *B. subtilis* spores. The electrochemical treatment exerted a low sporicidal effect on small numbers of *B. subtilis* spores, but it failed for large number of spores. These results indicated that the electrochemical treatment is an efficient method for water disinfection, but is not sufficient to remove disinfectant-resistant bacteria such as spore forming bacteria. This study showed that ethanol significantly increases the sporicidal efficiency of the electrochemical process.

Conclusion

The results obtained in this study show that *B. subtilis* spores were killed at 90 min by electrochemical water disinfection using ethanol. It was observed that increasing the operational time and adding ethanol to the electrochemical process improved the spore removal efficiency. Moreover, increasing the supporting electrolyte concentration in the solution reduces the specific electrical energy consumption.

Competing interests
The authors declare that they have no competing interests.

Authors' contributions
All authors read and approved the final manuscript.

Acknowledgements
The authors would like to thank the Tarbiat Modares University for funding and supporting this project (code: 1123803).

Author details
[1]Department of Environmental Health Engineering, Faculty of Medical Sciences, Tarbiat Modares University, Tehran, Iran. [2]Department of Environmental Health Engineering, School of Public Health, and Center for Water Quality Research, Institute for Environmental Research, Tehran University of Medical Sciences, Tehran, Iran. [3]Department of Pathobiology, School of Public Health, Tehran University of Medical Sciences, Tehran, Iran.

References
1. Zhou P, Giovanni GDD, Meschke GS, Dodd MC. Enhanced Inactivation of *Cryptosporidium parvum* Oocysts during Solar Photolysis of Free Available Chlorine. Environ Sci Technol Lett. 2014;1(11):453–8.
2. Forsyth JE, Zhou P, Mao Q, Asato SS, Meschke JS, Dodd MC. Enhanced Inactivation of Bacillus subtilis Spores during Solar Photolysis of Free Available Chlorine. Environ Sci Technol. 2013;47(22):12976–84.
3. Rezaee A, Kashi G, Jonidi-Jafari A, Khataee AR, Nili-Ahmadabadi A. Effect of Hydrogen peroxide on Baciluss Subtilis spore removal in an electrophotocatalytic system. Fresenius Environ Bull. 2011;20(10a):2750–5.
4. Kerwick MI, Reddy SM, Chamberlain AHL, Holt DM. Electrochemical disinfection, an environmentally acceptable method of drinking water disinfection? Electrochim Acta. 2005;50:5270–7.
5. Kraft A. Electrochemical water disinfection: A short review. Platinum Metals Rev. 2008;52(3):177–85.
6. Cho J, Choi H, Kim IS, Amy S. Chemical aspects and byproducts of electrolyser. Water Sci Technol. 2001;1(4):159–64.
7. Qin GF, Li ZY, Chen XD, Russell AB. An experimental study of an NaClO generator for anti-microbial applications in the food industry. J Food Eng. 2002;54:111–7.
8. Mart'nez-Huitle CA, Brillas E. Electrochemical alternatives for drinking water disinfection. Angew Chem Int Ed. 2008;47:2–10.
9. Bazrafshan E, Mahvi AH, Nasseri S, Shaieghi M. Performance evaluation of electrocoagulation process for diazinon removal from aqueous environments by using iron electrodes. Iranian J Environ Health Sci Eng. 2007;4(2):127–32.

10. Bazrafshan E, Ownagh KA, Mahvi AH. Application of electrocoagulation process using iron and aluminum electrodes for fluoride removal from aqueous environment. E J Chem. 2012;9(4):2297–308.

11. Chambers ST, Peddie B, Pithie A. Ethanol disinfection of plastic-adherent micro-organisms. J Hospital Infect. 2006;63:193–6.

12. Priscila GM, Angela FJ, Leticia C, Patricia M, Thereza CVP. Minimal inhibitory concentration (MIC) determination of disinfectant and/or sterilizing agents. Brazilian J Pharmaceutical Sci. 2009;45(2):241–8.

13. Wayne PA, National Committee for Clinical Laboratory Standards: Methods for dilution antimicrobial susceptibility tests for bacteria that grow aerobically. Approved Standard, M7-A2, National Committee for Clinical Laboratory Standards: USA; 1990.

14. Setlow B, Loshon CA, Genest PC, Cowan AE, Setlow C, Setlow P. Mechanisms of killing of spores of *Bacillus subtilis* by acid, alkali and ethanol. J Appl Microbiol. 2002;92:362–75.

15. Setlow P. Spores of Bacillus subtilis: their resistance to and killing by radiation, heat and chemicals. J Appl Microbiol. 2006;101:514–25.

16. Tennen R, Setlow B, Davis KL, Loshon CA, Setlow P. Mechanisms of killing of spores of *Bacillus subtilis* by iodine, glutaraldehyde and nitrous acid. J Appl Microbiol. 2000;89:330–8.

17. Anglada A, Urtiaga A, Ortiz I. Contributions of electrochemical oxidation to waste-water treatment: Fundamentals and review of applications. J Chem Technol Biotechnol. 2009;84:1747–55.

18. Jeong J, Kim C, Yoon J. The effect of electrode material on the generation of oxidants and microbial inactivation in the electrochemical disinfection processes. Water Res. 2009;43:895–901.

19. Francisco V, Selma F. Electrochemical disinfection: An efficient treatment to inactivate *Escherichia Coli* 0157:H7 in process wash water containing organic matter. Microbiol. 2010;30:146–56.

20. Kida N, Mochizuki Y, Taguchi F. Effects on the sporicidal activity by using various metal ions in the formulation combining ferric chloride, ethylenediaminetetraacetic acid and ethanol. Biocontrol Sci. 2004;9:29–32.

12

Optimization of sonochemical degradation of tetracycline in aqueous solution using sono-activated persulfate process

Gholam Hossein Safari[1], Simin Nasseri[1,2*], Amir Hossein Mahvi[1,3], Kamyar Yaghmaeian[1,3], Ramin Nabizadeh[1,4] and Mahmood Alimohammadi[1]

Abstract

Background: In this study, a central composite design (CCD) was used for modeling and optimizing the operation parameters such as pH, initial tetracycline and persulfate concentration and reaction time on the tetracycline degradation using sono-activated persulfate process. The effect of temperature, degradation kinetics and mineralization, were also investigated.

Results: The results from CCD indicated that a quadratic model was appropriate to fit the experimental data ($p < 0.0001$) and maximum degradation of 95.01 % was predicted at pH = 10, persulfate concentration = 4 mM, initial tetracycline concentration = 30.05 mg/L, and reaction time = 119.99 min. Analysis of response surface plots revealed a significant positive effect of pH, persulfate concentration and reaction time, a negative effect of tetracycline concentration. The degradation process followed the pseudo-first-order kinetic. The activation energy value of 32.01 kJ/mol was obtained for $US/S_2O_8^{2-}$ process. Under the optimum condition, the removal efficiency of COD and TOC reached to 72.8 % and 59.7 %, respectively. The changes of UV–Vis spectra during the process was investigated. The possible degradation pathway of tetracycline based on loses of N-methyl, hydroxyl, and amino groups was proposed.

Conclusions: This study indicated that sono-activated persulfate process was found to be a promising method for the degradation of tetracycline.

Keywords: Tetracycline degradation, Persulfate, Response surface methodology, Central composite design, Optimization

Background

Tetracycline (TC) is extensively used for the prevention and treatment of infectious diseases in human and veterinary medicine and as feed additives for promote growth in agriculture [1, 2]. Because of their extensive usage, their strongly hydrophilic feature, low volatility [2] and relatively long half-life [3], TC antibiotic has been frequently detected in different environmental matrices: surface waters (0.07-1.34 μg/L) [4], soils (86.2-198.7 μg/kg) [5], liquid manures (0.05-5.36 μg/kg) [5] and in 90 % of farm lagoon samples (>3 μg/L) [6]. In addition to environmental contamination, the occurrence of TC in the aquatic environments would also increase antibiotic resistance genes [7]. However, due to the antibacterial nature of TC, they cannot effectively be removed by conventional biological processes [8]. In wastewater treatment plants, the TC removal efficiency varied in the range of 12 % to 80 % [9, 10]. For example, concentrations of TC residues have been detected in values of 0.97 to 2.37 μg/L in the final effluent from wastewater treatment plants [11]. Hence, the effort to develop new processes to minimize the tetracycline residues discharges into the environment is become essential. Physicochemical processes such as membrane filtration and adsorption using activated carbon have been used to removal of TC. These processes are not efficient enough, transfer the pollutant from one phase to another [12, 13]. Advanced oxidation processes (AOPs) such as (O_3/H_2O_2, US/O_3, UV/O_3, UV/H_2O_2, H_2O_2/Fe^{2+}, $US-TiO_2$ and

* Correspondence: naserise@tums.ac.ir
[1]Department of Environmental Health Engineering, School of Public Health, Tehran University of Medical Sciences, Tehran, Iran
[2]Center for Water Quality Research, Institute for Environmental Research, Tehran University of Medical Sciences, Tehran, Iran
Full list of author information is available at the end of the article

UV-TiO$_2$) have been proposed as very effective alternatives to degrade tetracycline antibiotics. The primary of AOPs is production of hydroxyl radical in water, a much powerful oxidant in the degradation of a wide range of organic pollutants [12–15]. Recently, the application of sulfate radical-based advanced oxidation processes (SR-AOPs) to oxidation of biorefractory organics have attracted great interest [16, 17]. Persulfate (PS, S$_2$O$_8^{2-}$) is a powerful and stable oxidizing agent (E$_0$ = 2.01 V vs. NHE), which has high aqueous solubility and high stability at room temperature as compared to hydrogen peroxide (H$_2$O$_2$, E$_0$ = 1.77 V vs. NHE) [18, 19].

Sulfate radicles could be produced through the activation of persulfate (PS, S$_2$O$_8^{2-}$) with ultraviolet [20], heat [21, 22], microwave [23], sonolysis [24], base [25], granular activated carbon [26], quinones [27], phenols [28], soil minerals [29], radiolysis [30] and transition metals [31, 32]. Sulfate radicals are more effective than hydroxyl radical in the oxidation of organic contaminants. They have higher redox potentials, longer half-life and higher selectivity in the oxidation of organic contaminants (SO$_4^-$•, E$_0$ = 2.5-3.1, half-life = 30–40 μs) than hydroxyl radical (HO•, E$_0$ = 1.89–2.72 V, half-life = 10^{-3} μs) [33–39]. Hence, the organic pollutants could be oxidized entirely by SO$_4^-$•, especially benzene derivatives compounds [18]. Generally, sulfate radical reacts with organic contaminants predominantly through selective electron transfer, while hydroxyl radical mainly reacts through hydrogen abstraction and addition. Therefore, the possibility of sulfate radical scavenging by nontarget compounds is lower than hydroxyl radical [39–42].

Sonochemical treatment is an emerging and efficient process that applied pyrolytic cleavages to degradation of organic compounds [42, 43]. This process is a cleaner and safe technique compared with UV, ozonation, and has the ability of operation under ambient conditions [43, 44]. However, combination of ultrasound with various processes has been detected as an economical and successful alternative for the degradation and mineralization of some recalcitrant organic compounds in aqueous solution [42]. The combination of ultrasound and persulfate (US/S$_2$O$_8^{2-}$) has been effective for the degradation of compounds such as; methyl tert-butyl ether (MTBE) [45], nitric oxide [18], 1,4-dioxane [46], arsenic(III) [44], amoxicillin [47], tetracycline [48] and dinitrotoluenes [24]. In aqueous solutions, acoustic cavitation leading to produce plasma in water and free radicals and other reactive species such as HO• and H• radicals due to the thermal degradation of water according to Reaction (1) and (2). The HO• and H• radicals can also react with PS to production of more reactive SO$_4^-$• radicals according to Reactions (3) to (7) [42, 44, 49, 50].

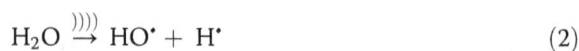

$$H_2O \xrightarrow{))))} H_2O \text{ plasma} \tag{1}$$

$$H_2O \xrightarrow{))))} HO^• + H^• \tag{2}$$

Where "))))))" refers to ultrasonication.
In the presence of S$_2$O$_8^{2-}$:

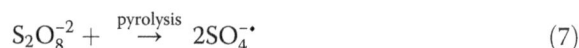

$$S_2O_8^{-2} + \xrightarrow{))))} 2SO_4^{-•} \tag{3}$$

$$SO_4^{-•} + H_2O \rightarrow SO_4^{2-} + HO^• + H^+ \tag{4}$$

$$S_2O_8^{-2} + HO^• \rightarrow HSO_4^- + SO_4^{-•} + \frac{1}{2} O_2 \tag{5}$$

$$S_2O_8^{-2} + H^• \rightarrow HSO_4^- + SO_4^{-•} \tag{6}$$

$$S_2O_8^{-2} \xrightarrow{pyrolysis} 2SO_4^{-•} \tag{7}$$

In aqueous solution, Hydroxyl radicals may be produced via the degradation of persulfate and/or ultrasonic irradiation. Ultrasonic irradiation could also lead to cavitation through the formation, growth and collapse of tiny gas bubbles in the water [51]. Moreover, during US irradiation, the collapse of cavitation bubbles leads to higher temperatures and pressures that produces free radicals and other reactive species and would also increase the number of collisions between free radicals and contaminants [42, 44, 49, 50].

The specific objectives of this study were to optimize the TC degradation in aqueous solution using US/S$_2$O$_8^{2-}$ process. Response surface methodology (RSM) is a reliable statistical technique for developing, improving and optimizing processes and can be used to assess the relative significance of several affecting factors with the least experiments [52–54]. Therefore, an experimental design methodology using RSM and CCD was used to evaluate the effect of operational parameters such as initial TC concentration, initial S$_2$O$_8^{2-}$ concentration, initial pH and reaction time on the sonochemical degradation of tetracycline. In addition, the effect of temperature, degradation kinetics, mineralization, changes of ultraviolet Visible (UV–Vis) spectra and the proposed degradation pathway of TC by the US/S$_2$O$_8^{2-}$ process were investigated. This study as part of a PhD dissertation of the first author was performed at Department of Environmental Health Engineering, School of Public Health, Tehran University of Medical Sciences in 2015.

Materials and methods
Materials
Tetracycline hydrochloride [C$_{22}$H$_{25}$N$_2$O$_8$Cl] (AR, 99 %), was provided from Sigma–Aldrich. Chemical properties of tetracycline hydrochloride are shown in Table 1 [2]. Sodium persulfate (Na$_2$S$_2$O$_8$, 98 %) was provided from Sigma–Aldrich. All other chemicals were of analytical

Table 1 Chemical properties of tetracycline hydrochloride

Molecule	Formula	Molecular weight (g/mol)	Solubility (mol/L)	pK$_{a1}$	pK$_{a2}$	pK$_{a3}$
TC	$C_{22}H_{24}O_8N_2.HCl$	480.9	0.041	3.2 ± 0.3	7.78 ± 0.05	9.6 ± 0.3

grade and were used without further purification. The water used in all experiments was purified by a Milli- Q system.

Procedure

Schematics of the experimental setup applied in this study is demonstrated in Fig. 1. A stock solution of tetracycline was daily prepared with distilled deionized water and diluted as required initial concentration. Sonochemical treatment was carried out with a fixed volume of 100 mL of TC solution in a glass vessel of 200 mL. The vessel was wrapped with tinfoil in order to avoid any photochemical effects. The pH adjustments were conducted with 1 m NaOH or 1 m HCL (Merck Co.) using a pH meter (E520, Metrohm, Tehran, Iran). Sonochemical treatment was performed with an ultrasonic generator at a frequency of 35 kHz and power of 500 W (Elma, Singen, Germany). The reactor was immersed into the ultrasonic bath and its location was always kept similarly. All experiments were conducted at constant temperature using cooling water and temperature controller. At pre-specified time intervals, 2 mL sample was withdrawn, filtered through 0.22 μm syringe filter and mixed with the same volume of methanol to quench the reaction before analysis [3].

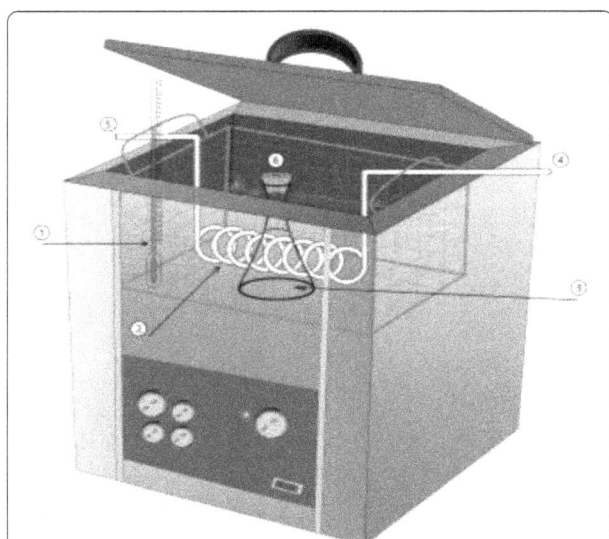

Fig 1 Schematic of the experimental used in this study; (1) temperature controller, (2) water-circulating (3) TC solution reactor (4) cooling water inlet, (5) cooling water outlet (6) sampling port

Analytical methodology

The pH was determined at room temperature using an S-20 pH meter, which was calibrated with pH 4.0 and 7 reference buffer solutions. The concentration of TC in aqueous solution was analyzed by HPLC, with a LC-20 AB pump, Shimadzu, Kyoto, Japan) with a reversed-phase column (VP-ODS-C18 4.6 mm × 250 mm, 5 μm, Shim-Pack, Kyoto, Japan), and UV detector (Shimadzu UV-1600 spectrophotometer). The injection volume was 20 μL; the mobile phase was acetonitrile 0.01 M, oxalic acid solution (31:69, v/v) with a flow rate of 1.0 mL min^{-1}. The detection wavelength and retention time of tetracycline were 360 nm and 2.38 min, respectively. In this study, limit of detection (LOD) were found to be 0.02-0.03 mg/L based on linear regression method.

Experimental design

A central composite statistical experiment design was used to evaluate the effects of four independent variables (initial solution pH (A), initial TC concentration (B), initial $S_2O_8^{-2}$ concentration (C) and reaction time (D)) on the TC degradation. The application of RSM provides a mathematical relationship between variables and experimental data can be fitted to an empirical second-order polynomial model as the following Eq. (8). [55–57].

$$
\begin{aligned}
Y = {} & \beta_0 + \beta_1\,A + \beta_2\,B + \beta_3\,C + \beta_4\,D + \beta_{12}\,AB \\
& + \beta_{13}\,AC + \beta_{14}\,AD + \beta_{23}BC + \beta_{24}\,BD \\
& + \beta_{34}CD + \beta_{11}\,A^2 + \beta_{22}\,B^2 + \beta_{33}C^2 \\
& + \beta_{44}\,D^2
\end{aligned}
\tag{8}
$$

Where, y (%) is the predicted response (TC degradation rate), β_0 is interception coefficient, β_1, β_2, β_3 and β_4 are the linear coefficients, β_{12}, β_{13}, β_{14}, β_{23}, β_{24} and β_{34} are interaction coefficients, β_{11}, β_{22}, β_{33} and β_{44} are the quadratic coefficients and A, B, C and D are the independent variables.

The natural and coded levels of independent variables based on the central composite design are shown in Table 2. The experimental values for each independent variables were chosen according to the results obtained from preliminary analysis. Table 3 indicates the four-factor, five-level CCD and the obtained and predicted values for the TC degradation rate (%) using the developed quadratic model. In RSM analysis, the approximation of y was proposed using the fitted second-order polynomial regression model which is called the quadratic model. A quadratic regression is the process of finding the equation of the parabola that fits best for a set of data [58].

Table 2 Natural and coded levels of independent variables based on the central composite design

Independent variable	Symbol	Coded levels				
		−2	**−1**	**0**	**+1**	**+2**
		Natural level				
pH	A	2.5	5	7.5	10	12.5
Tetracycline (mg/L)	B	10	30	50	70	90
Persulfate (Mm)	C	1	2	3	4	5
Reaction time (min)	D	30	60	90	120	150

Resulta and Disscusion
Analysis of variance (ANOVA)

The results of the analysis of variance test is summarized in Table 4. The probability > F for the model is less than 0.05 which implies that the model is significant and the terms in the model have significant effects on the response. In this case A, B, C, D, AB, AC, AD, BC, BD, CD, A2, B^2, C^2, D^2 are significant model terms at the 95 % confidence level (α =5 %). The model F-value of 1387.59 and P-value of < 0.0001 implies that the model is highly significant. Based on the ANOVA results, the values of R^2, Adjusted R^2 and Predicted R^2 were 0.9992, 0.9985 and 0.9971, respectively. This result suggests that the regression model is well interpreted the relationship

Table 3 Four-factor five-level central composite design for RSM

Run	Experimental conditions				TC degradation rate (%)	
	pH (A)	Tetracycline (mg/L) (B)	Persulfate (mM) (C)	Time (min) (D)	Observed (%)	Predicted (%)
1	7.5 (0)	50 (0)	3 (0)	90 (0)	51.06	49.82
2	2.5 (−2)	50 (0)	3 (0)	90 (0)	55.64	55.74
3	7.5 (0)	50 (0)	3 (0)	90 (0)	48.45	49.82
4	5 (−1)	70 (+1)	4 (+1)	60 (−1)	45.16	44.64
5	5 (−1)	30 (−1)	4 (+1)	120 (+1)	86.62	86.33
6	10 (+1)	70 (+1)	2 (−1)	120 (+1)	61.02	61.32
7	7.5 (0)	50 (0)	3 (0)	150 (+2)	81.85	81.17
8	5 (−1)	30 (−1)	2 (−1)	60 (−1)	34.55	35.25
9	10 (+1)	30 (−1)	2 (−1)	60 (−1)	47.25	46.91
10	10 (+1)	30 (−1)	4 (+1)	60 (−1)	70.44	69.95
11	5 (−1)	30 (−1)	2 (−1)	120 (+1)	61.85	61.79
12	7.5 (0)	50 (0)	1 (−2)	90 (0)	28.72	28.18
13	10 (+1)	70 (+1)	4 (+1)	120 (+1)	85.05	84.34
14	7.5 (0)	50 (0)	3 (0)	90 (0)	49.75	49.82
15	10 (+1)	30 (−1)	2 (−1)	120 (+1)	75.56	76.07
16	10 (+1)	30 (−1)	4 (+1)	120 (+1)	94.25	95.04
17	5 (−1)	70 (+1)	2 (−1)	60 (−1)	12.65	11.98
18	5 (−1)	30 (−1)	4 (+1)	60 (−1)	64.04	63.86
19	7.5 (0)	90 (+2)	3 (0)	90 (0)	41.15	41.31
20	10 (+1)	70 (+1)	4 (+1)	60 (−1)	54.88	55.05
21	7.5 (0)	50 (0)	3 (0)	90 (0)	50.55	49.82
22	7.5 (0)	50 (0)	3 (0)	90 (0)	49.75	49.82
23	7.5 (0)	50 (0)	5 (+2)	90 (0)	79.38	79.82
24	12.5 (+2)	50 (0)	3 (0)	90 (0)	80.62	80.42
25	7.5 (0)	10 (−2)	3 (0)	90 (0)	75.55	75.28
26	7.5 (0)	50 (0)	3 (0)	90 (0)	49.35	49.82
27	7.5 (0)	50 (0)	3 (0)	30 (−2)	24.75	25.33
28	5 (−1)	70 (+1)	2 (−1)	120 (+1)	42.25	42.72
29	10 (+1)	70 (+1)	2 (−1)	60 (−1)	27.68	27.96
30	5 (−1)	70 (+1)	4 (+1)	120 (+1)	70.85	71.31

Table 4 ANOVA results for Response Surface Quadratic Model

Source	Sum of squares	df	Mean square	F -Value	P- value
model	12060.48	1	861.46	1387.59	<0.0001
A-pH	914.15	1	914.15	1472.45	<0.0001
B-TCcon.	1730.94	1	1730.94	2788.08	<0.0001
C-PScon.	3999	1	3999	6441.32	<0.0001
D-Time	4676.04	1	4676.04	7531.85	<0.0001
AB	18.66	1	18.66	30.06	<0.0001
AC	30.97	1	30.97	49.88	<0.0001
AD	6.84	1	6.84	11.01	0.0047
BC	16.4	1	16.4	26.42	0.0001
BD	17.64	1	17.64	28.41	<0.0001
CD	16.61	1	16.61	26.75	0.0001
A^2	571.64	1	571.64	920.76	<0.0001
B^2	123.3	1	123.3	198.6	<0.0001
C^2	29.96	1	29.96	48.26	<0.0001
D^2	20.18	1	20.18	32.5	<0.0001
Residual	9.31	15	0.62		
Lack of Fit	5.13	10	0.51	0.61	0.7607
Pure Error	4.18	5	0.84		
Cor Total	12069.8	29			

$R^2 = 0.9992$; Adjusted $R^2 = 0.9985$ and Predicted $R^2 = 0.9971$

between the independent variables and the response. Furthermore, the adequate precision ratio of 149.08 in the study shows that this model could be applied to navigate the design space defined by the CCD.

Evaluation of model adequacy

There are many statistical techniques for the evaluation of model adequacy, but graphical residual analysis is the primary statistical method for assessment of model adequacy [59].

The normal probability plot indicates that the points on this plot are formed a nearly linear pattern (Fig. 2 (a)). Therefore, the normal distribution is a good model for this data set. Random scattering of the points of internally studentized residual (the residual divided by the estimated standard deviation of that residual) versus predicted values between −3 and +3 emphasizes highly accurate prediction of the experimental data through the derived quadratic model (Fig. 2 (b)).

The plot of predicted vs Actual values (Fig. 2 (c)) indicate a higher correlation and low differences between actual and predicted values. Hence, the predictions of the experimental data by developed quadratic models for the TC degradation is perfectly acceptable and this model fits the data better. Also, the random spread of

the residuals across the range of the data between −3 and +3 implies that there are no evident drift in this process and the model was a goodness fit (Fig. 2 (d)). The Box-cox plot is used for determine the suitability of a power low transformation for the selected data (Fig. 3 (a)). In this study, the best lambda values of 0.92 was obtained with low and high confidence interval 0.73 and 1.11, respectively. Therefore, recommend the standard transformation by the software is 'None'. The plot of points Leverage vs Run order is shown in Fig. 3 (b). The factorial and axial points have the most influence with a leverage of approximately 0.59, while the center points have the least effect with a leverage of 0.16.

Design matrix evaluation for response surface quadratic model

Design matrix evaluation implies that there are no aliases for the quadratic model. In general, a minimum of degrees of freedom 3 and 4 has been recommended for lack-of-fit and pure error, respectively. Therefore, degrees of freedom obtained in this study ensured a valid lack of fit test (Table 5).

The standard error (SE) used to measure the precision of the estimate of the coefficient. The smaller standard error implies the more accurate the estimate. The variables of A, B, C and D have a standard errors = 0.16. The interceptions of AB, AC, AD, BC, BD and CD have slightly high standard errors = 0.2, while A^2, B^2, C^2 and D^2 have standard errors = 0.15. An approximate 95 % confidence interval for the coefficient is given by the estimate plus and minus 2 times the standard error. For example, with 95 % confidence can be said that the value of the regression coefficient A is between 6.49 and 5.85 ($6.17 \pm 2 \times 0.16$).

The quadratic model coefficients for the CCD are shown in Table 6. This results suggested that the variables coefficients and their interactions are estimated adequately without multicollinearity. The low Ri-squared for independent variables and their interactions imply that the model is a good fit. In general, power should be approximately 80 % for detecting an effect [60]. In this study, there are more than 99 % chance of detecting a main effect while it is twice the background sigma.

Final equation and model graphs

The values of regression coefficients were determined and the experimental results of CCD were fitted with second order polynomial equation. The quadratic model for TC degradation rate in terms of coded were determined using as following Eq. (9):

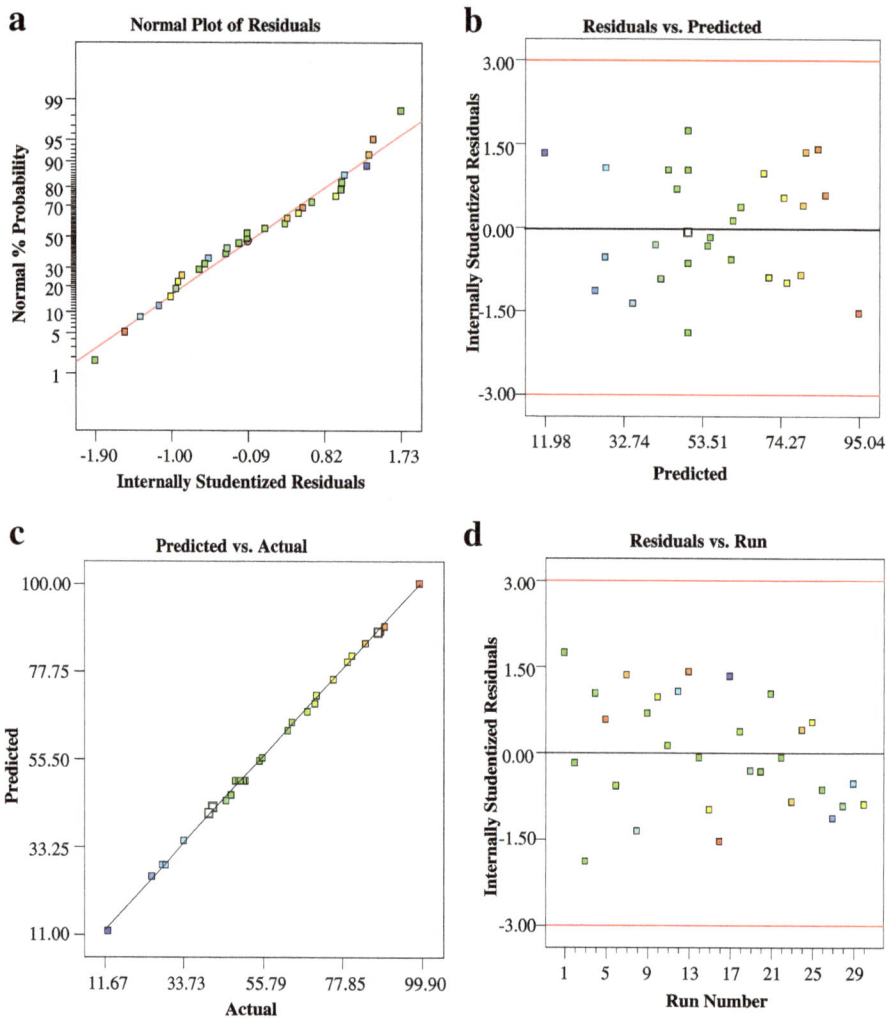

Fig 2 The plot of: (**a**) Normal Plot of Residuals; (**b**) residuals vs Predicted Response; (**c**) predicted vs Actual values; (**d**) Residuals vs Run Order

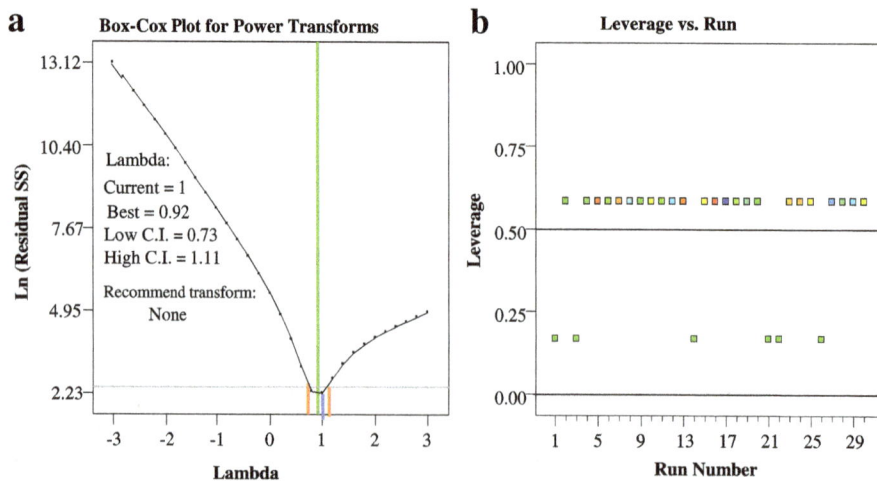

Fig 3 The plot of: (**a**) Box-Cox Plot for Power Transforms; (**b**) The points Leverage vs Run order for the CCD Design

Table 5 Degrees of freedom for evaluation

Model	14
Residuals	15
Lack of Fit	10
Pure Error	5

Final equation in terms of coded factors

$$Y = + 49.82 + 6.17 * A - 8.49 * B + 12.91 * C$$
$$+ 13.96 * D + 1.08 * A * B - 1.39 * A * C$$
$$+ 0.65 * A * D + 1.01 * B * C + 1.05 * B$$
$$* D - 1.02 * C * D + 4.57 * A^2 + 2.12 * B^2$$
$$+ 1.05 * C^2 + 0.86 * D^2 \qquad (9)$$

The factors in the quadratic equation were coded to produce the response surface with limiting the responses into a range of -1 to $+1$. The ramp function graph for the maximum TC degradation rate is shown in Fig. 4. The optimization of experimental conditions was conducted for maximize the TC degradation at defined criteria of the variable. The developed quadratic model for the TC degradation (Eq. (8)) was applied as an objective function to the optimization of operating conditions. Consequently, the optimum parameters were achieved using the numerical technology based on the predicted model and the variable in their critical range. The maximum degradation of 95.01 % was achieved at pH = 9.9, TC concentration = 30.19 mg/L, PS concentration = 3.97 mM and reaction time = 119.98 min. in order to evaluation of the model validity, the experiments were carried out under the optimal operating conditions. 93.45 % TC degradation was

obtained under the optimum operating conditions, which supported the results of the developed model.

The perturbation Plot of independent variables implies that reaction time (D) has the most significant effect (steepest slope) on the TC degradation rate, followed by $S_2O_8^{-2}$ concentration (C) and TC concentration (B), whereas pH (A) has the lowest effect on the TC degradation. (Fig. 5).

Interactive effect of independent variables on the TC degradation

Three-dimensional surfaces and contour plots are graphical representation of regression equation for the optimization of reaction Status. The results of the interactions between four independent variables and dependent variable are indicated in Figs. 6 and 7.

Figure 6(a) indicates the interaction effect of TC concentration and PS concentration on the TC degradation rate with reaction time of 120 min. with the increasing PS concentration, the TC degradation rate significantly enhanced. With increasing PS concentration from 2 to 4 mM, the TC degradation rate increased from 75.56 % to 94.25 % at TC concentration of 30 mg/L. These results suggest that with increasing PS concentration, more sulfate radicals are produced which leads to more quickly TC degradation [32].

Figure 6(b) indicates the interaction effect of initial TC concentration and reaction time on the TC degradation rate. The TC degradation rate strongly increased with increase of sonication time from 60 to 120 min. with increasing reaction time from 60 to 120 min, TC concentration of 30 and 70 mg/L, the TC degradation rate

Table 6 The Quadratic model coefficients for the CCD

Term	StdErr**	VIF	Ri-Squared	Power at 5 % SN = 0.5	Power at 5 % SN = 1	Power at 5 % SN = 2
A	0.16	1	0	20.90 %	63.00 %	99.50 %
B	0.16	1	0	20.90 %	63.00 %	99.50 %
C	0.16	1	0	20.90 %	63.00 %	99.50 %
D	0.16	1	0	20.90 %	63.00 %	99.50 %
AB	0.2	1	0	15.50 %	46.50 %	96.20 %
AC	0.2	1	0	15.50 %	46.50 %	96.20 %
AD	0.2	1	0	15.50 %	46.50 %	96.20 %
BC	0.2	1	0	15.50 %	46.50 %	96.20 %
BD	0.2	1	0	15.50 %	46.50 %	96.20 %
CD	0.2	1	0	15.50 %	46.50 %	96.20 %
A^2	0.15	1.05	0.0476	68.70 %	99.80 %	99.90 %
B^2	0.15	1.05	0.0476	68.70 %	99.980 %	99.90 %
C^2	0.15	1.05	0.0476	68.70 %	99.80 %	99.90 %
D^2	0.15	1.05	0.0476	68.70 %	99.80 %	99.90 %

**Basis Std. Dev. = 1.0

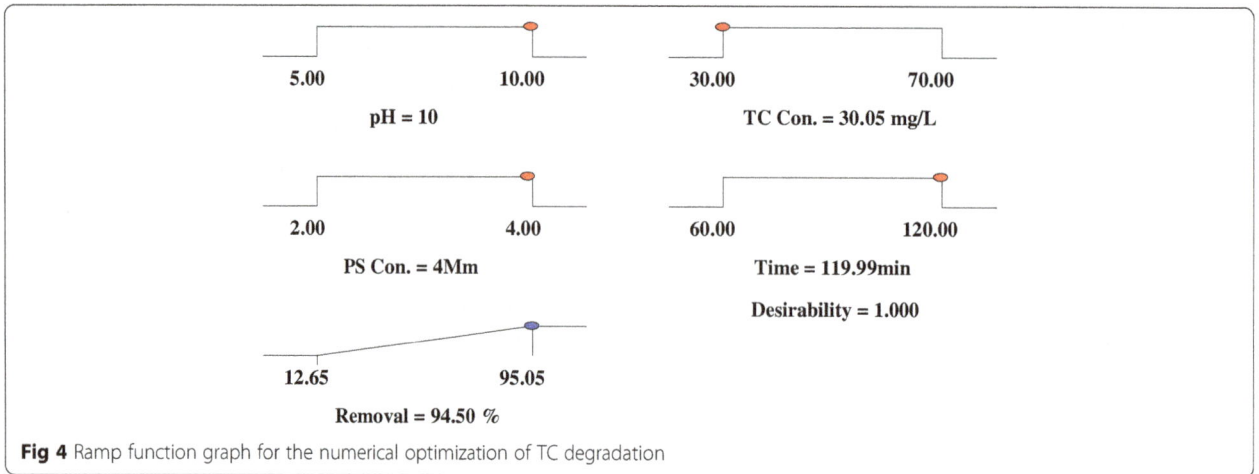

Fig 4 Ramp function graph for the numerical optimization of TC degradation

increased from 70.44 % to 94.25 % at TC concentration of 30 mg/L. With increasing the TC concentration from 30 to 70 mg/L, the TC degradation rate decreased from 94.25 % to 85.05 %. In the constant conditions, with the increasing TC concentration, possibility of reaction between TC molecules and reactive species were declined. Moreover, the higher concentration of TC may lead to the creation of resistant byproducts and consequently decreases the degradation rate of TC [14, 61]. However, the total amount of degraded TC increased with the increasing initial TC concentration. This results are in agreement with the results obtained by other researchers [50].

Figure 7 indicates the interaction influence of pH value and initial TC concentration on the TC degradation rate. With increasing pH from acidic (5) to natural (7.5), the degradation rate slightly decreased, whereas with increasing pH from neutral (7.5) to alkaline (10), the degradation rate significantly enhanced. The TC degradation rate increased from 86.62 % to 94.25 % with increasing pH from 5 to 10, at TC concentrations of 30 mg/L. Under alkaline conditions (pH ≥10), alkaline-activated persulfate is the primary responsible for the production of SO4$^{-\bullet}$, O$_2^{\bullet}$ and HO$^{\bullet}$ radicals as following equations: [62, 63].

$$S_2O_8^{2-} + 2H_2O \xrightarrow{OH^-} HO_2^- + 2SO_4^{2-} + 3H^+ \qquad (10)$$

$$HO_2^- + S_2O_8^{2-} \rightarrow SO_4^{-\bullet} + SO_4^{2-} + H^+ + O_2^{-\bullet} \qquad (11)$$

$$SO_4^{-\bullet} + OH^- \rightarrow SO_4^{2-} + HO^{\bullet} \qquad (12)$$

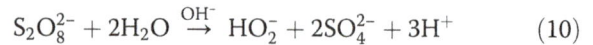

Also, at alkaline pH, sulfate radicals can react with hydroxyl anions to generate hydroxyl radicals (HO$^{\bullet}$) according to Eq. (3). In addition, a theory was introduced by other researchers that with increasing pH, the PS degradation into HO$^{\bullet}$ and SO$_4^{-\bullet}$ increased [64].

The SO$_4^{-\bullet}$ is the predominant radical responsible for TC degradation at acidic pH, whereas both SO$_4^{-\bullet}$ and OH$^{\bullet}$ are contributing in TC degradation at natural pH. Thus, three reactions compete with each other in natural pH: the reaction between SO$_4^{-\bullet}$ and HO$^{\bullet}$, the reaction between SO$_4^{-\bullet}$ and TC, and the reaction between HO$^{\bullet}$ and TC, the simultaneous occurrence of these reactions may reduce the TC degradation rate [37, 65].

Kinetics of tetracycline degradation

The sonochemical degradation process typically follows pseudo first-order kinetics as shown in the following Eqs. (13) and (14). Many studies have suggested that oxidation of organic pollutants by ultrasound follows pseudo first-order kinetics [42, 47, 52].

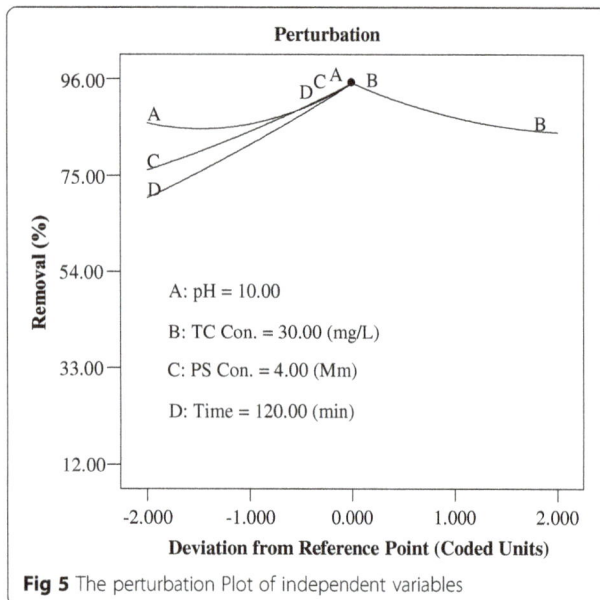

Fig 5 The perturbation Plot of independent variables

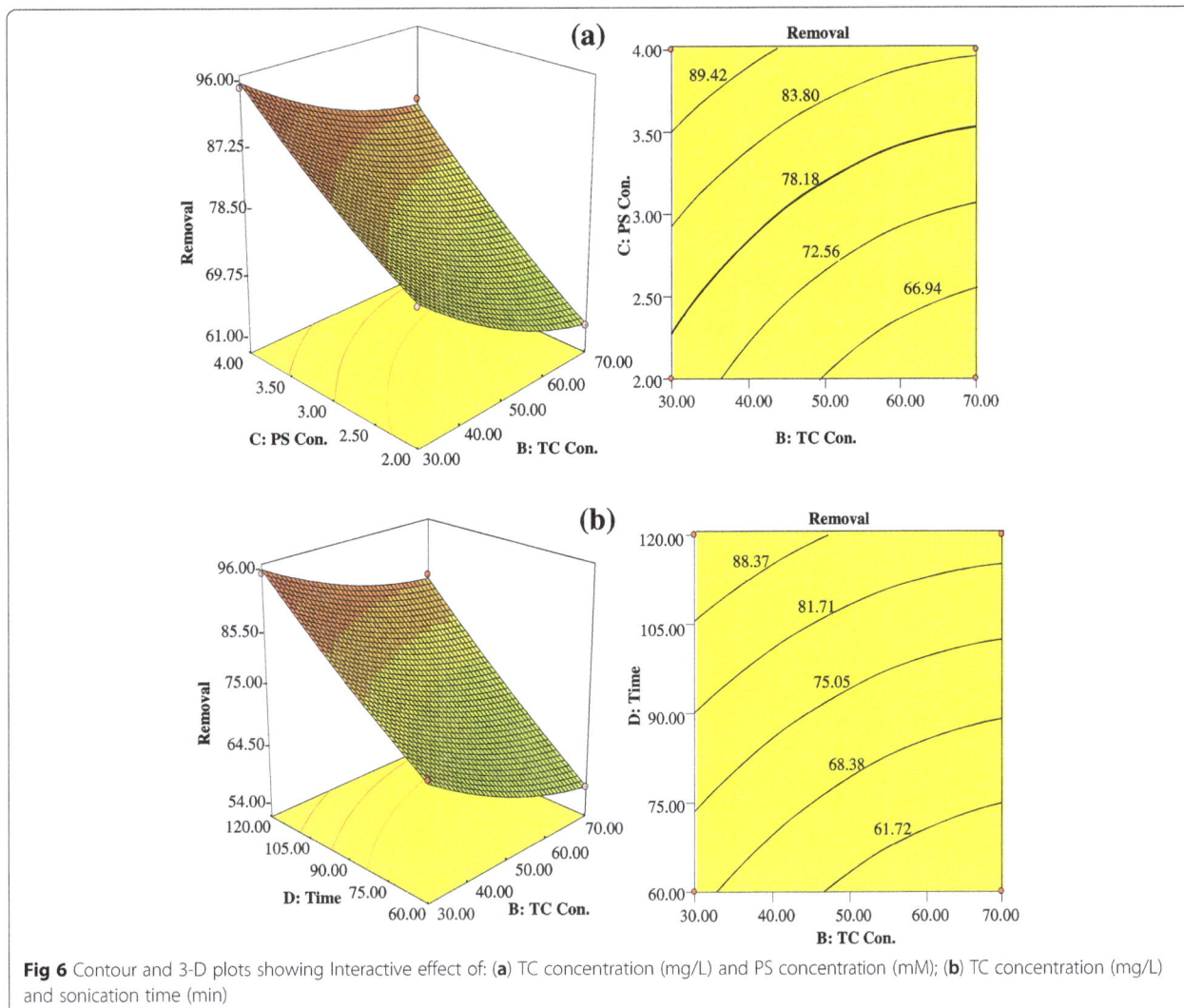

Fig 6 Contour and 3-D plots showing Interactive effect of: (**a**) TC concentration (mg/L) and PS concentration (mM); (**b**) TC concentration (mg/L) and sonication time (min)

$$-d[TC]/dt = k[TC] \qquad (13)$$

Eq. (13) can be rewritten as:

$$\ln C_i/C_t = kt \qquad (14)$$

Where C_i is the initial TC concentration, C_t is the TC concentration at time t, k is the pseudo first order reaction rate t is constant (min^{-1}) and the reaction time (min). To study the TC degradation by US/$S_2O_8^{2-}$ process, the data obtained was investigated using the pseudo first order kinetics. The effect of different parameters such as initial TC concentration, initial PS concentration, pH and temperature on the kinetic of TC degradation was evaluated. In all the experiments, TC degradation well-fitted to the using the pseudo first order kinetics with higher correlation coefficients (R^2). The values of kinetic rate constants

Table 7 Effect of operation parameters on the kinetics degradation of TC

parameter	Value	k_0 (min^{-1}) $\times 10^{-2}$	R^2	$t_{1/2}$ (min)
TC concentration (mg/L)	25	2.29	0.9973	30.2
	50	1.75	0.9952	39.6
	75	1.23	0.9956	56.3
PS concentration (mM)	2	1.15	0.9816	60.6
	3	1.52	0.9946	45.9
	4	2.29	0.9973	30.2
pH	5	1.62	0.9937	42.7
	7.5	1.12	0.9942	62.8
	10	2.29	0.9973	30.2
Temperature (°C)	25	2.29	0.9973	30.2
	45	5.70	0.9127	12.1
	55	7.87	0.921	8.8
	65	10.42	0.9824	6.6

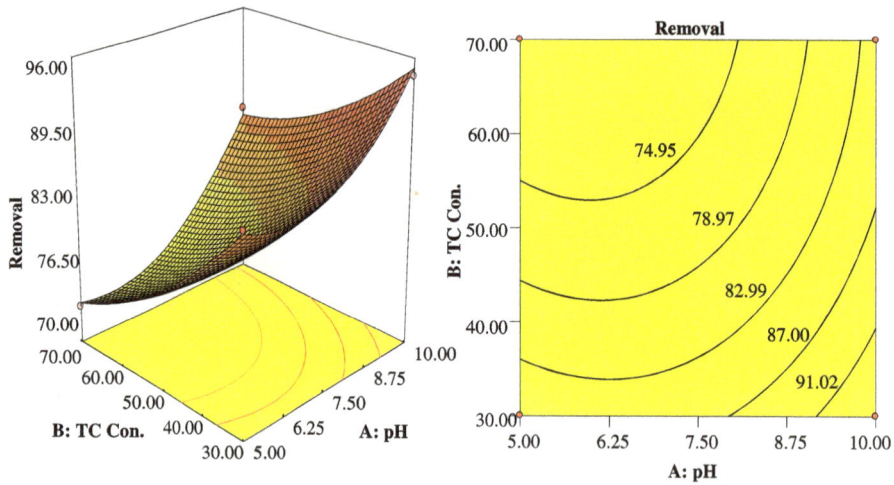

Fig 7 Contour and 3-D plots showing Interactive effect of pH and TC concentration (mg/L)

(k) related to the different parameters, with their regression coefficients R^2 are shown in Table 7.

The effect of temperature on the degradation of tetracycline

To investigate the effect of temperature on the TC degradation rate, experiments were done with various temperature varying from 25 to 65 °C. With increasing temperature from 25 to 65 °C, the degradation rate constant increased from 0.0229 to 0.1042 min^{-1}. Complete TC degradation occurs after 40, 60 and 75 min of reaction at 65, 55 and 45 °C respectively. The activation of $S_2O_8^{2-}$ can be done under heat to form $SO_4^{\bullet-}$ radical as following Eq. (15). Therefore, complete removal of TC by high temperature could be as a result of thermally activated

$S_2O_8^{2-}$ oxidation. Moreover, the increase of temperature significantly enhanced the cavitation activity and chemical effects, resulting in greater degradation rate of TC by US/$S_2O_8^{2-}$ process [22, 60].

$$S_2O_8^{-2} + \overset{\text{Termal–activation}}{\longrightarrow} 2SO_4^{-\bullet} \qquad (30^{\circ\circ}C < T < 99^{\circ\circ}C)$$

$$(15)$$

To investigate the effect of ultrasound on the process kinetics, significant parameters such as activation energy (Ea) play a remarkable role. The effect of temperature on the rate of the reaction and rate constant (k) is obtained by Arrhenius equation according with Eq. (16) [66].

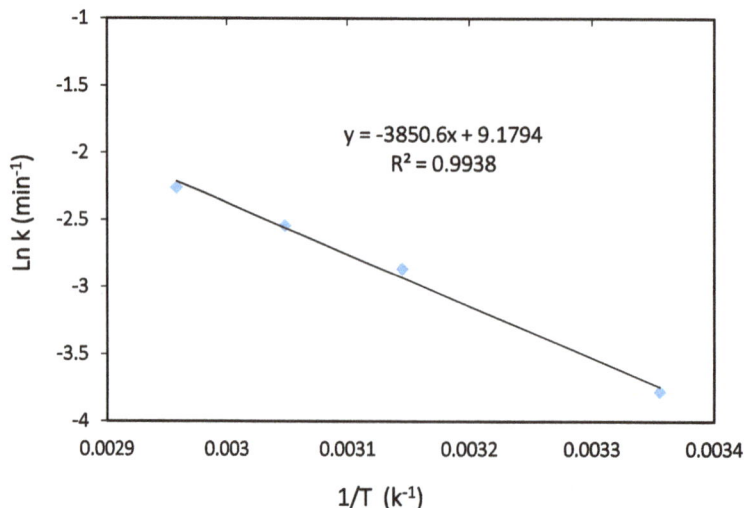

Fig 8 Arrhenius equation graph representation the temperature dependence on chemical reaction rate

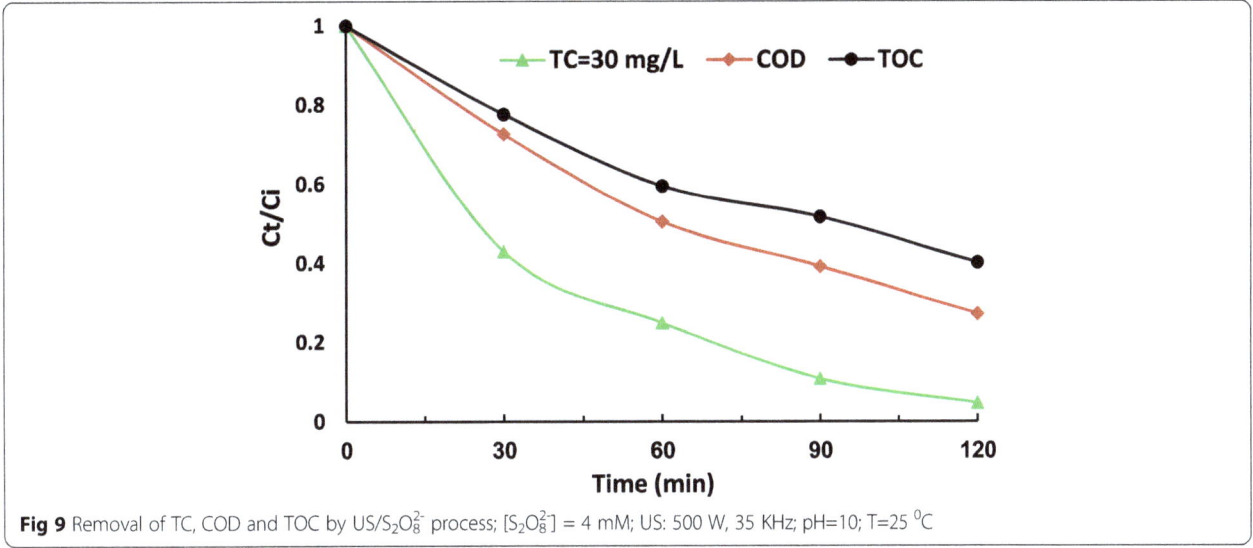

Fig 9 Removal of TC, COD and TOC by US/$S_2O_8^{2-}$ process; [$S_2O_8^{2-}$] = 4 mM; US: 500 W, 35 KHz; pH=10; T=25 ^{0}C

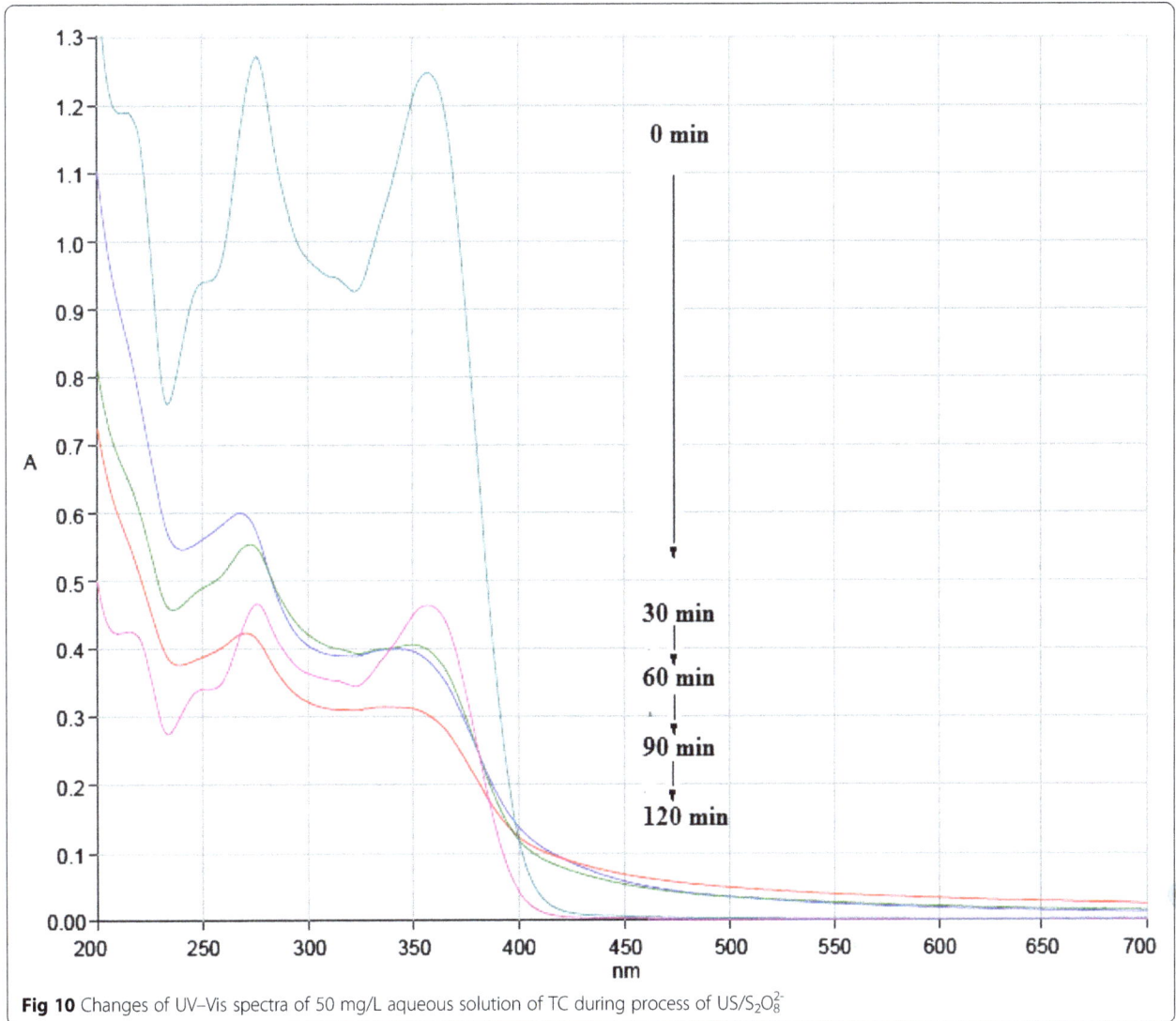

Fig 10 Changes of UV–Vis spectra of 50 mg/L aqueous solution of TC during process of US/$S_2O_8^{2-}$

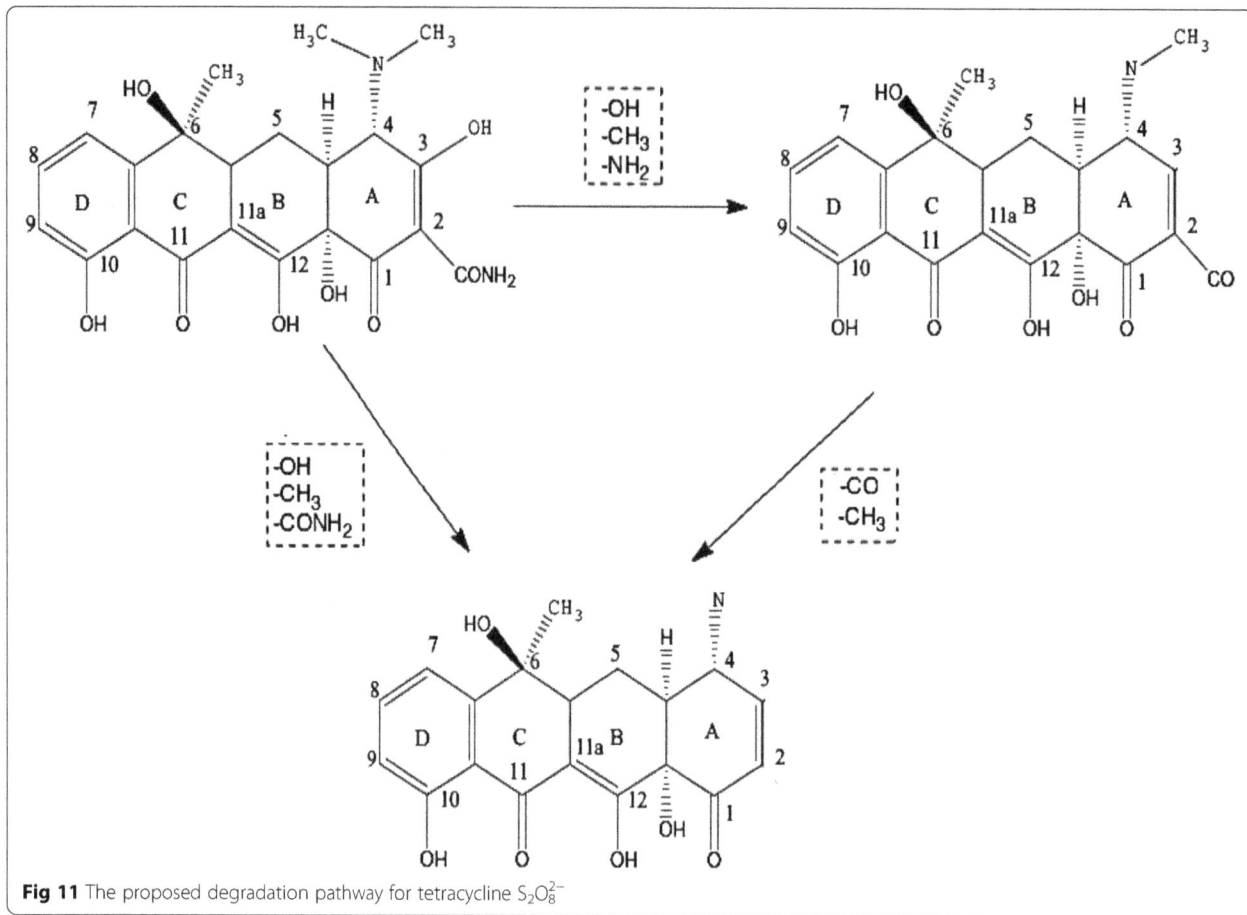

Fig 11 The proposed degradation pathway for tetracycline $S_2O_8^{2-}$

$$\mathrm{LnK} = A \exp\left(-\frac{Ea}{RT}\right) \qquad (16)$$

Arrhenius plot can be used to calculate the Activation Energy at various temperatures by graphing ln k (rate constant) versus 1/T (kelvin). The graph between ln k and 1/T is a straight line with an intercept of ln A and the slop of the graph is equal to $-E_a/R$, where R is a constant equal to 8.314 J/mol-K. According with Arrhenius plot (Fig. 8), the activation energy values of 32.01 (kJ/mol) obtained for degradation of TC by $S_2O_8^{2-}$/US process. It means that for a successful reaction, the colliding molecules must have a

total kinetic energy of 32.01 kJ/mol. The low activation energy indicates that the degradation of TC by $S_2O_8^{2-}$/US process is thermodynamically feasible.

Mineralization, Changes of ultraviolet Visible (UV–Vis) spectra and the proposed degradation pathway

Usually, sonochemical treatment lead to degradation of structure and ultimately mineralization of organic compounds [67]. While perfect mineralization for most antibiotics are difficult because of the structural stability [68]. Therefore, the changes of TOC and COD were evaluated during US/S_2O8^{2-} process and the result are shown

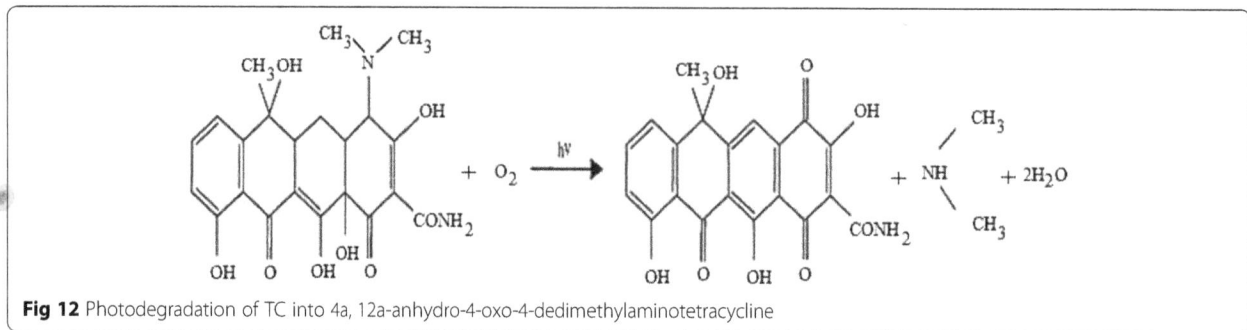

Fig 12 Photodegradation of TC into 4a, 12a-anhydro-4-oxo-4-dedimethylaminotetracycline

Table 8 Degradation of different types of organic pollutants in aqueous solutions using US/$S_2O_8^{2-}$ process

compound	Concentration (mg/L)	operating conditions	Summary of results	reference
Tetracycline	100	$[S_2O_8^{2-}]$ = 200 mM; US = 80 W, 20 KHz; pH = 3.7; T = ambient	More than 51 % degradation after 120 min.	Hou et al. [48]
Trichloroethane	50	$[S_2O_8^{2-}]$ = 0.94 mM US = 400 kHz, 100 W pH = 7; T = 20 ± 2 °C	100 % degradation after 120 min.	Li et al. [42]
Perfluorooctanoic acid	50	$[S_2O_8^{2-}]$ = 46 mM US = 150 W, 40 KHz pH = 4.3; T = 25 °C	More than 98 % degradation after 120 min.	Lin et al. [52]
2,4 Dichlorophenol	50	$[S_2O_8^{2-}]$ = 4 Mm US = 40 KHz, pH = 3; T = 30 °C	More than 95 % degradation after 60 min.	Seid Mohammai [74]
Acid Orange 7	30	$[S_2O_8^{2-}]$ = 1.25 Mm US = 100 W, 20 kHz pH = 5.8; T = ambient	More than 10 % degradation after 20 min.	Wang et al. [75]
Tetracycline	30	$[S_2O_8^{2-}]$ = 4 Mm US = 500 W, 35 KHz pH = 10; T = 25 ± 2 °C	More than 95 % degradation, COD and TOC removal of 72 % and 59 % after 120 min	Present study

in Fig. 9. After 120 min of reaction, TC, COD and TOC were removed approximate 95 %, 73 % and 60 %. The incomplete mineralization implies the potential formation of intermediate products and further identification of the degradation by-products is required.

To evaluate structural changes of TC, the UV–Vis spectra obtained before and after US/ $S_2O_8^{2-}$ process in various time are shown in Fig. 10. The UV–Vis spectra obtained before process indicates two main absorption bands at 275 and 360 nm. The absorption of TC in 360 nm is due to aromatic rings B–D, such as the developed chromophores [68, 69]. With increase of reaction time, the absorption band slightly decreased because of the fragmentation of phenolic groups attached to aromatic ring B [69, 70]. The generation of acylamino and hydroxyl groups led to reduction of absorbance at 270 nm band [70]. The absorption decay at 360 nm band faster than 275 nm. This implies that the ring containing the N-groups (responsible for the absorbance at 276 nm) hardly opened than the other rings, or the created intermediate products absorbed at this wavelength [71]. The proposed degradation pathway for tetracycline based on loses of N-methyl, hydroxyl, and amino groups is shown in Fig. 11. This possible pathway corresponded with conducted studies by other researchers [72]. In addition, TC has a naphthol ring with high stability, which remains unchanged in the reaction and is not easily mineralized. Also, the absorption decay at 360 nm band was found with a relatively small absorption in the visible region. This could be due to the forming of 4a,12a-anhydro-4-oxo-4-dedimethylaminotetracycline according to Fig. 12 [73].

Performance evaluation of US/$S_2O_8^{2-}$ process in the removal of different organics

An overview of performance of US/$S_2O_8^{2-}$ process for removal of different organics along with present study was presented in Table 8. The overview confirm that US/$S_2O_8^{2-}$ process is an attractive alternative technique for degradation of the wide range of organic compounds in aqueous solutions. This process could effectively decomposed organic pollutants in aqueous solution, and the degradation rate depends heavily on the operating conditions, such as physical and chemical characteristics and initial concentration of pollutant, $S_2O_8^{2-}$ concentration, initial pH, reaction time, ultrasound power, ultrasound frequency, and temperature of the medium. Therefore, the various experimental conditions could lead to the various removal efficiencies of organic compounds using US/$S_2O_8^{2-}$ process.

Conclusion

Sonochemical degradation of TC in the presence of $S_2O_8^{2-}$ was investigated with focusing on the optimizing of the operation parameters such as pH, $S_2O_8^{2-}$ concentration, initial TC concentration and reaction time. This study indicated that RSM was the suitable method to optimizing the best operating conditions to maximizing the TC degradation. The reaction time showed the highest effect on the TC degradation, followed by initial $S_2O_8^{2-}$ concentration, initial TC concentration and pH. Under optimal conditions, the TC degradation rate, COD and TOC removal efficiency were found to be 95.01 %, 72.8 % and 59.7 %, respectively. The degradation process followed the pseudo-first-order kinetics. The activation energy value of 31.71 kJ/mol implies that the degradation of TC by US/$S_2O_8^{2-}$ process is thermodynamically feasible. The ultraviolet visible spectra obtained before and after ultrasound irradiation in the presence of $S_2O_8^{2-}$ indicated that proposed degradation pathway for tetracycline was based on loses of N-methyl, hydroxyl, and amino groups. Overall, US/$S_2O_8^{2-}$ process was found to be a promising technology for TC degradation in aqueous solution.

Competing interests
The authors declare that they no competing interests.

Authors' contributions

GHS was the main investigator, collected the data, performed the statistical analysis, and drafted the manuscript. SN and AHM supervised the study and participated in the design of the study, final revised of manuscript and intellectual helping for analyzing of data. KY, RN, and MA were advisors of the study and carried out statistical and technical analysis of data, participated in design of study and manuscript preparation. All authors read and approved the final manuscript.

Acknowledgments

This research was part of a PhD dissertation of the first author and has been supported by Tehran University of Medical Sciences under grant no. 92-03-46-24084. The authors express their gratitude to all laboratory staff of the Department of Environmental Health Engineering.

Author details

[1]Department of Environmental Health Engineering, School of Public Health, Tehran University of Medical Sciences, Tehran, Iran. [2]Center for Water Quality Research, Institute for Environmental Research, Tehran University of Medical Sciences, Tehran, Iran. [3]Center for Solid Waste Research, Institute for Environmental Research, Tehran University of Medical Sciences, Tehran, Iran. [4]Center for Air Pollution Research, Institute for Environmental Research, Tehran University of Medical Sciences, Tehran, Iran.

References

1. Javid A, Nasseri S, Mesdaghinia A, Mahvi AH, Alimohammadi M, Aghdam RM, et al. Performance of photocatalytic oxidation of tetracycline in aqueous solution by TiO_2 nanofibers. J Environ Health Sci Eng. 2013;11:24.
2. Daghrir R, Drogui P. Tetracycline antibiotics in the environment: a review. Environ Chem Lett. 2013;11:209–27.
3. Jiang WT, Chang PH, Wang YS, Tsai Y, Jean JS, Li Z. Sorption and desorption of tetracycline on layered manganese dioxide birnessite. Int J Environ Sci Technol. 2015;12:1695–704.
4. Lindsey ME, Meyer M, Thurman E. Analysis of trace levels of sulfonamide and tetracycline antimicrobials in groundwater and surface water using solid-phase extraction and liquid chromatography mass spectrometry. Anal Chem. 2001;73:4640–6.
5. Alavi N, Babaei AA, Shirmardi M, Naimabadi A, Goudarzi G. Assessment of oxytetracycline and tetracycline antibiotics in manure samples in different cities of Khuzestan Province. Iran Environ Sci Pollut Res. 2015;1–7.
6. Zhu J, Snow DD, Cassada DA, Monson SJ, Spalding RF. Analysis of oxytetracycline, tetracycline and chlortetracycline in water using solid-phase extraction and liquid chromatography-tandem mass spectrometry. J Chromatogr A. 2001;928:177–86.
7. Guler UA, Sarioglu M. Removal of tetracycline from wastewater using pumice stone: equilibrium, kinetic and thermodynamic studies. J Environ Health Sci Eng. 2014;12:1.
8. Reyes C, Fernandez J, Freer J, Mondaca MA, Zaror C, Malato S, et al. Degradation and inactivation of tetracycline by TiO_2 photocatalysis. J Photochem Photobiol A. 2006;184:141–6.
9. Spongberg AL, Witter JD. Pharmaceutical compounds in the wastewater process stream in Northwest Ohio. Sci total environ. 2008;397:148–57.
10. Karthikeyan KG, Michael TM. Occurrence of antibiotics in wastewater treatment facilities in Wisconsin, USA. Sci total environ. 2006;361:196–207.
11. Deblonde T, Cossu-Leguille C, Hartemann P. Emerging pollutants in wastewater: a review of the literature. Int J Hyg Environ Eealth. 2011;214:442–8.
12. Homem V, Santos L. Degradation and removal methods of antibiotics from aqueous matrices-a review. J Environ Manag. 2011;92:2304–47.
13. Choi KJ, Kim SG, Kim SH. Removal of antibiotics by coagulation and granular activated carbon filtration. J Hazard Mater. 2008;151:38–43.
14. Safari GH, Hoseini M, Seyedsalehi M, Kamani H, Jaafari J, Mahvi AH. Photocatalytic degradation of tetracycline using nanosized titanium dioxide in aqueous solution. Int J Environ Sci Technol. 2014;12:603–16.
15. Kakavandi B, Takdastan A, Jaafarzadeh N, Azizi M, Mirzaei A, Azari A. Application of Fe_3O_4@C catalyzing heterogeneous UV-Fenton system for tetracycline removal with a focus on optimization by a response surface method. J Photochem Photobio A Chem. 2015;2016(314):178–88.
16. Tsitonaki A, Petri B, Crimi M, Mosbaek H, Siegrist RL, Bjerg PL. In situ chemical oxidation of contaminated soil and groundwater using persulfate: a review. Crit Rev Environ Sci Technol. 2010;40:55–91.
17. Li SX, Wei D, Mak NK, Cai ZW, Xu XR, Li HB, et al. Degradation of diphenylamine by persulfate: performance optimization, kinetics and mechanism. J Hazard Mater. 2009;164:26–31.
18. Adewuyi YG, Owusu SO. Ultrasound-induced aqueous removal of nitric oxide from flue gases: effects of sulfur dioxide, chloride, and chemical oxidant. J Phys Chem A. 2006;110:11098–107.
19. Weng CH, Tao H. Highly efficient persulfate oxidation process activated with Fe^0 aggregate for decolorization of reactive azo dye Remazol Golden Yellow. Arabian J Chem. 2015; doi.org/10.1016/j.arabjc.2015.05.012.
20. Fang JY, Shang C. Bromate formation from bromide oxidation by the UV/persulfate process. Environ Sci Technol. 2012;46:8976–83.
21. Johnson RL, Tratnyek PG, Johnson ROB. Persulfate persistence under thermal activation conditions. Environ Sci Technol. 2008;42:9350–6.
22. Mora VC, Rosso JA, Le C, Roux G, Roux GC, Martire DO, et al. Thermally activated peroxydisulfate in the presence of additives: a clean method for the degradation of pollutants. Chemosphere. 2009;75:1405–9.
23. Asgari G, Seidmohammadi AM, Chavoshani A. Pentachlorophenol removal from aqueous solutions by microwave/persulfate and microwave/H_2O_2: a comparative kinetic study. J Environ Health Sci Eng. 2014;12:94.
24. Chen WS, Su YC. Removal of dinitrotoluenes in wastewater by sono-activated persulfate. Ultrason Sonochem. 2012;19:921–7.
25. Furman OS, Teel AL, Watts RJ. Mechanism of base activation of persulfate. Environ Sci Technol. 2010;44:6423–8.
26. Yang S, Yang X, Shao X, Niu R, Wang L. Activated carbon catalyzed persulfate oxidation of azo dye acid orange 7 at ambient temperature. J Hazard Mater. 2011;86:659–66.
27. Fang G, Gao J, Dionysiou DD, Liu C, Zhou D. Activation of persulfate by quinones: Free radical reactions and implication for the degradation of PCBs. Environ Sci Technol. 2013;47:4605–11.
28. Liang C, Liang CP, Chen CC. pH dependence of persulfate activation by EDTA/Fe (III) for degradation of trichloroethylene. J Contam Hydrol. 2009;106:173–82.
29. Liang C, Guo Y, Chien Y, Wu Y. Oxidative degradation of MTBE by pyrite-activated persulfate: proposed reaction kinetics. J Contam Hydrol. 2010;49:8858–64.
30. Criquet J, Leitmer NKV. Electron beam irradiation of aqueous solution of persulfate ions. Chem Eng J. 2011;169:258–62.
31. Ahmad M, Teel AL, Watts RJ. Mechanism of persulfate activation by phenols. Environ Sci Technol. 2013;47:5864–71.
32. Xu XR, Li XZ. Degradation of azo dye Orange G in aqueous solutions by persulfate with ferrous ion. Sep Purif Technol. 2010;72:105–11.
33. Rivas FJ. Polycyclic aromatic hydrocarbons sorbed on soils: a short review of chemical oxidation based treatments. J Hazard Mater. 2006;138:234–51.
34. Anipsitakis GP, Dionysiou DD. Radical generation by the interaction of transition metals with common oxidants. Environ Sci Technol. 2004;38:3705–12.
35. Zhao J, Zhang Y, Quan X, Chen S. Enhanced oxidation of 4-chlorophenol using sulfate radicals generated from zero-valent iron and peroxydisulfate at ambient temperature. Sep Purif Technol. 2010;71:302–7.
36. Antoniou MG, Cruz AA, Dionysiou DD. Degradation of microcystin-LR using sulfate radicals generated through photolysis, thermolysis and e- transfer mechanisms. Appl Catal B Environ. 2010;96:290–8.
37. Lin YT, Liang CJ, Chen GH. Feasibility study of ultraviolet activated persulfate oxidation of phenol. Chemosphere. 2011;82:1168–72.
38. Liang HY, Zhang YQ, Huang SB, Hussain I. Oxidative degradation of p-chloroaniline by copper oxidate activated persulfate. Chem Eng J. 2013;218:384–91.
39. Olmez-Hanci T, Arslan-Alaton I. Comparison of sulfate and hydroxyl radical based advanced oxidation of phenol. Chem Eng J. 2013;224:469–74.
40. Ji Y, Dong C, Kong D, Lu J, Zhou Q. Heat-activated persulfate oxidation of atrazine: Implications for remediation of groundwater contaminated by herbicides. Chem Eng J. 2015;263:45–54.
41. Zhou D, Chen L, Zhang C, Yu Y, Zhang L. novel photochemical system of ferrous sulfite complex: Kinetics and mechanisms of rapid decolorization of Acid Orange 7 in aqueous solutions. Water Res. 2014;57:85–97.
42. Li B, Li L, Lin K, Zhang W, Lu S, Luo Q. Removal of 1,1,1-trichloroethane from aqueous solution by a sono-activated persulfate process. Ultrason Sonochem. 2013;20:855–63.

43. Sivakumar R, Muthukumar K. Sonochemical degradation of pharmaceutical wastewater. Clean Soil Air Water. 2011;39:136–41.

44. Neppolian B, Doronila A, Ashokkumar M. Sonochemical oxidation of arsenic (III) to arsenic (V) using potassium peroxydisulfate as an oxidizing agent. Water Res. 2010;44:3687–95.

45. Neppolian B, Jung H, Choi H, Lee JH, Kang JW. Sonolytic degradation of methyl tert-butyl ether: the role of coupled Fenton process and persulphate ion. Water Res. 2002;36:4699–708.

46. Son HS, Choi SB, Khan E, Zoh KD. Removal of 1, 4-dioxane from water using sonication: effect of adding oxidants on the degradation kinetics. Water Res. 2006;40:692–8.

47. Su S, Guo W, Yi C, Leng Y, Ma Z. Degradation of amoxicillin in aqueous solution using sulphate radicals under ultrasound irrdidation. Ultrason Sonochem. 2012;19:469–74.

48. Hou L, Zhang H, Xue X. Ultrasound enhanced heterogeneous activation of peroxydisulfate by magnetite catalyst for the degradation of tetracycline in water. Sep Purif Technol. 2012;84:147–52.

49. Gayathri P, Dorathi RPJ, Palanivelu K. Sonochemical degradation of textile dyes in aqueous solution using sulphate radicals activated by immobilized cobalt ions. Ultrason Sonochem. 2010;17:566–71.

50. Kwon M, Kim S, Yoon Y, Jung Y, Hwang TM, Lee J, et al. Comparative evaluation of ibuprofen removal by UV/H_2O_2 and $UV/S_2O_8^{2-}$ processes for wastewater treatment. Chem Eng J. 2015;269:379–90.

51. Chen WS, Huang YL. Removal of dinitrotoluenes and trinitrotoluene from industrial wastewater by ultrasound enhanced with titanium dioxide. Ultrason sonochem. 2011;18:1232–40.

52. Lin JC, Lo SL, Hu CY, Lee YC, Kuo J. Enhanced sonochemical degradation of perfluorooctanoic acid by sulfate ions. Ultrason Sonochem. 2014;22:542–7.

53. Ghafoori S, Mowla A, Jahani R, Mehrvar M, Chan PK. Sonophotolytic degradation of synthetic pharmaceutical wastewater: Statistical experimental design and modeling. J Environ Manage. 2015;150:128–37.

54. Rezaee R, Maleki A, Jafari A, Mazloomi S, Zandsalimi Y, Mahvi AH. Application of response surface methodology for optimization of natural organic matter degradation by UV/H_2O_2 advanced oxidation process. J Environ Health Sci Eng. 2014;12:1.

55. Montgomery DC. Design and Analysis of Experiments. 6th ed. New York: John Wiley & Sons; 2005.

56. Affam AC, Chaudhuri M. Optimization of Fenton treatment of amoxicillin and cloxacillin antibiotic aqueous solution. Desal Water Treat. 2014;52:1878–84.

57. Zuorro A, Fidaleo M, Fidaleo M, Lavecchia R. Degradation and antibiotic activity reduction of chloramphenicol in aqueous solution by UV/H_2O_2 process. J Environ Manage. 2014;133:302–8.

58. Morshedi A, Akbarian M. Application of response surface methodology: design of experiments and optimization: a mini review. In J Fund Appl Life Sci. 2014;54:2434–9.

59. Dominguez JR, Munoz MJ, Palo P, Gonzalez T, Peres JA, Cuerda-Correa EM. Fenton advanced oxidation of emerging pollutants: parabens. Int J Energy Environ Eng. 2014;5:1–10.

60. Xu M, Du H, Gu X, Lu S, Qiu Z, Sui Q. Generation and intensity of active oxygen species in thermally activated persulfate systems for the degradation of trichloroethylene. RSC Advances. 2014;76:40511–7.

61. Hoseini M, Safari GH, Kamani H, Jaafari J, Ghanbarain M, Mahvi AH. Sonocatalytic degradation of tetracycline antibiotic in aqueous solution by sonocatalysis. Toxicol Environ Chem. 2014;95:1680–9.

62. Saeed W. The effectiveness of persulfate in the oxidation of petroleum contaminants in saline environment at elevated groundwater temperature. Ontario, Canada: Waterloo; 2011.

63. Zhao D, Liao X, Xiulan Y, Huling SG, Chai T, Huan T. Effect and mechanism of persulfate activated by different methods for PAHs removal in Soi. J Hazard Mater. 2013;254:228–35.

64. Guan YH, Ma J, Li XC, Fang JY, Chen LW. Effect of pH on the formation of sulfate and hydroxyl radicals in the UV/peroxymonosulfate system. Environ Sci Technol. 2011;45:9308–14.

65. Liang C, Su HW. Identification of sulfate and hydroxyl radicals in thermally activated persulfate. Ind Eng Chem Res. 2009;48:5558–62.

66. Ji YF, Ferronato C, Salvador A, Yang X, Chovelon JM. Degradation of ciprofloxacin and sulfamethoxazole by ferrous-activated persulfate: Implications for remediation of groundwater contaminated by antibiotics. Sci Total Environ. 2014;472:800–8 [38].

67. Jiao S, Zheng S, Yin D, Wang L, Chen L. Aqueous photolysis of tetracycline and toxicity of photolytic products to luminescent bacteria. Chemosphere. 2008;73:377–82.

68. Shaojun JIAO, Zheng S, Daqiang YIN, Lianhong WANG, Liangyan CHEN. Aqueous oxytetracycline degradation and the toxicity change of degradation compounds in photoirradiation process. J Environ Sci. 2008;20:806–13.

69. Wang Y, Zhang H, Zhang J, Lu C, Huang Q, Wu J, et al. Degradation of tetracycline in aqueous media by ozonation in an internal loop-lift reactor. J Hazard Mater. 2011;192:35–43.

70. Zhu XD, Wang YJ, Sun RJ, Zhou DM. Photocatalytic degradation of tetracycline in aqueous solution by nanosized TiO_2. Chemosphere. 2013;92:925–32.

71. Dalmazio I, Almeida MO, Augusti R. Monitoring the degradation of tetracycline by ozone in aqueous medium via atmospheric pressure ionization mass spectrometry. J Am Soc Mass Spectrom. 2007;18:679–87.

72. Ma Y, Gao N, Li C. Degradation and pathway of tetracycline hydrochloride in aqueous solution by potassium ferrate. Environ Eng Sci. 2012;29:357–62.

73. Addamo M, Augugliaro V, Di Paola A, Garcia-Lopez E, Loddo V, Marci G, et al. Removal of drugs in aqueous systems by photoassisted degradation. J Appl Electrochem. 2005;35:765–74.

74. Seid Mohammadi A, Asgari G, Almasi H. Removal of 2,4 Di-Chlorophenol Using Persulfate Activated with Ultrasound from Aqueous Solutions. J Environ Eng. 2014;4:260–68.

75. Wang X, Wang L, Li J, Qiu J, Cai C, Zhang H. Degradation of Acid Orange 7 by persulfate activated with zero valent iron in the presence of ultrasonic irradiation. Sep Pur Technol. 2014;122:41–46.

13

Cadmium removal from aqueous solution by green synthesis iron oxide nanoparticles with tangerine peel extract

Mohammad Hassan Ehrampoush[1], Mohammad Miria[1], Mohammad Hossien Salmani[1*] and Amir Hossein Mahvi[2]

Abstract

Background: The adsorption process by metal oxide nanoparticles has been investigated an effective agent for removing organic and inorganic contaminants from water and wastewater. In this study, iron oxide nanoparticles were synthesized in the presence of tangerine peel extract as adsorbent for cadmium ions removal from contaminated solution. Iron oxide nanoparticles prepared by co-precipitation method and tangerine peel extract was used to prevent accumulation and reduce the diameter of the particles. Effect of various parameters such as contact time, pH, metal concentration and adsorbent dosage was determined on the removal efficiency.

Results: The different concentrations of tangerine peel had an impact on the size of nanoparticles. As, increasing the concentration of tangerine peel extract from 2 to 6 % the average size of synthesized iron oxide nanoparticles decreased 200 nm to 50 nm. The maximum removal of cadmium ions (90 %) occurred at pH of 4 and adsorbent dose of 0.4 g/100 ml. Adsorption of cadmium ions by synthesized iron oxide nanoparticles followed Freundlich adsorption model and pseudo-second-order equation.

Conclusion: The cadmium ions are usually soluble in acidic pH and the maximum removal of cadmium by green synthesis iron oxide nanoparticles was obtained in the pH of 4, so these nanoparticles can be a good adsorbent for the removal of cadmium from wastewater.

Keywords: Iron oxide nanoparticles, Cadmium removal, Tangerine peel extract

Background

Cadmium usually is a trace ion in the ground water and surface water. It may be hydrated ions, complexes - such as inorganic carbonate, hydroxide, chloride, sulfate, or as organic complexes with humic acids in water [1]. The high levels of cadmium in water may be existed in the plating and coating of pipes and fittings, soldering with silver-coated tubes places [2]. Cadmium is toxic for animals and human, and recently, much attention has been on the long-term risk factors for general population health that are affected by cadmium. Cadmium can be introduced to body via inhalation, ingestion or absorption through the skin and affects on the body organs. The effects of short-term exposure on cadmium include: nausea, vomiting, diarrhea, muscle cramps, dry mouth, impaired senses, liver damage, convulsions, shock and renal defects, influences on the lung and the cardiovascular system, liver, and nervous system [3, 4]. Long-term effects of exposure on cadmium include the impacts on the lungs, kidneys, bones, growth and carcinogenic. Plants, fruits and vegetables are grown in the fertile soil with superphosphate fertilizer relation to the soil has higher levels of cadmium. The daily intake cadmium for the average person weighing 70 kg was estimated 25–60 µg by The United States of America and Europe [5]. The WHO is recommended maximum allow of 10 µg/L cadmium ion in drinking water sources [6]. Therefore, in recent years, considerable attention has been devoted to the study of effective methods to remove low concentrations of cadmium in contaminated water.

When cadmium concentration in contaminated water is less than 0.5 mg/L, treatment of water by conventional methods such as precipitation with lime and treatment with alum and ferric sulfate is not suitable to remove it

* Correspondence: mhsn06@yahoo.com
[1]Department of Environmental Health, School of Public Health, Shahid Sadoughi University of Medical Science, Yazd, Iran
Full list of author information is available at the end of the article

and also is less effective [7]. Most researchers has proposed ion exchange, reverse osmosis, coagulation-filtration and adsorption processes as the best available technology for effective removal of cadmium to the 0.005 mg/L or 5 ppb [3, 8]. However, most of these methods have disadvantages such as expensive devices, production of toxic sludge or other waste materials, space and high energy requirements. Among these methods, adsorption has attracted the attention of many researches because it is a simple, low-cost, and effective for the removal of heavy metal ions in low and medium concentrations [9]. A number of minerals, clays, and waste materials were regularly used to remove heavy metals from water and wastewater [10]. Recently, it emphasized that the nanoparticles and nanostructure sorbents can be used as an efficient and convenient replacement instead of conventional adsorbents [11, 12]. Adsorption of metal ions on iron nanoparticles is an environmentally friendly technology that it has been studied as an effective agent for removing organic contaminants and heavy metal ions from water and wastewater [13–21].

Number various methods studied to synthesis of nanoparticles such as: co-precipitation, sol-gel synthesis, micro emulsion, and the oxidation of nanoparticles [22]. The chemical co-precipitation of ferric and ferrous ions in the alkaline solution is a commonly and large scale amount for synthesis of iron oxide nanoparticles [23]. During the separation of particles from bulk of solution, nanoparticle will lead to aggregation under Vander Waals force. The synthesis nanoparticles to agglomeration which converts a greater particles size in diameter. For prevention of agglomeration, it often used a surface modification using inorganic or organic material [24]. Most of water soluble polymers or surfactants can often be used a good dispersion or capping agent, but they are toxic for environment. In this study, tangerine peel extract a green material was applied to synthesis of iron oxide nanoparticles by co-precipitation methods and then these nanoparticles were used to remove cadmium ions from aqueous solutions. This method dose not required organic solvent for prevention of agglomeration that nanoparticles with an average diameter size of 50 nm can be produced in an aqueous peel extract solution. Also, this approach is entirely friendly environment and is not an environment pollutant.

Methods

Tangerine Peel Extract

Tangerine peel was collected from the local market at Yazd city. These washed with distillated water to clean the surface pollutants and dried in the air and absence of sunlight at the Lab. temperature (27 ± 2 °C). The dried peels were milled by electric milling and sieved to

powders. 50 g of powders were introduced to 500 ml of distilled water in the beaker and heated to 80 °C for 15 min. After cooling to Lab. temperature filtered with 0.45 μ filter paper. This solution as capping agent was maintained in the refrigerator for further use.

Preparation of nanoparticles

The co-precipitation method was used to prepare nanoparticles [25, 26]. For this purpose, 5.35 g of $FeCl_3$ and 8.10 g of $FeCl_2.4H_2O$ were separately dissolved in 500 ml of different concentration of extract peel to form a solution with the concentration of 0.1 M for Fe(III) and Fe(II) in presence of extract peel as surfactant and stabilizer. Then two solutions were mixed a ratio of 40–60 ml and raised it to 80 °C using an electric heater. Finally, NH_4OH solution 25 % was dropped to this solution with extremely string at least 20 min until the final pH was came over than 9. The brown precipitate collected by magnetic piece and washed three times with distilled water. The treatment precipitant was dried at Lab. temperature at 1 week. Scanning electron microscopy (SEM) and dynamic light scattering (DLS) were used to specify nanoparticles size. These iron oxide nanoparticles were used to study of removal cadmium ions from aqueous solutions.

Adsorption process

To obtain different concentrations of working and standard solution of cadmium ions, the stock solution of 1 g/l of cadmium ions was prepared with pure cadmium nitrate Cd $(NO_3)_2.4H_2O$ and distilled water. Experiments were performed on four periods (15, 30, 60, 90 min), three levels of pH (2, 4, 6.5), four different concentrations of cadmium ions (5, 10, 15, 20 mg/l), and three levels of absorbent dose (0.2, 0.4, 0.6 gr/100 ml). Then cadmium ions solution contacted with the adsorbent, shake at specified time on the 120 rpm orbital shaker, separated by a piece of magnet, and centrifuged in the 5000 rpm. In the end of each experiment, the residual cadmium concentration in solution was determinate.

Data analysis

The remaining concentration of cadmium in clear solution was determined using of atomic absorption spectrometer (spectra model AA-20). For this purpose, different concentration (0.0, 2.0, 5.0, 8.0 and 10.0 mg/l) standard solutions of Cd^{2+} were prepared from stock solution of 100 mg/l and used for calibration of AA. Absorbance was recorded three fold at wavelength 228.8 nm and 0.5 nm in a standard mode and concentration of cadmium was calculated from Beer-Lambert equation. The removal efficiency of nanoparticles and adsorption capacity was calculated from experimental data as eqs. (1 and 2).

$$\% \text{ RE} = \frac{(C_0 - C)}{C_0} \times 100 \qquad (1)$$

$$q_e = \frac{(C_0 - C)}{m} \times V \qquad (2)$$

After determining of adsorption data, using linear equations related to isotherm and kinetic models, modeling of Cd^{2+} adsorption study on the synthesized iron oxide nanoparticles were determined.

Results and Discussion
Characterization of synthesized iron oxide nanoparticles
Iron oxide nanoparticles were prepared in the presence of various concentrations of tangerine peel extract (2, 4, 6, 8, and 10 %) that the size of them was controlled using scanning electron microscopy (SEM) and also the range size of produced particles was determined by dynamic light scattering (DLS). Average size of nanoparticles was about 200 nm at 2 % concentration of tangerine

peel extract. As well as, the same average was obtained when tangerine peel extract increased to concentration of 4 %. While the tangerine peel extract reached to a concentration of 6 %, a significant decrease in the size of the iron oxide nanoparticles were appeared, so an average size was about 50 nm. Tangerine peel extract concentration of 10 % indicates severe aggregation and significant increasing in the size of the nanoparticles that the average size of nanoparticle was reached to 1 μ. Figure 1a, b shows the SEM and DLS of prepared nanoparticles at optimal conditions that was 6 % of tangerine peel extract.

Tangerine peel extract was used to prepare iron oxide nanoparticles as well as to reduce the impact on average nanoparticles size for increasing of its removal efficiency. The results showed that the concentration of tangerine peel extract can affect the nanoparticles size. So that with increasing concentration of tangerine peel extract from 2 to 6 %, the average size of nanoparticle reach out about 200 nm to 50 nm and the increasing

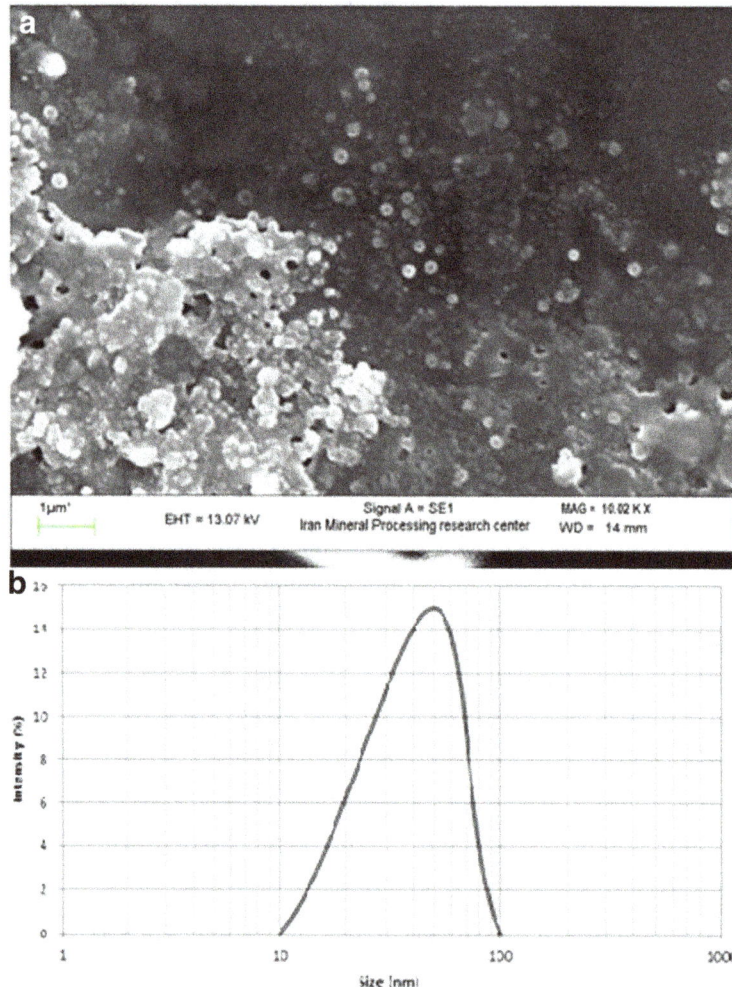

Fig. 1 Characterization of synthesized iron oxide nanoparticles **a)** SEM image **b)** Size distribution by DLS

concentration of tangerine peel extract continued up to 10 % the size of nanoparticles rose again due to the strong adhesion of nanoparticles. The tangerine peel extract concentration of 6 % is optimal for the production of iron oxide nanoparticles. Over the last few years, several synthetic methods have been focused on produce of controlled size nanoparticles and enhance to produce regular shape [27]. From SEM image predicted that iron oxide nanoparticles synthesized by tangerine peel extract of 6 % yield relatively uniform spherical shape in average range 50 nm that calculated from DLS micrograph. So, it seems that different agglomeration can occur depending on the used concentration of peel extract, which directly influences the size of particles.

Also, the synthesized iron oxide nanoparticles in presence of tangerine peel extract applied to remove of Cd^{2+} in aqueous solution. For obtaining the best condition, several experiments were done by variation of effective parameters such as contact time, pH, initial concentration and adsorbent dosage in a batch system.

The effect of contact time

The effect of contact time on cadmium removal efficiency by iron oxide nanoparticles synthesized in the presence of tangerine peel extract was investigated in an initial concentration of 5 mg/l of cadmium ions at 15, 30, 60 and 90 min. During this phase, adsorbent mass was 0.4 g/100 ml and the solution pH adjusted to 4. The effect of contact time is shown in Fig. 2. As observed, the removal rate increases with time so the optimal time is about 90 min for the best removal and this time was chosen to subsequent experiments.

According to Fig. 2, with increasing contact time, cadmium ion removal efficiency increases, because of cadmium ions are more opportunities for contact with the adsorbent surface when time increases. As can be seen from Fig. 2, the rate of cadmium ions removal was fast in the beginning times (first 15 min) due to the larger surface area of the adsorbent available. As time increases to 90 min, more amount of cadmium ions adsorb onto the surface of the adsorbent by attraction forces and

cause a complete removal of cadmium ions. Therefor, the equilibrium time for this absorbent is about 90 min. Kosa et al. (2012) have been studied as heavy metals removal using carbon nanotubes modified with 8-hydroxyl-quinoline. Their results showed that removal efficiency was increased by enhancing the contact time between the adsorbent and cadmium ions. The maximum absorption occurred in the first 10 min that the results of this study are in line with the present study results [28].

The effect of solution pH on Cd^{2+} removal

pH is the most important factor in the absorption process. In this study, the effect of pH on cadmium removal efficiency by prepared iron oxide nanoparticles was determined in the absorption process at pH 2, 4, and 6.5. For each experiment, the 0.4 g of adsorbent was added into the 100 ml initial cadmium concentration of 5 mg/l at each above adjusted initial pH. After a contact time of 90 min, residual concentrations of cadmium ions in the solution were measured and the removal efficiency was calculated based on the results. The results are presented in Fig. 3. The optimum pH for removal efficiency of cadmium was 4 that were used in the following experiments.

Figure 3 revealed that the removal efficiency is lower at high acidic pH. It seems that the positive charge on the adsorbent is created in the acidic pH. So, there is an electrostatic repulsion between the adsorbent and cadmium ions in solution. The hydrogen ions instead of cadmium ions are placed in to the adsorbent sites when the amount of hydrogen ions increases in solution, and so the removal efficiency decreases. This study showed that changing of pH from 2 to 4, the removal efficiency increases but it has a decreasing trend. The similar observation was obtained in our previous study [29] . The zinc oxide nanoparticles were used for the removal of cadmium at high ionic strength solution that with a change in pH from 4 to 5, the removal efficiency of cadmium increased in a great deal, and when the pH increased to 7, the removal efficiency of cadmium has had

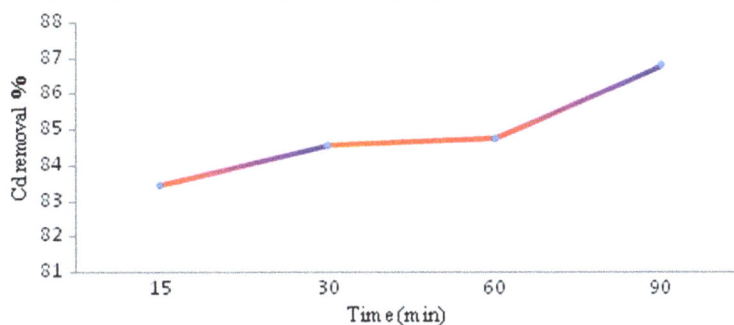

Fig. 2 The effect of contact time on the Cd removal efficiency

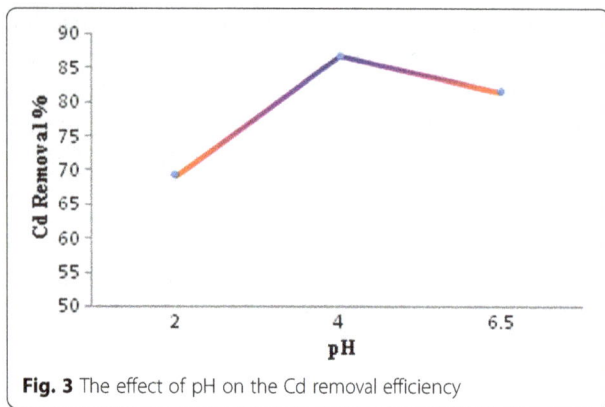

Fig. 3 The effect of pH on the Cd removal efficiency

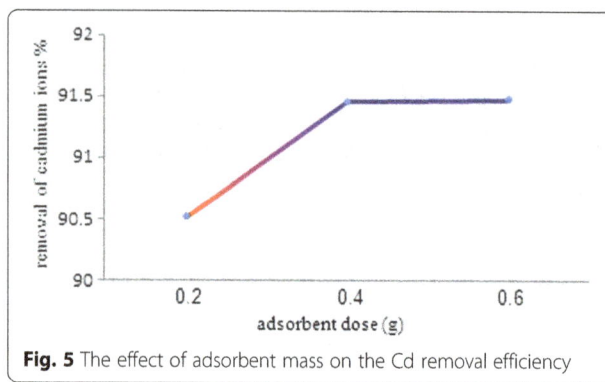

Fig. 5 The effect of adsorbent mass on the Cd removal efficiency

a gradual increase. Thus, the highest removal efficiency gained in pH = 7 that was 89.6 %, and the lowest efficiency measured as 38 % in pH = 4.

A study conducted by Afkhami et al. (2010), The alumina nanoparticles modified with 2, 4-di-nitro phenyl hydrazine was used to remove heavy metals such as Pb (II), Cd (II), Cr (III), Co (II), Ni (II) and Mn (II). The specific effect of pH was determined to change of pH 1.5–5.5 ranges. The results showed that removal efficiency increases up to pH of 5 and then decreases with excess of pH up to 5. These results are in line with the present study [30]. The obtained results of the present study are in line with the mentioned research findings.

The effect of initial concentration on Cd^{2+} removal

Effect of initial cadmium concentration on the removal efficiency was investigated in the pH = 4, the mass of nanoparticles 0.4 g/100 ml solution with a concentration 5, 10, 15, and 20 mg/L of cadmium ions. The results are shown in Fig. 4. As observed, with increasing cadmium concentration of 5–20 mg/L, the removal efficiency is increased from 87 to 88.7 %.

An increased ratio of initial number of cadmium ions to the available surface area resulted in high concentration; hence fractional adsorption dependent on initial concentration. For a given adsorbent dose the total

number of available adsorption active sites is constant thereby adsorbing almost the same amount of adsorbate, thus a decrease in the removal of adsorbate was resulted to an increase in initial concentration of cadmium. Similar results were also reported [31].

The effect of adsorbent mass on Cd^{2+} removal

In this study, the effect of iron oxide nanoparticles mass on the removal of cadmium was examined by changing of the mass of 0.2, 0.4, and 0.6 g at the 100 ml initial cadmium concentration of 15 mg/l. All suspensions were shacked 90 min, separated of adsorbent, measured the residual cadmium ions in solution and calculated the removal efficiency. The obtained results are presented in Fig. 5. The results showed the increasing of adsorbent mass from 0.2 to 0.6 g/ 100 ml the removal efficiency increased up to mass of 0.4 g and then remained constant. Hence value of 0.4 g/100 ml of adsorbent dosage was conducted as optimum value.

The influence of adsorbent dose on adsorption of cadmium ions was studied for obtaining the right adsorbent mass. Figure 5 shows, at the constant cadmium ions concentration, the adsorption percent of cadmium ions increases with increasing mass of the adsorbents. This is due to the existence of more available sites on surface of adsorbents at higher doses. When adsorbent dose was increased from 0.4 to 0.6 % w/v, the removal of cadmium ions had a low increasing. This effect may be due to the fact that some adsorption sites remain unsaturated during the batch adsorption process. This drop in adsorbed amount per unit mass of adsorbent is a routine behavior in the batch process.

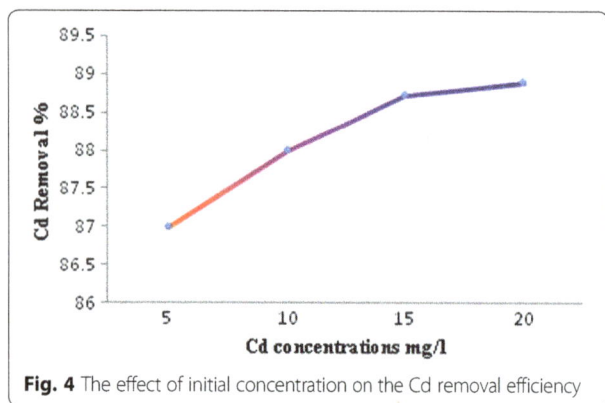

Fig. 4 The effect of initial concentration on the Cd removal efficiency

Table 1 Linear equation and adsorption isotherm parameters for Cd^{2+} removal

Isotherm Linear equation	Langmuir			Freundlich		
	$\frac{1}{q_e} = \frac{1}{q_{max} K_i C_e} + \frac{1}{q_{max}}$			$\ln q_e = \ln K_F + \frac{1}{n} \ln C_e$		
Parameters	R^2	b	q_{max}	R^2	n	k_F
Obtained values	0.993	0.117	15.5	0.997	0.866	1.789

Table 2 Kinetic equation and obtained results for Cd^{2+} removal

	pseudo-first order		pseudo-second order	
model	$\log(q_e - q_t) = \log q_e - \frac{k_1}{2.303} t$		$\frac{t}{q_t} = \frac{1}{k_2 q_e^2} + \frac{1}{q_e} t$	
Equation	$Y = -0.0352x - 2.3298$	$R^2 = 0.798$	$Y = 0.9163\,x + 0.9007$	$R^2 = 0.999$
Obtained Value	$K_1 = 0.081$	$q_e = 4.7$	$K_2 = 1.119$	$q_e = 10.9$

Kinetic and isotherm adsorption

The adsorption data are usually described by adsorption equilibrium isotherms that indicate the effect of the initial concentration of pollutant on the adsorption quantity. In present study, the equilibrium adsorption of cadmium by iron oxide nanoparticles was explored at the contact time of 90 min., the adsorbent dose of 0.4 g/100 ml, cadmium initial concentrations in range of 5, 10, 15, 20 mg/L and the temperature of 25 °C. The *Langmuir* and *Freundlich* models were applied to determine the adsorption isotherm for cadmium removal by synthesized iron nanoparticles that the obtained results are presented in Table 1.

One of the most salient factors to design an adsorption system (to determine the optimum contact time) is anticipating the speed of the adsorption process that is controlled by isotherm and kinetic system. The common isotherm models are Freundlich and Langmuir that the linear equations are presented in 1. According to regression coefficient (Table 1), experiments data of cadmium adsorption on synthesized iron oxide nanoparticles followed Freundlich models better than those for the Langmuir model. Freundlich equation is an empirical adsorption isotherm equation that can be applied in case of low and intermediate concentration ranges. Freundlich isotherm gives the parameters, n, indicative of bond energies between metal ion and the adsorbent and K_F, related to bond strength [32]. The slope n and K_F from the linear equation was 0.866 and 1.789, respectively, satisfying of the condition for favorable adsorption.

In order to ascertain cadmium adsorption kinetic on iron oxide nanoparticles, the experiments were carried out under the optimized conditions and four different contact times as 15, 30, 60, and 90 min that were selected to obtain adsorption kinetic. *Pseudo-first order* and *pseudo- second order* equation were considered to determine the adsorption kinetic. The obtained results are presented in Table 2.

The pseudo-first-order and pseudo-second-order equations were applied to assess the suitability of the rate equation for the experiment data. The obtained rate equations were compiled in Table 2. The results show that the correlation coefficients for second order rate equations (0.999) were higher than those for the first order rate equations. Hence, the pseudo second order rate equation is more suitable to explain the cadmium adsorption on the iron oxide nanoparticles. In case of pseudo second order kinetics, the plot of (t/q_t) versus t gives a linear relationship that allows computation of q_e and K_2. The second order rate constant (k_2) value and equilibrium adsorption capacity (q_e) was 1.119 $g.mg^{-1}.min^{-1}$ and 10.9 $mg.g^{-1}$ for cadmium adsorption by synthesized iron nanoparticles in presence of tangerine peel extract.

Conclusion

The use of peel extracts as stabilizer agent for preparation of metal oxide nanoparticles is inexpensive, and eco-friendly. It is especially for preparation of nanoparticles that have been free of toxic contaminations. The peel extract can controlled the size and morphology of nanoparticles during synthesis process. Simple green synthesis of iron oxide nanoparticles through co-precipitation in alkali condition can exhibit an excellent adsorption for the cadmium ions that followed Freundlich adsorption model and pseudo-second-order equation. The obtained results of this investigation indicated that the synthesized adsorbent was more able to remove cadmium ions.

Competing interests
The authors declare that they have no competing interests.

Authors' contributions
The study was directed by MM who performed all the experiments. MHE is the first author read the manuscript. MHS (Corresponding author) advised the experimental methods and drafted the manuscript. AHM has given consultancy for the experiments and the English edition of manuscript. All authors have read and approved the final manuscript.

Acknowledgments
This project was financially supported by the Faculty of Public Health, Shahid Sadoughi University of Medical Sciences. The authors are grateful to the head of Environmental Chemistry Laboratory for his help.

Author details
[1]Department of Environmental Health, School of Public Health, Shahid Sadoughi University of Medical Science, Yazd, Iran. [2]Center for solid Waste Research, Institute for Environmental Research, Tehran University of Medical Science, Tehran, Iran.

References
1. Pan B, Qiu H, Pan B, Nie G, Xiao L, Lv L, et al. Highly efficient removal of heavy metals by polymer-supported nanosized hydrated Fe (III) oxides: behavior and XPS study. Water Res. 2010;44(3):815–24.
2. Mondiale B. Pollution prevention and abatement handbook-Toward cleaner production. Washington, DC: The World Bank; 1998. p. 457.
3. Mahvi A. Application of agricultural fibers in pollution removal from aqueous solution. Inter Environ Sci Tech. 2008;5(2):275–85.

4. Mbarek S, Saidi T, Mansour HB, Stéphane MP, Rostene W, Chaouacha-Chekir RB. Effect of cadmium on water metabolism regulation by Meriones shawi (Rodentia, Muridae). Environm Eng Sci. 2011;28(3):237–48.

5. Shils ME, Shike M. Modern nutrition in health and disease. Baltimore: Lippincott Williams & Wilkins; 2006.

6. WHO. Guidelines for drinking-water quality: recommendations. Geneva: World Health Organization; 2004.

7. Eddy M, Tchobanoglous G, Burton F, Stensel H. Theory and practice of water and wastewater treatment. New York: J. Wiley; 2003.

8. Ehrampoush MH, Masoudi H, Mahvi AH, Salmani MH. Prevalent kinetic model for Cd (II) adsorption from aqueous solution on barley straw Fresenius. Environ Bulletin. 2013;22(8):2314–8.

9. Fu F, Wang Q. Removal of heavy metal ions from wastewaters: a review. J Environ Manag. 2011;92(3):407–18.

10. Engates KE, Shipley HJ. Adsorption of Pb, Cd, Cu, Zn, and Ni to titanium dioxide nanoparticles: effect of particle size, solid concentration, and exhaustion. Environm Sci and Pollution Res. 2011;18(3):386–95.

11. Sheela T, Nayaka YA. Kinetics and thermodynamics of cadmium and lead ions adsorption on NiO nanoparticles. Chem Eng J. 2012;191:123–31.

12. Salmani MH, Ehrampoush MH, Aboueian-Jahromi M, Askarishahi M. Comparison between Ag (I) and Ni (II) removal from synthetic nuclear power plant coolant water by iron oxide nanoparticles. J Environ Health Sci Eng. 2013;11(21):1.

13. Xu P, Zeng GM, Huang DL, Feng CL, Hu S, Zhao MH, et al. Use of iron oxide nanomaterials in wastewater treatment: a review. Sci Total Environment. 2012;424:1–10.

14. Gong J, Wang X, Shao X, Yuan S, Yang C, Hu X. Adsorption of heavy metal ions by hierarchically structured magnetite-carbonaceous spheres. Talanta. 2012;101:45–52.

15. Mosaferi M, Nemati S, Khataee A, Nasseri S, Hashemi AA. Removal of Arsenic (III, V) from aqueous solution by nanoscale zero-valent iron stabilized with starch and carboxymethyl cellulose. J Environ Health Sci Eng. 2014;12(1):74.

16. Rafati L, Mahvi A, Asgari A, Hosseini S. Removal of chromium (VI) from aqueous solutions using Lewatit FO36 nano ion exchange resin. Inter J Environm Sci Tech. 2010;7(1):147–56.

17. Bazrafshan E, Mahvi AH, Naseri S, Mesdaghinia AR. Performance evaluation of electrocoagulation process for removal of chromium (VI) from synthetic chromium solutions using iron and aluminum electrodes. Turk J Eng Environ Sci. 2008;32(2):59–66.

18. Nouri J, Mahvi A, Bazrafshan E. Application of electrocoagulation process in removal of zinc and copper from aqueous solutions by aluminum electrodes. Inter J Environ Res. 2010;4(2):201–8.

19. Ebrahimi R, Maleki A, Shahmoradi B, Daraei H, Mahvi AH, Barati AH, et al. Elimination of arsenic contamination from water using chemically modified wheat straw. Desalin Water Treat. 2013;51(10–12):2306–16.

20. Mahvi A, Gholami F, Nazmara S. Cadmium biosorption from wastewater by Ulmus leaves and their ash. European J Sci Res. 2008;23(2):197–203.

21. Mahvi AH, Ebrahimi SJA-d, Mesdaghinia A, Gharibi H, Sowlat MH. Performance evaluation of a continuous bipolar electrocoagulation/electrooxidation–electroflotation (ECEO–EF) reactor designed for simultaneous removal of ammonia and phosphate from wastewater effluent. J Hazard Mater. 2011;192(3):1267–74.

22. Lu AH, Salabas EL, Schüth F. Magnetic nanoparticles: synthesis, protection, functionalization, and application. Angew Chem Int Ed. 2007;46(8):1222–44.

23. Nedkov I, Merodiiska T, Slavov L, Vandenberghe R, Kusano Y, Takada J. Surface oxidation, size and shape of nano-sized magnetite obtained by co-precipitation. J Mag and Mag Mater. 2006;300(2):358–67.

24. He F, Zhao D. Preparation and characterization of a new class of starch-stabilized bimetallic nanoparticles for degradation of chlorinated hydrocarbons in water. J Environ Sci & Tech. 2005;39(9):3314–20.

25. Berger P, Adelman NB, Beckman KJ, Campbell DJ, Ellis AB, Lisensky GC. Preparation and properties of an aqueous ferrofluid. J Chemical Edu. 1999;76(7):943.

26. Nassar NN. Rapid removal and recovery of Pb (II) from wastewater by magnetic nanoadsorbents. J Hazard Mater. 2010;184(1):538–46.

27. Duan H, Wang D, Li Y. Green chemistry for nanoparticle synthesis. Chem Soc Rev. 2015;44:5778–92.

28. Kosa SA, Al-Zhrani G, Salam MA. Removal of heavy metals from aqueous solutions by multi-walled carbon nanotubes modified with 8-hydroxyquinoline. Chem Eng J. 2012;181:159–68.

29. Salmani M, Zarei S, Ehrampoush M, Danaie S. Evaluations of pH and high ionic strength solution effect in cadmium removal by zinc oxide nanoparticles. J Appl Sci Environ Manag. 2014;191:123–31.

30. Afkhami A, Saber-Tehrani M, Bagheri H. Simultaneous removal of heavy-metal ions in wastewater samples using nano-alumina modified with 2, 4-dinitrophenylhydrazine. J Hazard Mater. 2010;181(1):836–44.

31. Kakavandi B, Jonidi AJ, Rezaei RK, Nasseri S, Ameri A, Esrafily A. Synthesis and properties of FeO-activated carbon magnetic nanoparticles for removal of aniline from aqueous solution: equilibrium, kinetic and thermodynamic studies. J Environ Health Sci Eng. 2013;10(1):19. doi:10.1186/1735-2746-10-19.

32. Li X, Li A, Long M, Tian X. Equilibrium and kinetic studies of copper biosorption by dead Ceriporia lacerata biomass isolated from the litter of an invasive plant in China. J Environ Health Sci Eng. 2015;13:37. doi:10.1186/s40201-015-0191-1.

14

Enzymatic catalysis treatment method of meat industry wastewater using lacasse

K. Thirugnanasambandham and V. Sivakumar[*]

Abstract

Background: The process of meat industry produces in a large amount of wastewater that contains high levels of colour and chemical oxygen demand (COD). So they must be pretreated before their discharge into the ecological system.

Methods: In this paper, enzymatic catalysis (EC) was adopted to treat the meat wastewater.

Results: Box-Behnken design (BBD), an experimental design for response surface methodology (RSM), was used to create a set of 29 experimental runs needed for optimizing of the operating conditions. Quadratic regression models with estimated coefficients were developed to describe the colour and COD removals.

Conclusions: The experimental results show that EC could effectively reduce colour (95 %) and COD (86 %) at the optimum conditions of enzyme dose of 110 U/L, incubation time of 100 min, pH of 7 and temperature of 40 °C. RSM could be effectively adopted to optimize the operating multifactors in complex EC process.

Keywords: Enzymatic catalysis, Meat wastewater, Colour removal, Box-Behnken design, Model development, Process optimization

Background

Meat industry is the world's fastest growing sector due to ever increasing demand of its products. Meat processing industries use approximately 62 Mm^3/y of fresh water from river and canals [1]. Meat-based products have become an essential part of every day's life and its high demand has resulted in a large quantity of meat wastewater that needs to be treated in order to protect the environment and aquatic life [2]. The meat wastewater contains higher level of suspended solids and organic materials and these particles cannot be easily separated. For these reasons many attempts have been made to treat meat wastewater using conventional wastewater treatment methods [3]. There are a number of processes available for wastewater treatment such as chemical coagulation, electro coagulation, sedimentation precipitation, ozonation, evaporation, membrane filtration, adsorption, ion-exchange, oxidation and advanced oxidation, incineration, bio-degradation and biological treatment. Moreover, these conventional methods are also usually expensive and treatment efficiency is

inadequate because of the large variability of the composition of meat wastewater [4].

Enzymatic catalysis using Laccase (EC) is one of the most practiced technologies extensively used on industrial scale wastewater treatment. Meanwhile, the EC treatment can be simpler and more efficient than the traditional physical-chemical treatments [5]. Laccase has the advantages over conventional chemical or microbial catalysts such as biodegradable, high level of catalytic efficiency, high degree of specificity, easily removed from contaminated streams, easily standardized in commercial preparations and absence of side-reactions [6]. These characteristics provide substantial process energy savings and reduced manufacturing costs. Nevertheless, the efficiency of EC process depends on several factors including the enzyme dose, incubation time, pH and temperature. The optimization of these factors may significantly increase the process efficiency [7].

Traditionally, optimization in wastewater treatment has been carried out by monitoring the manipulate of one factor at a time on an experimental response. While only one factor is changed, others are kept at a constant level. This optimization technique is called one-variable-at-a-time. Its major disadvantage is that it does not include the

* Correspondence: drvsivakumar@yahoo.com
Department of Chemical Engineering, AC Tech Campus, Anna University, Chennai 600 025TN, India

combined effects among the variables studied [8]. As a consequence, this technique does not depict the complete effects of the parameter on the response. Another disadvantage of the one-factor optimization is the increase in the number of experiments necessary to conduct the research, which leads to an increase of time, man power and operating cost [9]. In order to overcome this problem, nowadays optimization has been carried out by using multivariate statistic techniques. Among the most relevant multivariate techniques used in optimization process wastewater treatment is response surface methodology (RSM). RSM is a collection of mathematical and statistical techniques based on the fit of a polynomial equation to the experimental data, which must portray the performance of a data set with the aim of making statistical previsions. It can be well applied when a response or a set of responses of interest are influenced by several variables. In RSM, Box–Behnken design (BBD) is a statistical technique for designing experiments, building models, evaluating the effects of several factors, and searching optimum conditions for desirable responses. The main advantage of this method of other statistical experimental design methods is the reduced number of experiments trials needed to evaluate multiple parameters and their interactions [10].

An extensive literature survey shows that there is lack of knowledge regarding the optimization of EC paramerets to treat meat industry wastewater using RSM. Hence, in this present study an attempt was made to investigate the optimize the EC process parameters such as enzyme dose, incubation time, pH and temperature on the colour and COD removals from meat industry wastewater using four factors three level Box-Behnken design (BBD). The results will obtain shows the treatment efficiency of EC and its possibility to implement in industrial scale level by analyzing removal efficiencies of colour and COD.

Methods

Raw materials and chemicals

Wastewater used in this study was collected from a meat industry located in Erode, Tamilnadu, India and its physico-chemical properties was determined and shown in Table 1. In this present study, analytical reagent grade

Table 1 Characteristics of meat industry wastewater

Characteristics	Value	Permissible values
pH	5.6	6–8
Colour (CU$_s$(Pt-Co)	223	5
COD (mg/l)	4658	500
Turbidity (NTU)	1568	10
Conductivity (mS/cm)	1.78	0.5
BOD (mg/l)	1685	100

chemicals such as citrate hydrogen phosphate and disodium hydrogen phosphate were used to adjust the pH, which was supplied by Merck chemicals, Chennai. Commercial laccase formulation (DeniLite® IIS; 120 U/g) produced from genetically modified *Aspergillus oryzae* was purchased from local suppliers, Erode.

Experimental setup

Jar-test experiments were conducted on meat wastewaters in different graduated glass beakers. A Jar containing 100 ml wastewater were tested at pH range between 5 and 9 with different enzyme dose (80–120 U/L). After each enzyme dose, sample was rapidly mixed at 180 rpm during 3 min and incubated (60–120 min) with different temperature (25–45 °C). After treatment, samples were centrifuged at 10,000 rpm for 15 min and analyzed for colour intensity and for COD.

Analytical method

Colour, COD, and conductivity were determined according to the standard methods described by American Public Health Association (APHA). Colour measurement was determined with standard dilution multiple method and by comparing absorbance to a calibration curve. Colour removal was determined by monitoring the decrease in the absorbance peak at the maximum wavelength (678 nm). Double beam UV-visible spectrophotometer (Shimadzu UV 1650 PC) was used in all experiments. COD was determined by open reflux method. The sample was refluxed in an acidic medium with a known excess of potassium dichromate ($K_2Cr_2O_7$) and the remaining dichromate was titrated with ferrous ammonium sulphate (FAS). Turbidity was measured using a turbidity meter (ELIKO, 456) in accordance with standard method via display. The conductivity was determined by conductivity meter (Lil120) via digital display. The removal efficiency (RE) of colur and COD was calculated by using the following equation [11]

$$RE = \left(\frac{c_0 - c_e}{c_0} \right) \times 100 \qquad (1)$$

where, c_0 and c_e is the initial and final concentrations of colour and COD respectively.

Stastical experimental design

In this present study, Box-Behnken response surface experimental design (BBD) with four factors at three levels was used to optimize and investigate the influence of process variables such as enzyme dose (A), incubation time (B), pH (C) and temperature (D) on enzymatic catalysis process to reduce colour (Y_1) and COD (Y_2) from from meat wastewater. Process variables and their ranges (Table 2) were determined based on the preliminary

Table 2 Process variables and their ranges

Process variables	Level		
	−1	0	1
A (U/L)	80	100	120
B (min)	60	90	120
C	5	7	9
D (°C)	25	35	45

Table 3 BBD experimental design with results

Run	A	B	C	D	Y_1	Y_2
1	120	90	5	35	85.48	69.83
2	100	90	7	35	89.42	79.85
3	120	120	7	35	89.36	79.79
4	120	90	7	25	84.42	74.85
5	80	90	5	35	59.42	55.85
6	120	90	7	45	90.36	80.79
7	100	60	5	35	42.12	39.55
8	100	60	7	25	34.56	24.99
9	100	90	7	35	89.42	79.85
10	80	90	7	25	49.72	48.15
11	100	90	9	25	55.66	50.09
12	100	60	9	35	44.74	39.17
13	100	90	5	45	73.03	67.46
14	100	90	9	45	96.42	86.85
15	100	120	9	35	74.42	67.85
16	80	60	7	35	31.94	22.37
17	100	90	7	35	89.42	79.85
18	100	120	7	45	73.36	63.79
19	100	90	7	35	89.42	79.85
20	100	90	7	35	89.42	79.85
21	80	120	7	35	63.92	50.35
22	100	60	7	45	64.46	54.89
23	100	120	5	35	77.72	66.15
24	100	90	5	25	68.72	59.15
25	80	90	7	45	74.42	66.85
26	120	90	9	35	77.42	67.85
27	120	60	7	35	59.38	49.73
28	100	120	7	25	75.42	65.85
29	80	90	9	35	64.42	50.85

studies. Preliminary studies were carried out using Placket Burmann (PB) design. BBD design consists of 29 experiments with five centre points were designed and the data was analyzed by multiple regression analysis (Sequential sum of squares and model summary statistics) in order to study the ability of various mathematical models to express the enzymatic catalysis process [12].

All the statistical analyses were done with the help of Stat ease Design Expert 8.0.7.1 statistical software package (Stat-Ease Inc., Minneapolis, USA). Then the adequacy of mathematical model was analysed with various statistical analysis such as determination coefficient (R^2), adjusted determination of coefficient (R_a^2), predicted determination of coefficient (R_p^2), adequate precision (AP) and coefficient of variation (CV). Then, the individual and combined effects of process parameters on responses were studied by constructing three dimensional (3D) response surface plots from polynomial model [13].

Optimization of process variables for maximum colour and COD was carried out by derringer's desired function methodology. In this present study, goals of the operating conditions were selected as in a range and the responses goal was selected as maximize. After optimization, adequacy of the model equation for predicting the optimum response value was validated [14]. Triplicate verification experiments were performed under the optimal conditions and the average value of the experiments was compared with the predicted value of the developed model equation [15].

Results and discussions

In this present study, removal of colour and COD from meat industry wastewater are investigated using enzymatic catalysis method under various operating conditions such as enzyme dose, incubation time, pH and temperature. Four factors with three levels BBD response surface design (BBD) is used to examine and optimize the process variables. BBD experimental design consists of 29 experiments are carried out and the results are shown in Table 3.

Mathematical modelling

In order to select the suitable mathematical among various models such as linear, interactive, quadratic and cubic, BBD experimental data is analyzed by multi regression analysis namely the sequential model sum of squares and model summary statistics (Table 4). The obtained results indicates that, linear and interactive (2FI) models were exhibited lower R^2, adjusted R^2, predicted R^2 and also having high p-values, when compared with quadratic model. Cubic model was found to be aliased [16]. Therefore the quadratic model ischosen to describe the effects of process variables on the enzymatic catalysis treatment method. Mean while, second-order polynomial equation has been developed by fitting coefficient of process variables (individual and interactive) to the

Table 4 Sequential model sum of squares and model summary statistics for responses

Source	Sum of squares	DF	Mean Square	F Value	Prob > F	Remarks
Sequential model sum of squares for colour removal (%)						
Mean	146045.62	1	146045.62			
Linear	5202.01	4	1300.50	8.09	0.0003	
2FI	727.90	6	121.32	0.70	0.6548	
Quadratic	2998.97	4	749.74	80.57	<0.0001	Suggested
Cubic	129.16	8	16.15	86.70	<0.0001	Aliased
Residual	1.12	6	0.19			
Total	155104.77	29	5348.44			
Sequential model sum of squares for COD removal (%)						
Mean	112009.84	1	112009.84			
Linear	4385.38	4	1096.35	7.14	0.0006	
2FI	502.86	6	83.81	0.47	0.8186	
Quadratic	2934.27	4	733.57	41.62	<0.0001	Suggested
Cubic	191.83	8	23.98	2.62	0.1285	Aliased
Residual	54.92	6	9.15			
Total	120079.10	29	4140.66			
Source	Std.Dev.	R^2	Adjusted R^2	Predicted R^2	PRESS	Remarks
Model summary statistics for colour removal (%)						
Linear	12.6773	0.5742	0.5033	0.4257	5202.4046	
2FI	13.1851	0.6546	0.4627	0.2757	6561.6618	
Quadratic	3.0505	0.9856	0.9712	0.9172	750.4137	Suggested
Cubic	0.4315	0.9999	0.9994	0.9822	160.8996	Aliased
Model summary statistics for COD removal (%)						
Linear	12.3893	0.5435	0.4674	0.3806	4998.1928	
2FI	13.2937	0.6058	0.3868	0.1276	7039.4443	
Quadratic	4.1982	0.9694	0.9388	0.8239	1421.2648	Suggested
Cubic	3.0253	0.9932	0.9682	0.0200	7907.8116	Aliased

generalized quadratic model and the final model obtained in terms of coded factors are given below

$$Y_1 = 89.42 + 11.88A + 14.75B + 0.55C + 8.63D - 0.50AB - 3.27AC - 4.69AD \\ -1.48BC - 7.99BD + 9.11CD - 8.05A^2 - 20.41B^2 - 9.38C^2 - 6.76D^2 \tag{2}$$

$$Y_2 = 79.85 + 10.70A + 13.59B + 0.39C + 8.13D + 0.52AB + 0.75AC - 3.19AD \\ +0.52BC - 7.99BD + 7.11CD - 8.73A^2 - 20.33B^2 - 8.30C^2 - 5.42D^2 \tag{3}$$

Where, Y_1 and Y_2 are colour and COD removal (%) respectively; A, B, C and D are enzyme dose, incubation time, pH and temperature respectively.

Suitability of developed mathematical models

The predicting ability of the developed mathematical model are evaluated by constructing analytical plots (Fig. 1) such as residuals vs run, predicted versus actual

plot to find out the connection between predicted and experimental values and also to know the fitness of the model. From the Fig. 1, it is observed that, residuals for the prediction of each response is minimum and it indicated a good adequate agreement between experimental data and the data predicted by model. This result confirms the normal distribution of the observed data and adequacy of the developed model [17]. Statistical significance of the developed mathematical model is examined using pareto analysis of variance (ANOVA) with corresponding F and p-values of process variables, which is shown in Table 5. The higher model F value and lower p-values $(p < 0.0001)$ confirmed that, the developed model is significant. The goodness of fit of the model is also evaluated by the determination co-efficient of variance (CV), adequate precision (AP) and PRESS, which clearly stated that, the deviations between experimental and predicted values are low and confirms the reliability of the conducted experiments [18].

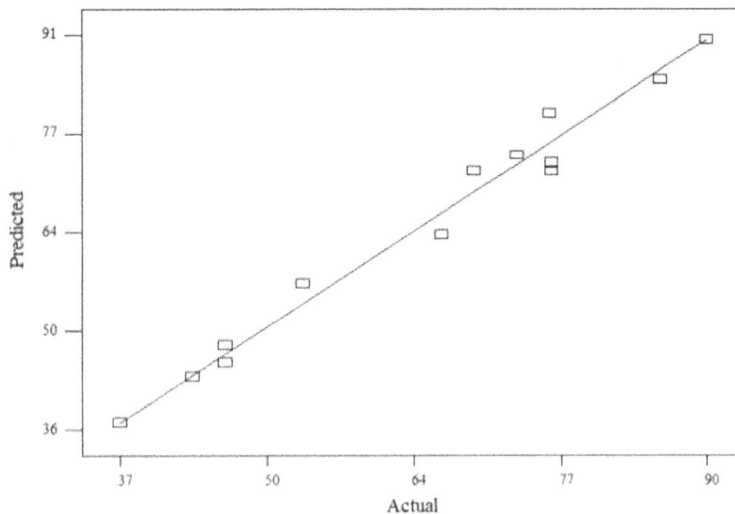

Fig. 1 Model adequacy plot

Influence of process parameters

Three dimensional (3D) response surface plots are plotted from the developed mathematical models in order to study the individual and combined effect of process variables on the responses to treat meat industry wastewater. In this present study, the model has more than two factors. So, the 3D plots are drawn by maintaining one factor at a constant level (in turn at its central level),

whereas the other two factors were varied in their range, which are shown in Fig. 2.

Effect of enzyme dose

Enzyme dose is one of the crucial parameter, which affects the performance of the enzymatic catalysis for treating meat wastewater significally. So that, experiments were carried out to study the effect of enzyme dose (80, 100 and 120 U/L) over the colour and COD removal and the results are shown in Fig. 2a–c. From the experimental results, it is observed that, the colour and COD removals were increased linearly with increasing enzyme dose upto 110 U/L. This is mainly due to the fact that by increasing the enzyme dose, an increase in the number of active sites takes place. At higher concentration of the enzyme the inhibitors will fall short. More active sites will reduce the colur and COD in the given period of time thus treatment efficiency is enhanced [19]. However, it is noticed that beyond enzyme dose of 110 U/L shows negligible effect on treatment efficiency.

Effect of incubation time

Incubation time is one of the important factor for the treatment of meat industry wastewater using enzymatic catalysis method. In order to investigate the effect of incubation time, experiments were carried out various incubation time (60, 90 and 120 min) and results are shown in Fig. 2a–d. From the results, it could be found that, the colour and COD removals were increased linearly with increasing incubation time upto 100 min. This is mainly due to the fact that, increase in enzyme dose would increase the reaction kinetic; this happens because free activation centers of the enzyme bind to free substrates thus removal efficiency are increased

Table 5 ANOVA results for responses

Source	Colour removal (%)		COD removal (%)	
	F-value	P value	F-value	P value
Model	68.54	<0.0001	31.70	<0.0001
A	182.05	<0.0001	77.98	<0.0001
B	280.55	<0.0001	125.75	<0.0001
C	0.39	0.5429	0.10	0.7529
D	96.02	<0.0001	44.99	<0.0001
AB	0.11	0.7479	0.06	0.8079
AC	4.58	0.0504	0.13	0.7245
AD	9.45	0.0082	2.31	0.1508
BC	0.94	0.3484	0.06	0.8079
BD	27.44	0.0001	14.49	0.0019
CD	35.69	<0.0001	11.48	0.0044
A^2	45.15	<0.0001	28.04	0.0001
B^2	290.24	<0.0001	152.04	<0.0001
C^2	61.38	<0.0001	25.38	0.0002
D^2	31.87	<0.0001	10.82	0.0054
C.V. %	4.30		4.85	
PRESS	754.08		596.54	
AP	27.65		31.54	

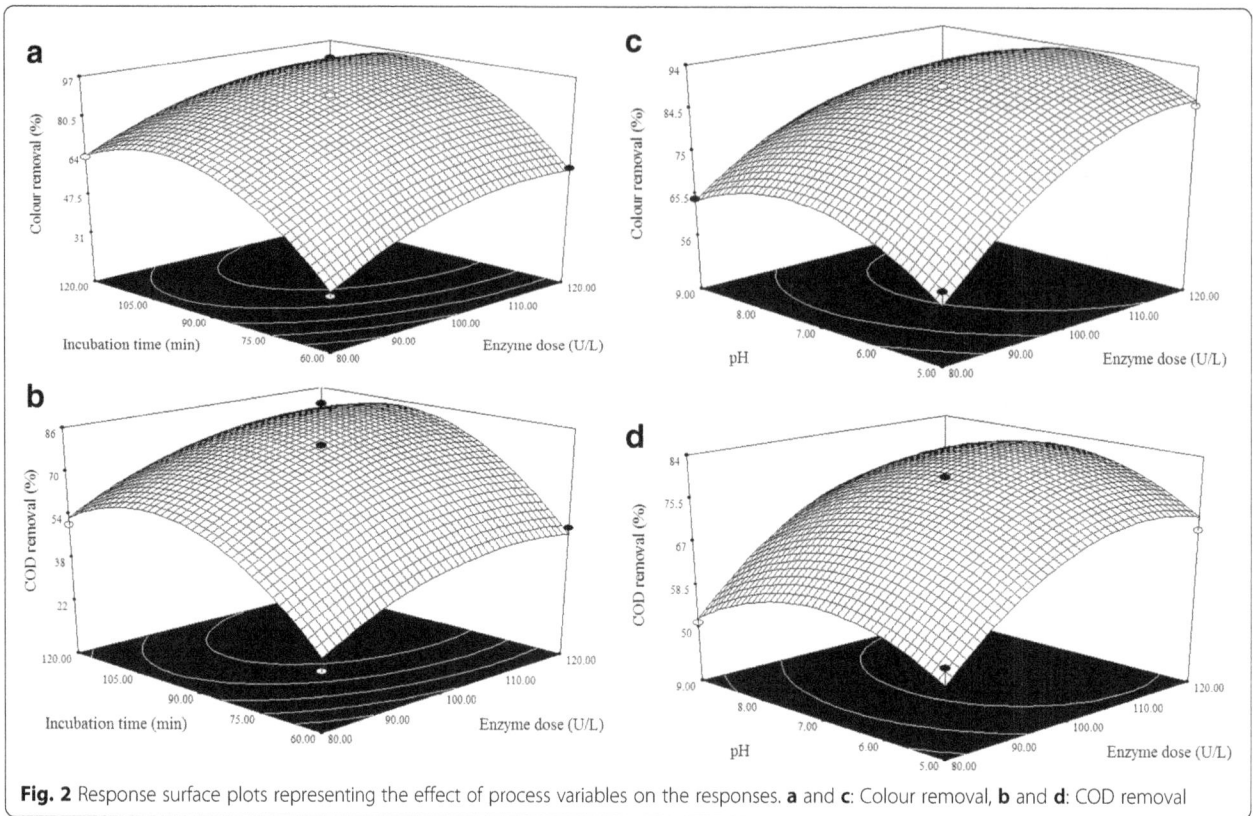

Fig. 2 Response surface plots representing the effect of process variables on the responses. **a** and **c**: Colour removal, **b** and **d**: COD removal

[20]. Thereafter, there is a negligible effect on the colour and COD removal efficiencies.

Effect of pH

The pH is also an important factor influences the treatment of meat wastewater using enzymatic catalysis method. The concentration of hydrogen ions in wastewater affects the enzyme activity. Each enzyme has maximal efficiency under an optimum pH otherwise there will be no enzymatic activity due to denaturation of enzymes. Therefore, in this study influence of pH on the colour and COD removal efficiencies was examined by varying its range (5, 7 and 9) and the results are illustrated in Fig. 3a–c. From the results, it is observed that, colour and COD removal efficiencies was increased with the increasing pH upto 8, due to the increase in the activity of enzyme and enhanced the treatment efficiency [21]. Beyond pH of 8 shows the negligible effect on removal efficiency of colour and COD.

Effect of temperature

Removal efficiency of colour and COD from meat wastewater using enzymatic catalysis method is highly affectd by temperature and its influence on treatment efficiency is investigated by varying temperature (25, 35 and 45 °C) and the results are depicted in Fig. 3a–d. From the results, it is found that, colour and COD removal efficiencies were increased with the increasing temperature upto 40 °C and it can be explained the fact that, there are distinct temperature ranges under which enzymes operate and there is a specific temperature levels (optimum temperature) in which enzymes have maximum efficiency. Therefore temperature variations affect enzymatic activity and the kinetic of the reactions they catalyze. In addition, enzymes can be denatured under extreme temperatures and loses their catalytic activity [22]. These results indicates the key role of temperature on enzymatic catalysis method process for colour and COD removals.

Optimization and validation

For optimization, simultaneous optimization of the multiple responses is carried out using Derringer's desired function methodology in order to find out the optimum operating conditions for maximum removal efficiencies of colour and COD. This numerical optimization technique evaluates a point that maximizes the desirability function and optimum operating conditions were found to be as follows: enzyme dose of 110 U/L, incubation time of 100 min, pH of 7 and temperature of 40 °C. Under these conditions, the experimental results show that EC could effectively reduce the colour (95 %) and COD (86 %). Then, the suitability of optimum

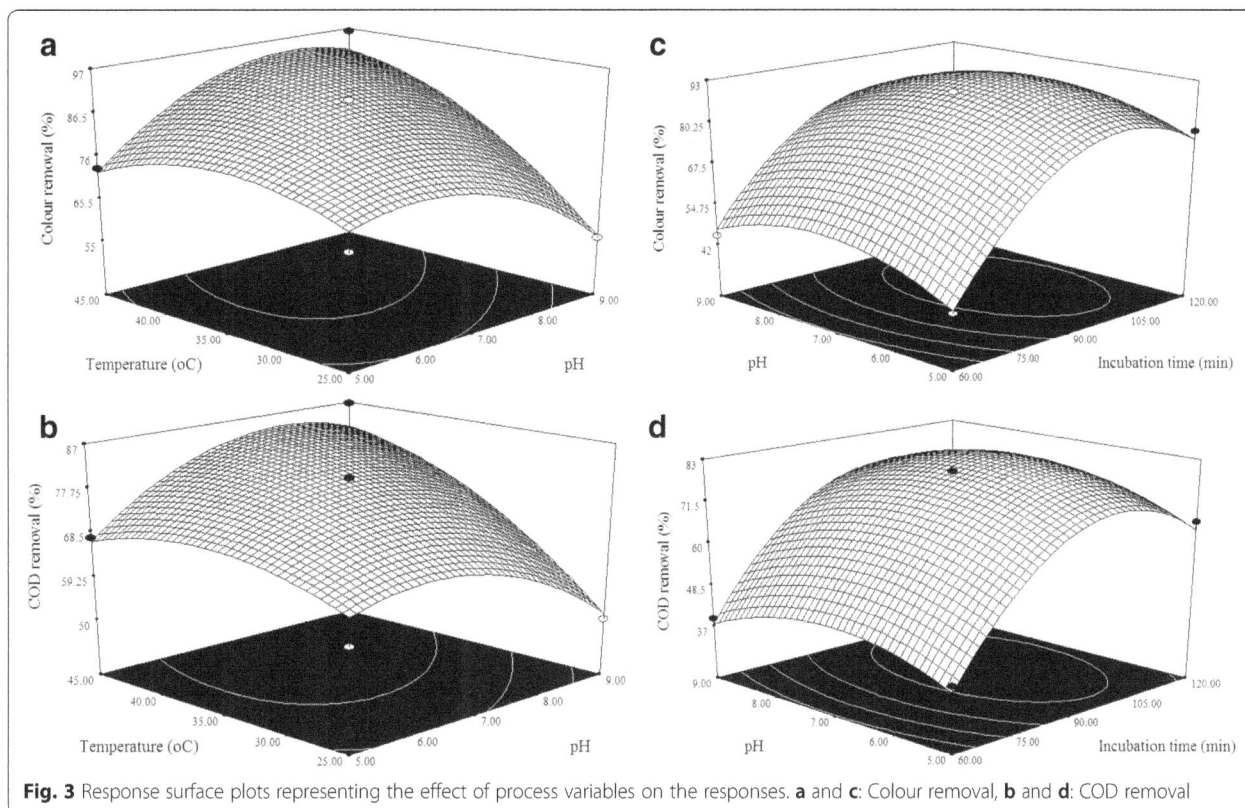

Fig. 3 Response surface plots representing the effect of process variables on the responses. **a** and **c**: Colour removal, **b** and **d**: COD removal

conditions for predicting optimum response value is tested based on above mentioned conditions. Triplicate experiments were performed under the optimized conditions and the mean value (95.35, 85.68 % for colour and COD removal respectively) obtained from real experiments, demonstrated the validation of the optimized conditions.

Conclusion

In this study, BBD was employed to study and optimize the process variables such as enzyme dose, incubation time, pH and temperature on the removal of colour and COD from meat wastewater using enzymatic catalysis method. From the results, it was observed that, all the process variables have significant effects on the treatment efficiency and quadratic model were developed for predicting the responses. Optimum set of the independent variables was obtained by derringer's desired function methodology in order to find out the maximum colour (95 %) and COD (86 %) removal efficiencies and it was found to be: enzyme dose of 110 U/L, incubation time of 100 min, pH of 7 and temperature of 40 °C. These results indicates that the proposed enzymatic catalysis process is an effective and economically viable method to treat meat industry wastewater.

Competing interests
The authors declare that they have no competing interests.

Authors' contributions
Both authors read and approved the final manuscript.

Acknowledgments
The authors are thankful to University Grant Commission (F.No:39-853/2010), Government of India, for financial support to fabricate and use the experimental setup.

References
1. Yi Jing C, Mei Fong C, Chung Lim L, Hassell DG. A review on anaerobic–aerobic treatment of industrial and municipal wastewater. Chem Eng J. 2009;155:1–18.
2. Eriksson E, Auffarth K, Eilersen AM, Henze M, Ledin A. Household chemicals and personal care products as sources for xenobiotic organic compounds in grey wastewater. Water SA. 2003;29:135–46.
3. Elmitwalli TA, Shalabi M, Wendland C, Otterpohl R. Grey water treatment in UASB reactor at ambient temperature. Water Sci Technol. 2007;55:173–80.
4. Shin HS, Lee SM, Seo IS, Kim GO, Lim KH, Song JS. Pilot-scale SBR and MF operation for the removal of organic and nitrogen compounds from greywater. Water Sci Technol. 1998;38:79–88.
5. Lesjean B, Gnirss R. Greywater treatment with a membrane bioreactor operated at low SRT and low HRT. Desalination. 2006;199:432–4.
6. Li Z, Gulyas H, Jahn M, Gajurel DR, Otterpohl R. Greywater treatment by constructed wetland in combination with TiO$_2$-based photocatalytic oxidation for suburban and rural areas without sewer system. Water Sci Technol. 2003;48:101–6.
7. Bektas N, Akbulut H, Inan H, Dimoglo A. Removal of phosphate from aqueous solutions by electro-coagulation. J Hazard Mater. 2004;106:101–5.
8. Adhoum N, Monser L, Bellakhal N, Belgaied JE. Treatment of electroplating wastewater containing Cu^{2+}, Zn^{2+} and Cr (VI) by electrocoagulation. J Hazard Mater. 2004;B112:207–13.
9. Sridhar R, Sivakumar V, Prince Immanuel V, Prakash Maran J. Treatment of pulp and paper industry bleaching effluent by electrocoagulation process. J Hazard Mater. 2011;186:1495–502.

10. Lai CL, Lin SH. Treatment of chemical mechanical polishing wastewater by electrocoagulation: system performances and sludge settling characteristics. Chemosphere. 2004;54:235–42.

11. Kobya M, Can OT, Bayramoğlu M. Treatment of textile wastewaters by electrocoagulation using iron and aluminum electrodes. J Hazard Mater. 2003;B100:163–78.

12. İnan H, Dimoglo A, Simşek H, Karpuzcu M. Olive oil mill wastewater treatment by means of electro-coagulation. Sep Purif Technol. 2004;36:23–31.

13. Fathollah GB, Amir HM, Simin N, Mohammad AF, Ramin N, Mahmood A. Application of immobilized horseradish peroxidase for removal and detoxification of azo dye from aqueous solution. Res J Chem Environ. 2011;15:73–8.

14. Olmez T. The optimization of Cr(VI) reduction and removal by electrocoagulation using response surface methodology. J Hazard Mater. 2009;162:1371–8.

15. Fathollah GB, Amir HM, Simin N, Mohammad AF, Ramin N, Mahmood A. Enzymatic treatment and detoxification of acid orange 7 from textile wastewater. Appl Biochem Biotechnol. 2011;165:1274–84.

16. Behbahani M, Alavi Moghaddam MR, Arami M. Techno-economical evaluation of fluoride removal by electrocoagulation process: Optimization through response surface methodology. Desalination. 2011;271:209–18.

17. Prakash Maran J, Manikandan S, Thirugnanasambandham K, Nivetha CV, Dinesh R. Box-Behnken design based statistical modeling for ultrasound-assisted extraction of corn silk polysaccharide. Carbohyd Polym. 2013;92:604–11.

18. Sridhar R, Sivakumar V, Prince Immanuel V, Prakash Maran J. Development of model for treatment of pulp and paper industry bleaching effluent using response surface methodology. Environl Prog Sustain Energy. 2012;31:558–64.

19. Fathollah GB, Mohammad AF, Fatemeh NB, Amir HM. Oxidative degradation and detoxification of textile azo dye by horseradish peroxidase enzyme. Fresenius Environ Bull. 2013;22:1865–73.

20. Prakash Maran J, Sivakumar V, Sridhar R, Thirgananasambandham K. Development of model for barrier and optical properties of tapioca starch based films. Carbohyd Polym. 2013;92:1335–47.

21. Mahesh S, Prasad B, Mall ID, Mishra IM. Electrochemical degradation of pulp and paper mill waste water COD and color removal. Ind Eng Chem Res. 2006;45:2830–9.

22. Mayank K, Infant Anto Ponselvan F, Jodha Ram M, Vimal Chandra S, Indra Deo M. Treatment of bio-digester effluent by electrocoagulation using iron electrodes. J Hazard Mater. 2009;165:345–52.

15

Sonocatalytic degradation of humic acid by N-doped TiO$_2$ nano-particle in aqueous solution

Hossein Kamani[1], Simin Nasseri[1,2], Mehdi Khoobi[3], Ramin Nabizadeh Nodehi[1] and Amir Hossein Mahvi[1,4,5*]

Abstract

Background: Un-doped and N-doped TiO$_2$ nano-particles with different nitrogen contents were successfully synthesized by a simple sol–gel method, and were characterized by X-ray diffraction, field emission scanning electron microscopy, Energy dispersive X-ray analysis and UV–visible diffuse reflectance spectra techniques. Then enhancement of sonocatalytic degradation of humic acid by un-doped and N-doped TiO$_2$ nano-particles in aqueous environment was investigated. The effects of various parameters such as initial concentration of humic acid, N-doping, and the degradation kinetics were investigated.

Results: The results of characterization techniques affirmed that the synthesis of un-doped and N-doped TiO$_2$ nano-particles was successful. Degradation of humic acid by using different nano-particles obeyed the first-order kinetic. Among various nano-particles, N-doped TiO$_2$ with molar doping ratio of 6 % and band gap of 2.92 eV, exhibited the highest sonocatalytic degradation with an apparent-first-order rate constant of 1.56×10^{-2} min^{-1}.

Conclusions: The high degradation rate was associated with the lower band gap energy and well-formed anatase phase. The addition of nano-catalysts could enhance the degradation efficiency of humic acid as well as N-doped TiO$_2$ with a molar ratio of 6 %N/Ti was found the best nano-catalyst among the investigated catalysts. The sonocatalytic degradation with nitrogen doped semiconductors could be a suitable oxidation process for removal of refractory pollutants such as humic acid from aqueous solution.

Keywords: Humic acid, N-doped TiO$_2$, Sonocatalytic degradation

Background

Humic substances, as part of natural organic matters, have been a major issue in water treatment plants due to their non-biodegradability and their water-soluble formation [1, 2]. These substances can affect the water quality such as odor, taste and color. It has been also confirmed that these substances act as precursors to form disinfection by-products when water treated with chlorine [1, 3, 4]. Hence, removal of humic substances has been widely investigated for the protection of public health. In water treatment plants, portion of these substances are removed from raw water by conventional methods such as; coagulation, precipitation, filtration and adsorption [5–7]. Wang et al. reported that the removal of humic substances by using conventional processes is only 5-50 % [8].

In addition, application of high coagulant dosage isn't reasonable due to high cost operation and problem in sludge disposal. Besides, the presence of humic substances in water may reduce the efficiency of water treatment processes when membranes or microporous adsorbents are applied.

Chemical degradation is one of the best technologies that have been widely accepted for removal of humic substances [3, 9, 10]. Recently, sonolysis process attracted considerable attention as an advanced oxidation process (AOP) for degradation of pollutants in water [11–14]. However, this method consumes considerable energy and its efficiency is low compared to other methods. In order

* Correspondence: ahmahvi@yahoo.com
[1]Department of Environmental Health Engineering, School of Public Health, Tehran University of Medical Sciences, Tehran, Iran
[4]Center for Solid Waste Research, Institute for Environmental Research, Tehran University of Medical Sciences, Tehran, Iran
Full list of author information is available at the end of the article

to increase the degradation efficiency semiconductors have been added to the sonolysis processes [15, 16].

In recent years, application of heterogeneous sono-catalysis using TiO2 has become an environmentally sustainable treatment and cost-effective option for degradation of pollutants. Moreover, TiO2 is the most suitable photocatalyst for water treatment due to its high photocatalytic activity, long-time stability, relative low cost and non-toxicity [17–19]. It is well known that mechanism of sonocatalysis is similar to the photocatalysis [20, 21]. Thus, various techniques, including dye sensitization, semiconductor coupling and doping with metal and non-metal elements may enhance the sonoactivity of TiO2. According to previous studies, the doping of TiO2 with non-metal has been verified to be the most feasible method to improve photocatalytic activity of this catalyst [22]. It is also important to mention that the doping with nitrogen may be more effective than other non-metals because of its comparable atomic size with oxygen and small ionizing energy [23].

In the present study, un-doped and N-doped TiO$_2$ nano-particles with different nitrogen contents were successfully synthesized by a simple sol–gel method and were characterized by X-ray diffraction (XRD), field emission scanning electron microscopy (FE-SEM), energy dispersive X-ray analysis (EDX) and UV–visible diffuse reflectance spectra (UV-vis DRS) techniques.

The sonocatalytic activity of the as-synthesized TiO$_2$ for degradation of humic acid was investigated under ultrasonic irradiation with respect to the effects of nitrogen doping content, the initial concentration of humic acid and the addition of doped nanocatalyst into sonolysis process. Furthermore, the possible mechanism of sonocatalysis of N-doped TiO$_2$ was proposed.

Methods
Materials
Titanium tetraisopropoxide (TTIP, Ti(OC$_3$H$_7$)$_4$), Ethanol (EtOH), triethylamine, nitric acid (HNO$_3$), Hydrochloric acid (HCl) and sodium hydroxide (NaOH) were purchased from Merck Company, Germany, as analytical grade and were used without further purifying. Humic acid was purchased from Aldrich Company as sodium salt, and it was used after preparation. The stock solution of humic acid was prepared according to the methods [24]. The humic acid solution was prepared by addition of humic acid powder into deionized water and was heated up to 60 in order to accelerate the dissolution of humic acid. Then, the humic acid suspension cooled down to room temperature and was filtered through a 0.45-μm Milipore syringe filter. The residue of humic acid on the filter was dried in an oven at 105 until stable weight. The humic acid in filtered solution was

calculated by gravimetric method and stored as a stock solution for experimental use.

Synthesis of N-doped TiO$_2$
All catalyst samples were synthesized using a sol–gel method. To synthesize N-doped TiO$_2$ with a nominal molar doping of the dopant, 3 % "TN1", 6 % "TN2" and 12 % "TN3", 3 mL Titanium tetraisopropoxide and a certain amount of triethylamine was dissolved in 20 mL of ethanol, and the solution was stirred for 15 min (solution A). 2 mL deionized water was added into 10 mL of ethanol that contained nitric acid, this solution was also stirred for 15 min (solution B). Solution B was added drop wise to the solution A under magnetic stirring. After constantly stirring for 30 min, the semitransparent sol was obtained. Subsequently, the obtained semitransparent sol was set for 5 h at room temperature and then dried at 80 °C for 24 h in an oven. The dried powder was ground and calcinated under air at 500 °C for 1 h with a heating rate of 16 °C min^{-1}. For comparison, un-doped TiO$_2$ was also synthesized without the addition of dopant under the same conditions.

Characterization of N-doped TiO$_2$
In order to determine the effect of N-doping on the nano-particle structure, the analysis by X-ray diffraction (XRD), surface morphology, elemental analysis and photophysical properties were carried out. A Philips X'Pert X-ray Diffractometer with a diffraction angle range $2\theta =$ 10–70° using Cu Kα radiation ($\lambda = 1.5418$A) was used to collect XRD diffractograms. The accelerating voltage and emission current were 40 kV and 30 mA, respectively. The average crystallite size was determined according to the Scherrer equation using the full-width at half-maximum (FWHM) of the (1 0 1) peak. The UV–visible diffuse reflectance spectra (UV-DRS) were recorded using a UV–vis spectrophotometer (Avaspec-2048-TEC, Avantes, Netherland) with BaSO$_4$ as the reflectance standard. Then, the recorded data were converted to the absorbance units by using the Kubelka–Munk theory. The surface morphology and shape of the as-synthesized N-doped TiO$_2$ was observed through a field emission scanning electron microscope (FE-SEM, TESCAN) by gold-coated samples. Energy dispersive X-ray analysis (EDX) in the FE-SEM was also taken for the elemental analysis of the doped samples.

Sonocatalytic activity
Each suspension was prepared by adding 20 mg of each synthesized catalyst into a 100 mL of humic acid solution at concentrations 5, 10, and 20 mg L^{-1} in a reaction vessel. Prior to ultrasonic irradiation, the suspension was stirred using magnetic stirrer for 30 min in darkness to ensure a good dispersion and also to

complete adsorption/desorption equilibrium of humic acid on the catalyst surface. All experiments were carried out in laboratory scale and in batch system. The ultrasonic irradiation was generated by an Elma ultrasonic bath (model TI-H5) which was operated at a frequency of 130 kHz and a maximum output power of 100 W. During the sonocatalytic processes, the solution temperature was maintained at 25 ± 2 °C using a water cooling system in ultrasonic bath. After the desired reaction time, 5 mL aliquots were withdrawn at certain interval and centrifuged at 6000 rpm for 20 min to separate the catalysts by a centrifuge (Hettich, Germany, model D-78532). The residual humic acid concentration in supernatant solution was determined by UV-vis spectrophotometer (Perkin Elmer, USA) at 254 nm. For comparison of reaction rate among different condition, the kinetic model was used.

Results and discussion

X-ray diffraction pattern

An X-ray diffraction pattern was used to investigate the type of crystalline in material and also to know if any change was occurred after doping of TiO_2. Figure 1a shows the XRD patterns of the un-doped and N-doped TiO_2 samples. As shown in the XRD pattern, all synthesized samples had a sharp diffraction peak indicating a good characteristic crystal. The distinctive peaks at $2\theta = 25.49°$, $37.14°$, $37.99°$, $38.76°$, $48.35°$, $54.12°$, $55.33°$, 62, $90°$ and $68.95°$; correspond to the anatase (JCPDF Card No. 20-0387) were observed. The patterns also showed that the anatase was the main phase in un-doped and N-doped TiO_2 under all synthesis conditions.

These results revealed that the peak positions were nearly the same and no detectable dopant-related peaks

were observed, implying that the structure of TiO_2 has not been changed and also suggesting that nitrogen dopants do not react with TiO_2 to form new crystalline [25, 26]. It is noteworthy, that many documents have also reported that doping with the nitrogen ions have not exhibited additional phase except anatase [22, 27]. The pure anatase phase in N-doped TiO_2 could be due to the fact that the nitrogen dopants are so low and they have also moved into either the interstitial positions or into the substitution sites of the TiO_2 crystal structure [25, 28]. Compared to the un-doped TiO_2, the peak of N-doped TiO_2 samples exhibited a slight shift toward the lower angle corresponding to (1 0 1) plane of anatase (Fig. 1b), indicating a lattice distortion of the N-doped TiO_2. These defects and disorderly state in the particles caused by nitrogen dopants are reported as key factor for absorption edge shift towards the visible-light region [25, 27].

The average crystallite size of un-doped TiO_2 and N-doped TiO_2 were calculated according to the Debye-Scherrer formula as the following:

$$D = \frac{k\lambda}{\beta cos\theta} \qquad (1)$$

where:
D = the average crystallite size,
k = a dimensionless shape factor (usually = 0.9),
λ = the wave length of the X-ray radiation (0.15418 nm for Cu Ka),
β = the full width at half-maximum of the diffraction, and
θ = the corresponding diffraction angle in degree [21].

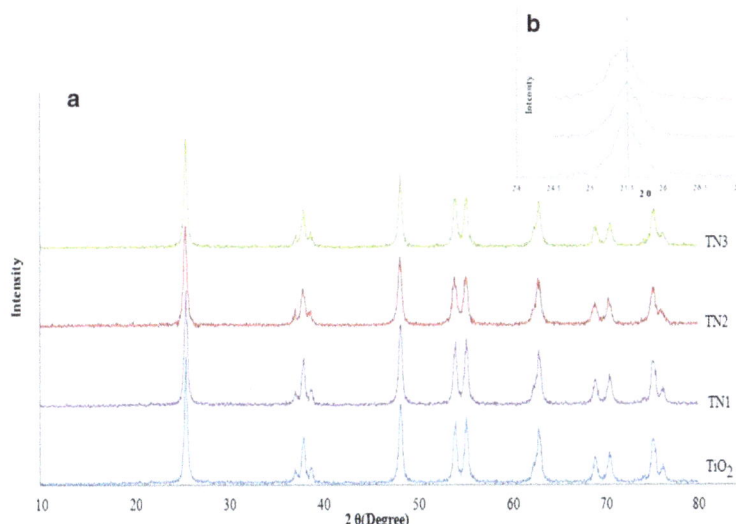

Fig. 1 XRD patterns (**a**) and lattice distortion (**b**) of un-doped and N-doped nano-particles

The calculated results were 30, 30, 26 and 34 nm for un-doped TiO_2, NT1, NT2 and NT3 nano-particles, respectively.

FE-SEM and EDX

The FE-SEM was used to show the shape and morphology of un-doped and *N*-doped TiO_2 particles (Fig. 2). The prepared nano-particles were found to be fine, irregular shape, slightly smooth surface and tend to agglomerate to form larger irregular grains. The diameter of particles was found to be 30-40 nm, which is in a good agreement with the crystal size obtained by XRD indicating that both un-doped and *N*-doped particle is nano-sized particles (Additional file 1: Figure S1). The energy dispersive X-ray Spectroscopy (EDX) of *N*-doped TiO_2 for different points of sample shows the appearance peaks of N, O and Ti atoms, which indicating that *N*-doped TiO_2 are mainly composed of these elements and confirm the N-doping process [29, 30]. The EDX spectra and the EDX elemental mapping (Additional file 1: Figure S2) also indicate no impurities in the samples and a good uniform distribution of N ions.

UV-vis diffuse reflectance spectra (UV-vis-DRS)

UV-visible diffuse reflectance spectra are the easiest and most convenient method to have a rough measure of the influence of doping [31]. As shown in Fig. 3a, doping of TiO_2 with nitrogen ion is clearly indicated by the reflectance spectra in the range of 300–700 nm. It is confirmed by various studies that N-doping has positive effect on the activity of the TiO_2 photocatalyst [31, 32]. As expected, N-doping caused a red shift from UV to the visible-light region. This red shift led to a better light absorption and consequently high radical generation and degradation efficiency.

Changing toward higher light absorption and red shift of absorption edge, which is in consistent with the yellowish color of nano-particles, can be attributed to narrowing of the band gap of synthesized nano-sized particles [26].

The band gap energies (E_g) of nano-sized particles can be determined according to the following equation [33]:

$$(\alpha h \nu) = A\left(h\nu - E_g\right)^r \tag{2}$$

where α is the absorption coefficient, h is Planck's constant, ν is the frequency of light, A is the absorption constant, E_g is the optical energy gap of the nano-sized particle and r is a number for characterizing the transition process, which is equal to 2 for indirect transition and 0.5 for direct transition. Therefore, the band gap energy of un-doped and *N*-doped TiO_2 can be determined from plots of the square root of $(\alpha h\nu)^{0.5}$ versus photon energy (Fig. 3 b).

The calculated optical band gaps were 3.02, 2.92, 2.91 and 3.09 eV for the TN1, TN2, TN3 and un-doped TiO_2 nano-particles, respectively. In all synthesized nano-particles the optical band gaps were lower than the band gap of commercial TiO_2 (3.2 eV) that is reported in various literatures [34]. This narrower band gap enhances transition of electrons from the valence band to the conduction band in the doped TiO_2 under ultrasonic irradiation and therefore it can increase sonocatalytic activities [34].

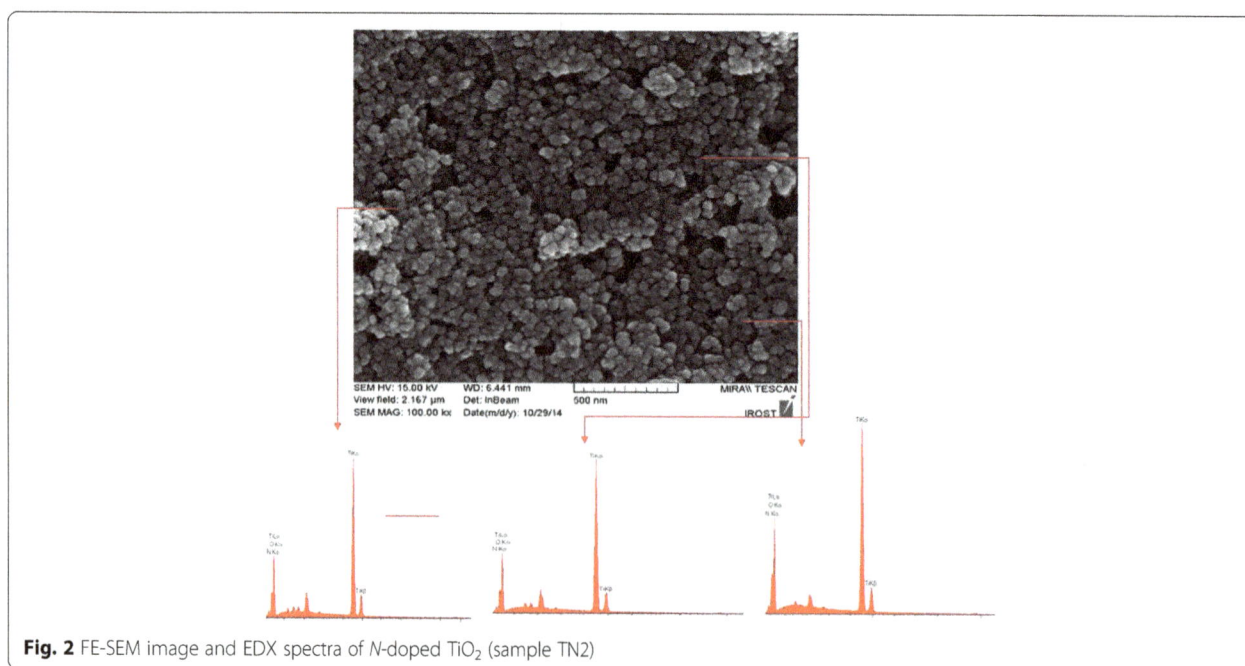

Fig. 2 FE-SEM image and EDX spectra of *N*-doped TiO_2 (sample TN2)

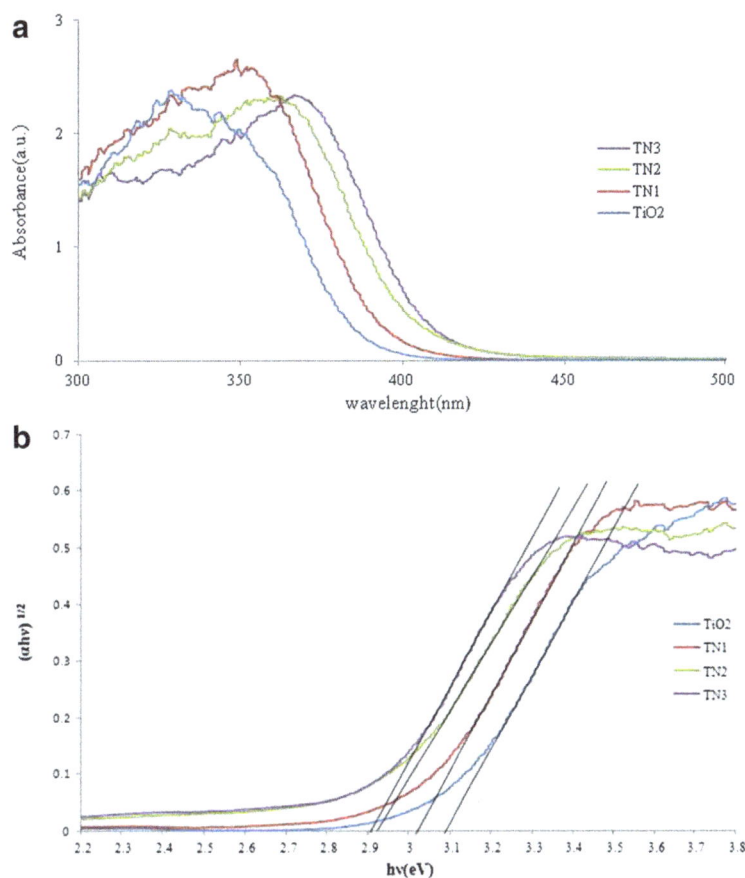

Fig. 3 a UV-Vis Diffuse Reflectance Spectra and (**b**) energy band gap of un-doped and *N*-TiO$_2$

The decrease in the band gap of *N*-doped TiO$_2$ can be attributed to the localized *N* 2p states in the structure of TiO$_2$ lattice in the form of substitutional and/or interstitial N states. It has been reported that substitutional N doping decreases the band gap by mixing of the O 2p and N 2p orbitals, while interstitial doping creates an additional state between the valence band and conduction band [22, 34].

Sonocatalytic performance of various sonocatalysts

The degradation of humic acid was studied using sonolysis, sonocatalysis with TiO$_2$, and sonocatalysis with different nitrogen contents doped in TiO$_2$. Figure 4 shows the degradation of humic acid under using different sonocatalysts at the neutral pH. The amount of adsorption for humic acid on the surface of the nano-particles was less than 3 % in darkness without ultrasonic irradiation; therefore it was negligible for un-doped and N-doped TiO$_2$.

As shown in Fig. 4, only 32 % of humic acid was degraded under ultrasonic irradiation after 90 min (without sonocatalyst), while the degradation efficiency of TiO$_2$, TN1, TN2 and TN3 sonocatalysts were 49.0, 55.0,

72.0 and 60.0.%, respectively. These results indicate that presence of sonocatalyst increases the degradation efficiency. This improvement could be due to this fact that the added sonocatalysts act as nuclei for bubble formation in aqueous solution as well as formation of oxygen vacancies in N-doped TiO$_2$ crystallite [15, 21]. These oxygen vacancies act as electron-trapping sites and prevent the recombination of hole-electron pairs, while, the additional amount of surface defect such as oxygen vacancies could increase the recombination of hole-electron pairs [21, 23].

As shown in Fig. 4, the highest sonocatalytic activity was achieved by TN2 with 72.0 % for humic acid degradation after 90 min of ultrasonic irradiation. According to the reported studies, the sonocatalytic activity of doped TiO$_2$ under ultrasonic irradiation is affected by different parameters such as surface area, phase of catalyst, oxygen vacancies, crystalinity of nano-particles and band gap energy [21, 23]. Therefore, the high sonocatalytic activity of TN2 could be attributed to the band gap narrowing resulting from doping of nitrogen and well-formed anatase phase. Figure 4 also indicates that the

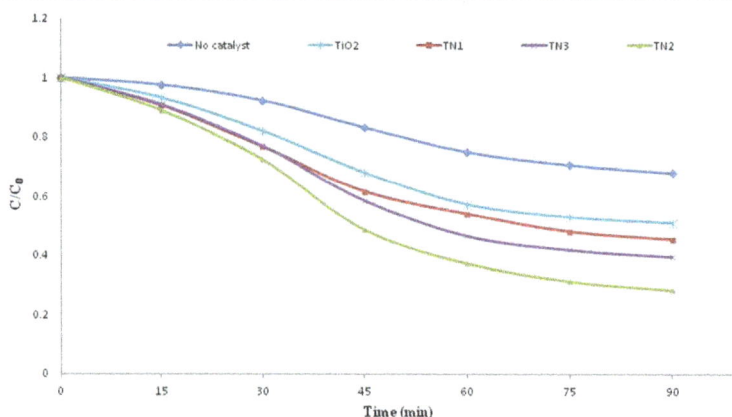

Fig. 4 Sonocatalytic degradation of humic acid for TiO_2 and different N-doped TiO_2 nano-particles

sonocatalytic activity of N-doped TiO_2 initially increased with the increase of N dopant but further increasing of dopant decreased the activity. Therefore for improvement of sonocatalytic activity of TiO_2, optimum amount of dopant is essential.

Kinetic study

The sonocatalytic degradation of humic acid can be well explained by a pseudo-first-order reaction and its kinetics can be expressed with the following equation:

$$\ln\left(\frac{C_0}{C}\right) = k_{app}t \qquad (3)$$

where k_{obs} is the apparent reaction rate constant, C_0 and C are the humic acid concentrations at initial and at time t, respectively. The k_{obs} were determined from the slopes of straight lines obtained by plotting $ln(C_0/C)$ versus irradiation time.

The values of apparent reaction rate constants (k_{app}) related to the various synthesized nano-sized particles are presented in Table 1. The correlation coefficients above 0.98 indicated the sonodegradation of humic acid by un-doped and N-doped TiO_2 suspensions obey the first-order kinetic model in solution. These results also indicated that reaction rate of humic acid can be improved by doping of nitrogen into the TiO_2 structure. The apparent reaction rate constant of un-doped TiO_2 was 0.84×10^{-2} min^{-1}, while the apparent reaction rate constant of TN2 was 1.56×10^{-2} min^{-1}. In addition, enhancement of the reaction rate constants of TN1, TN2 and TN3 were 1.98, 3.25 and 2.40 times higher than the reaction rate constant of sonolysis without catalyst, respectively. These results are in accordance with the those reported by Huang et al. [35] and Wu et al. [33] who studied the degradation of organic pollutants by un-doped TiO_2 and ion- doped TiO_2.

Effect of initial humic acid concentration

The initial concentration of solute in aqueous environment is a key factor on sonocatalytic degradation. As shown in Fig. 5, the degradation efficiency of humic acid increased with decrease in its initial concentration. Sonocatalytic degradation of humic acid with the initial concentrations of 5, 10, and 20 mg L^{-1} for 90 min lead to the conversion of 82.0, 76.0 and 68.0 % of humic acid, respectively. This result indicates that the high degradation efficiency could be obtained at lower humic acid concentration. Our results are in good agreement with the results reported in literature [36]. This result can be due to this fact that under the same conditions, the amount of formed radicals during the sonocatalytic reaction was equal in the entire volume of the solution; therefore, the reaction of humic acid molecules with radicals becomes more likely at lower humic acid concentrations [15].

Langmuir–Hinshelwood model is widely used for analysis of heterogeneous sonocatalytic degradation kinetics as well as to realize the dependence of observed initial reaction rate on the initial concentration of solute in the aqueous environment [9, 29, 37, 38]. The L-H kinetic model is defined as the following equation:

Table 1 Results of kinetic constant, k_{app}, relative increase and removal efficiency of different N-doped TiO_2

Catalyst	$k_{app}.10^{-2}$(min^{-1})	Relative increase	R^2	Removal efficiency after 90 min
Absent of catalyst	0.48	1.00	0.9868	32.0
TiO_2	0.84	1.75	0.9851	49.0
TN1	0.95	1.98	0.9895	55.0
TN2	1. 56	3.25	0.9869	72.0
TN3	1. 15	2.40	0.9846	60.0

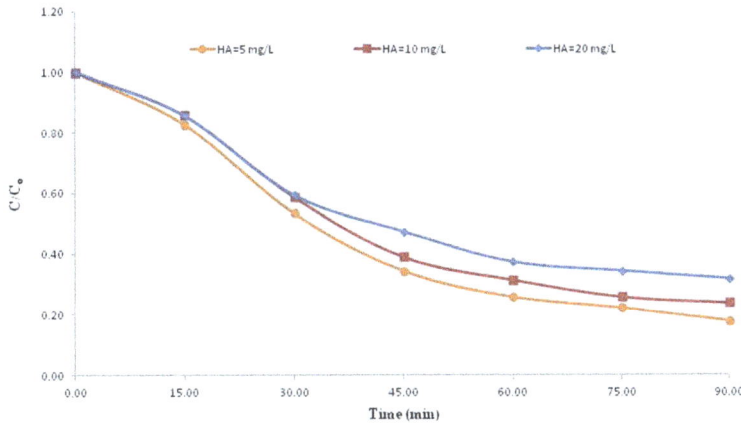

Fig. 5 Effect of initial humic acid concentration on sonocatalytic degradation of humic acid by *N*-doped TiO$_2$ (TN2) (catalyst concentration: 100 mg L^{-1})

$$r = -\frac{dc}{dt} = k_r\theta_x = \frac{k_rKC}{1+KC} \qquad (4)$$

where r is the reaction rate (mg L^{-1} min^{-1}), C is the concentration of solute at any time (mg L^{-1}), t is the reaction time (min), k_r is the Langmuir-Hinshelwood reaction rate constant, related to the limiting rate of reaction at maximum coverage for the experimental condition (mg L^{-1} min^{-1}) and K is the Langmuir adsorption constant reflecting the proportion of solute molecules which adhere to the catalyst surface (L mg^{-1}) and θ is the fraction of the surface of TiO$_2$ covered by solute. A linear expression of L-H model can be obtained by linearzing the Eq. (4) as follows:

$$\frac{1}{r_0} = \frac{1}{k_r} + \frac{1}{k_rKC_0} \qquad (5)$$

The parameters k_r and K which were calculated by plotting the reciprocal initial rate against the reciprocal initial concentration were 0.62 mg L^{-1} min^{-1} and 0.04 L mg^{-1},

respectively (Fig. 6). As shown in Fig. 6, from the correlation coefficient above 0.98 it could be observed that the experimental data are in good agreement with L-H model. According to the L-H model, the reaction is first order at low concentration and zero order at high concentration.

Possible mechanism

In sonolysis process, the sono-luminescene and localized hot-spots with high temperatures up to 5000 K and high pressures (approximately 1800 atm) caused by acoustic cavitation and collapse of micro-scale bubbles will occur [11, 12, 39]. These hot spots can pyrolysis water molecules to OH$^{'}$ and H$^{'}$ radicals as below Eq. (6):

$$H_2O +)))\rightarrow OH^{'} + H^{'} \qquad (6)$$

In addition, the sono-luminescene could induce the formation of flash light/energy which equals or exceeds the band gap energy of TiO$_2$ to excite the all synthesized nano-sized particles. The electron excitation from the

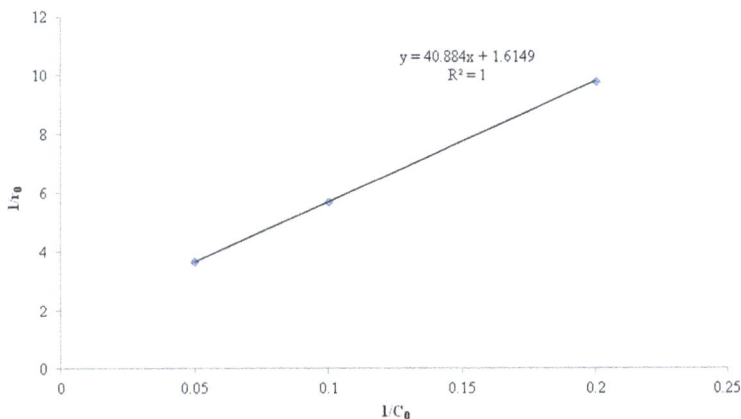

Fig. 6 Variation of reciprocal initial rate versus the reciprocal initial concentration of humic acid

local state of N 2p result in the generation of conduction band electrons (e–) and valence band holes (h$^{+)}$ as shown by Eqs. (7) and (8):

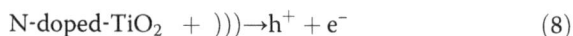

$$))) \rightarrow \text{light or energy} \qquad (7)$$

$$\text{N-doped-TiO}_2 \; + \;))) \rightarrow h^+ + e^- \qquad (8)$$

These charges migrate to the surface and finally react with a suitable electron donor and acceptor. The electrons are captured by Ti^{4+} to form Ti^{3+} states. Subsequently, the 3d orbital of Ti^{3+} ions are localized at 0.75–1.18 eV below the bottom of the conduction band. Ti^{3+} is known to be the most reactive site for oxidation process because it may cause more oxygen vacancy sites, as well as oxygen molecule is more easily adsorbed on TiO$_2$ surface. Besides, the electrons will react with these surface adsorbed oxygen molecules (O$_2$) to form superoxide radical anion (O$_2^{\prime}$) (Eq. 3) and is transformed further to hydroxyl radical (OH$^{\prime}$) as shown in Eqs. (9) – (14).

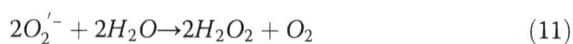

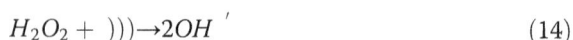

$$e^- + Ti^{4+} \rightarrow Ti^{3+} \qquad (9)$$

$$e^- + O_{2(ads)} \rightarrow O_2^{\prime -} \qquad (10)$$

$$2O_2^{\prime -} + 2H_2O \rightarrow 2H_2O_2 + O_2 \qquad (11)$$

$$O_2^{\prime -} + H^+ \rightarrow HOO^{\prime} \qquad (12)$$

$$HOO^{\prime} + H_2O \rightarrow H_2O_2 + OH^{\prime} \qquad (13)$$

$$H_2O_2 +))) \rightarrow 2OH^{\prime} \qquad (14)$$

The holes migrate to the surface and react with water molecules or chemisorbed OH$^-$ on the surface of N–doped TiO$_2$ and result in formation of OH$^{\prime}$ radicals (Eqs. (15) and (16)). Besides, the holes can directly oxidize organic substances adsorbed on the surface of catalyst (Eq. (17))

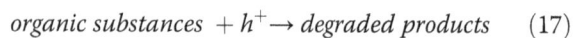

$$h^+ + Ti \cdot OH^- \rightarrow Ti \cdot OH^{\prime} \qquad (15)$$

$$H_2O + h^+ \rightarrow OH^{\prime} + H^+ \qquad (16)$$

$$organic\ substances + h^+ \rightarrow degraded\ products \qquad (17)$$

where ")))" denotes to the ultrasonic irradiation. It is widely accepted that O$_2^-$ and $^{OH^{\prime}}$ have strong oxidative degradation potential. Wu et al. found that the amounts of the produced $^{OH^{\prime}}$ radicals increase with doping of TiO$_2$ [33] . In this study, from degradation efficiency it can be understand that the highest amount of radicals is generated on the surface of TN2 because narrower band gap of TN2 facilitates the transition of electron from the valence band to the conduction band and eventually increases sonocatalytic activity. Thus, optimum amount of

nitrogen dopant play an important role in improving sonocatalytic activity.

Conclusions

In this study, a simple sol-gel method was used to synthesize of un-doped and N-dope TiO$_2$ for activity enhancement of sonolysis and sonocatalysis processes. The characterization of synthesized nano-particles was carried out by XRD, FE-SEM, EDX and UV-vis spectra. The characterization experiments confirmed that nitrogen doping has been successfully done in the TiO$_2$ structure.

The degradation of humic acid was used to evaluate the sonocatalytic activity of synthesized nano-particles. On the basis of the above results and discussion, addition of nano-catalysts could enhance the degradation efficiency of humic acid as well as N-doped TiO$_2$ with a molar ratio of N/Ti as 0.06 was found the best nano-catalyst among the investigated catalysts. The synthesized N-doped TiO$_2$ showed about 1.86 times higher sonocatalytic activity for humic acid degradation compared to the un-doped TiO$_2$.

The sonocatalytic degradation of humic acid with different catalysts followed the first-order kinetic model. L-H model confirmed the dependence of initial reaction rate on the initial humic acid concentrations and showed that the degradation efficiency decrease with the increase of initial humic acid concentrations. As a general conclusion, the results indicated that sonocatalytic degradation with nitrogen doped semiconductors could be a suitable oxidation process for removal of refractory pollutants such as humic acid from aqueous solution.

Competing interests
The authors declare that they have no competing interests.

Authors' contributions
HK was the main investigator, synthesized the nano-particles and drafted the manuscript. AM and SN supervised the study. RNN and MK were advisors of the study. RNN also contributed in analyzing of data. All authors read and approved the final manuscript.

Acknowledgments
This paper is a part of the results a PhD research thesis. The authors would like to thank the Center for Water Quality Research, Institute for Environmental Research, Tehran University of Medical Science for the financial support of this study (grant no. 94-33-61-20515). Authors also thank Mrs., Sheikhi and Mrs. Hoseini, the technical staffs in the chemical laboratory, for their cooperation.

Author details
[1]Department of Environmental Health Engineering, School of Public Health, Tehran University of Medical Sciences, Tehran, Iran. [2]Center for Water Quality Research, Institute for Environmental Research, Tehran University of Medical Sciences, Tehran, Iran. [3]Department of Medicinal Chemistry, Faculty of

Pharmacy and Pharmaceutical Sciences Research Center, Tehran University of Medical Sciences, Tehran, Iran. [4]Center for Solid Waste Research, Institute for Environmental Research, Tehran University of Medical Sciences, Tehran, Iran. [5]National Institute of Health Research, Tehran University of Medical Sciences, Tehran, Iran.

References

1. Rezaee R, Maleki A, Jafari A, Mazloomi S, Zandsalimi Y, Mahvi AH. Application of response surface methodology for optimization of natural organic matter degradation by UV/H_2O_2 advanced oxidation process. J Environ Health Sci Eng. 2014;12:67.

2. Mahvi A, Maleki A, Rezaee R, Safari M. Reduction of humic substances in water by application of ultrasound waves and ultraviolet irradiation. J Environ Health Sci Eng. 2010;6:233–40.

3. Sun DD, Lee PF. TiO_2 microsphere for the removal of humic acid from water: complex surface adsorption mechanisms. Sep Purif Technol. 2012;91:30–7.

4. Bazrafshan E, Biglari H, Mahvi AH. Humic acid removal from aqueous environments by electrocoagulation process using iron electrodes. J Chem. 2012;9:2453–61.

5. Bazrafshan E, Mostafapour FK, Hosseini AR, Raksh Khorshid A, Mahvi AH. Decolorisation of reactive red 120 dye by using single-walled carbon nanotubes in aqueous solutions. J Chem. 2013;1-8.

6. Alipour V, Nasseri S, Nodehi RN, Mahvi AH, Rashidi A. Preparation and application of oyster shell supported zero valent nano scale iron for removal of natural organic matter from aqueous solutions. J Environ Health Sci Eng. 2014;12:146.

7. Ashrafi S, Kamani H, Mahvi A. The optimization study of direct red 81 and methylene blue adsorption on NaOH-modified rice husk. Desalin Water Treat. 2016;57:738-746

8. Wang G-S, Liao C-H, Wu F-J. Photodegradation of humic acids in the presence of hydrogen peroxide. Chemosphere. 2001;42:379–87.

9. Safari G, Hoseini M, Seyedsalehi M, Kamani H, Jaafari J, Mahvi A. Photocatalytic degradation of tetracycline using nanosized titanium dioxide in aqueous solution. Int J Environ Sci Technol. 2015;12:603–16.

10. Mahvi AH, Ebrahimi SJA-d, Mesdaghinia A, Gharibi H, Sowlat MH. Performance evaluation of a continuous bipolar electrocoagulation/ electrooxidation–electroflotation (ECEO–EF) reactor designed for simultaneous removal of ammonia and phosphate from wastewater effluent. J Hazard Mater. 2011;192:1267–74.

11. Pang YL, Abdullah AZ. Comparative study on the process behavior and reaction kinetics in sonocatalytic degradation of organic dyes by powder and nanotubes TiO $_2$. Ultrason Sonochem. 2012;19:642–51.

12. Mahvi AH, Roodbari AA, Nodehi RN, Nasseri S, Dehghani MH, Alimohammadi M. Improvement of Landfill Leachate Biodegradability with Ultrasonic Process. PLoS One. 2012;7:e27571.

13. Dobaradaran S, Nabizadeh R, Mahvi A, Mesdaghinia A, Naddafi K, Yunesian M, et al. Survey on degradation rates of trichloroethylene in aqueous solutions by ultrasound. Iranian J Environ Health Sci Eng. 2010;7:307–12.

14. Maleki A, Mahvi AH, Ebrahimi R, Zandsalimi Y. Study of photochemical and sonochemical processes efficiency for degradation of dyes in aqueous solution. Korean J Che Eng. 2010;27:1805–10.

15. Hoseini M, Safari GH, Kamani H, Jaafari J, Ghanbarain M, Mahvi AH. Sonocatalytic degradation of tetracycline antibiotic in aqueous solution by sonocatalysis. Toxicol Environ Chem. 2013;95:1680–9.

16. Daraei H, Maleki A, Mahvi AH, Zandsalimi Y, Alaei L, Gharibi F. Synthesis of ZnO nano-sono-catalyst for degradation of reactive dye focusing on energy consumption: operational parameters influence, modeling, and optimization. Desalin Water Treat. 2014;52:6745–55.

17. Mahvi A, Ghanbarian M, Nasseri S, Khairi A. Mineralization and discoloration of textile wastewater by TiO_2 nanoparticles. Desalination. 2009;239:309–16.

18. Borji SH, Nasseri S, Mahvi AH, Nabizadeh R, Javadi AH. Investigation of photocatalytic degradation of phenol by Fe (III)-doped TiO_2 and TiO_2 nanoparticles. J Environ Health Sci Eng. 2014;12:101.

19. Javid A, Nasseri S, Mesdaghinia A, Hossein Mahvi A, Alimohammadi M, Aghdam RM, et al. Performance of photocatalytic oxidation of tetracycline in aqueous solution by TiO_2 nanofibers. J Environ Health Sci Eng. 2013;11:24.

20. Pang YL, Abdullah AZ. Effect of low Fe^{3+} doping on characteristics, sonocatalytic activity and reusability of TiO_2 nanotubes catalysts for removal of Rhodamine B from water. J Hazard Mater. 2012;235:326–35.

21. Zhang S. Synergistic effects of C–Cr codoping in TiO_2 and enhanced sonocatalytic activity under ultrasonic irradiation. Ultrason Sonochem. 2012;19:767–71.

22. Cheng X, Yu X, Xing Z, Wan J. Enhanced photocatalytic activity of nitrogen doped TiO_2 anatase nano-particle under simulated sunlight irradiation. Energy Procedia. 2012;16:598–605.

23. Pang YL, Abdullah AZ. Effect of carbon and nitrogen co-doping on characteristics and sonocatalytic activity of TiO_2 nanotubes catalyst for degradation of Rhodamine B in water. Chem Eng J. 2012;214:129–38.

24. Li X, Fan C, Sun Y. Enhancement of photocatalytic oxidation of humic acid in TiO_2 suspensions by increasing cation strength. Chemosphere. 2002;48:453–60.

25. Gurkan YY, Turkten N, Hatipoglu A, Cinar Z. Photocatalytic degradation of cefazolin over N-doped TiO_2 under UV and sunlight irradiation: prediction of the reaction paths via conceptual DFT. Chem Eng J. 2012;184:113–24.

26. Kuo Y-L, Su T-L, Kung F-C, Wu T-J. A study of parameter setting and characterization of visible-light driven nitrogen-modified commercial TiO_2 photocatalysts. J Hazard Mater. 2011;190:938–44.

27. Xie Y, Li Y, Zhao X. Low-temperature preparation and visible-light-induced catalytic activity of anatase F–N-codoped TiO_2. J Mol Catal A Chem. 2007;277:119–26.

28. Lin X, Rong F, Fu D, Yuan C. Enhanced photocatalytic activity of fluorine doped TiO_2 by loaded with Ag for degradation of organic pollutants. Powder Technol. 2012;219:173–8.

29. Wang P, Yap PS, Lim TT. C–N–S tridoped TiO_2 for photocatalytic degradation of tetracycline under visible-light irradiation. Appl Catal A Gen. 2011;399:252–61.

30. Song L, Chen C, Zhang S, Wei Q. Sonocatalytic degradation of amaranth catalyzed by La^{3+} doped TiO_2 under ultrasonic irradiation. Ultrason Sonochem. 2011;18:1057–61.

31. Livraghi S, Elghniji K, Czoska A, Paganini M, Giamello E, Ksibi M. Nitrogen-doped and nitrogen–fluorine-codoped titanium dioxide. Nature and concentration of the photoactive species and their role in determining the photocatalytic activity under visible light. J Photochem Photobiol A Chem. 2009;205:93–7.

32. Farzadkia M, Bazrafshan E, Esrafili A, Yang J-K, Shirzad-Siboni M. Photocatalytic degradation of Metronidazole with illuminated TiO2 nanoparticles. J Environ Health Sci Eng. 2015;13:35

33. Wu Y, Xing M, Tian B, Zhang J, Chen F. Preparation of nitrogen and fluorine co-doped mesoporous TiO_2 microsphere and photodegradation of acid orange 7 under visible light. Chem Eng J. 2010;162:710–7.

34. Ananpattarachai J, Kajitvichyanukul P, Seraphin S. Visible light absorption ability and photocatalytic oxidation activity of various interstitial N-doped TiO_2 prepared from different nitrogen dopants. J Hazard Mater. 2009;168:253–61.

35. Huang DG, Liao SJ, Liu JM, Dang Z, Petrik L. Preparation of visible-light responsive N–F-codoped TiO_2 photocatalyst by a sol–gel-solvothermal method. J Photochem Photobiol A Chem. 2006;184:282–8.

36. Kaur S, Singh V. Visible light induced sonophotocatalytic degradation of Reactive Red dye 198 using dye sensitized TiO $_2$. Ultrason Sonochem. 2007;14:531–7.

37. Saha S, Wang J, Pal A. Nano silver impregnation on commercial TiO_2 and a comparative photocatalytic account to degrade malachite green. Sep Purif Technol. 2012;89:147–59.

38. Jaafari J, Mesdaghinia A, Nabizadeh R, Hoseini M, Mahvi AH. Influence of upflow velocity on performance and biofilm characteristics of Anaerobic Fluidized Bed Reactor (AFBR) in treating high-strength wastewater. J Environ Health Sci Eng. 2014;12:139.

39. Salavati H, Tangestaninejad S, Moghadam M, Mirkhani V, Mohammadpoor-Baltork I. Sonocatalytic oxidation of olefins catalyzed by heteropolyanion–montmorillonite nanocomposite. Ultrason Sonoche. 2010;17:145–52.

16

A novel technique for detoxification of phenol from wastewater: Nanoparticle Assisted Nano Filtration (NANF)

L. D. Naidu[1*], S. Saravanan[1], Mukesh Goel[2], S. Periasamy[3] and Pieter Stroeve[4]

Abstract

Background: Phenol is one of the most versatile and important organic compound. It is also a growing concern as water pollutants due to its high persistence and toxicity. Removal of Phenol from wastewaters was investigated using a novel nanoparticle adsorption and nanofiltration technique named as Nanoparticle Assisted Nano Filtration (NANF).

Methods: The nanoparticle used for NANF study were silver nanoparticles and synthesized to three distinct average particle sizes of 10 nm, 40 nm and 70 nm. The effect of nanoparticle size, their concentrations and their tri and diparticle combinations upon phenol removal were studied.

Results: Total surface areas (TSA) for various particle size and concentrations have been calculated and the highest was 4710×10^{12} nm^2 for 10 nm particles and 180 ppm concentration while the lowest was for 2461×10^{11} for 70 nm and 60 ppm concentrations. Tri and diparticle studies showed more phenol removal % than that of their individual particles, particularly for using small particles on large membrane pore size and large particles at low concentrations. These results have also been confirmed with COD and toxicity removal studies.

Conclusions: The combination of nanoparticles adsorption and nanofiltration results in high phenol removal and mineralization, leading to the conclusion that NANF has very high potential for treating toxic chemical wastewaters.

Keywords: Phenol, Nanoparticle, Nanofiltration, NANF, COD, Nanoporous membranes

Background

Water is the central element of all vital social and economic processes. Because of the development of consumer society, harmful chemicals are being generated in huge quantities throughout the world. The problems derived from the toxicological effects of these organic compounds must be resolved for the benefits of entire society. The problem is certainly complex and it is imperative that novel procedures are required to deal with this extensive range of tribulations. There are ample treatment technologies for sewage, distillery effluents and so on, which contain biodegradable organics, but not so much for toxic effluents containing xenobiotic compounds, which are often non-biodegradable or only partially biodegradable. Biological treatment though promises much in this regard is handicapped by its slow oxidation characteristics and incomplete mineralization of toxic chemicals [1–5].

Nanotechnology has also attracted the versatile membrane filtration process for water treatment. Nanofiltration and Reverse Osmosis (RO), are two common membrane filtration processes for toxic chemical removal from wastewaters and have been successfully applied in removing BOD/COD from many wastewaters [6–9]. The powerful membrane technology is however, limited by high costs associated with it and other operational considerations. One of the significant ways to overcome this limitation is to couple membrane separation with another technology. An interesting solution is proposed by some researchers; polymer enhanced ultrafiltration (PEUF) [10, 11]. This was especially found to be more relevant in removing heavy metals from

* Correspondence: ldnaidu@gmail.com
[1]Department of Chemical Engineering, National Institute of Technology, Tiruchirapalli 15, India
Full list of author information is available at the end of the article

aqueous solutions. It combines UF with metal complexation using water-soluble polymers. The formed complexes are sufficiently large to be retained by a UF membrane. There has been several reports on the use of PEUF for treatment of wastewaters. This is however, limited to metal removal and has not found applications in removing toxic chemicals from water and wastewater [12, 13].

A modified version of this technology is called Nanoparticle Assisted Nanofiltration (NANF); nanofiltration in conjunction with adsorption using nanoparticles. The large surface area of nanoparticles makes them a potent adsorbent for toxic chemical removal. The first step in the process is adsorption of toxic organic chemicals on nanoparticles, followed by filtration with a NF membrane, which permits the passage of water and other smaller compounds, but retaining the toxic organic compounds. For polluted water, the full treatment steps conventionally adopted in practice includes preoxidation, enhanced coagulation, sedimentation, sand filtration, main oxidation, GAC filtration, etc. Such a treatment chain is however too long to be afforded for developing countries. Therefore NANF, using nanoadsorbents, is a suitable alternative toconventional wastewater treatment plants in dealing with toxic chemicals. In this work, silver nano-particles have been tested for their ability to adsorb organic compounds, which are then retained with a NF membranes. Phenol is used as the model organic compound for these experiments because it is one of the major waste-products in a wide range of manufacturing industries, e.g., chemical and pharmaceutical, paintsand textiles, paper and pulp, plastics and polymer, oil and gasoline as well as coking ovens and metallurgical furnaces [14, 15]. Three different nanoparticles sizes, 10, 40 and 70 nm, and four different NF membranes of diameter, 10, 30, 50 and 80 nm were used for the experiments.

Methods
Adsorbent
Silver nanoparticles of different sizes were synthesized using a chemical reduction method [16]. Dynamic light scattering (DLS, Brookhaven Instruments Corporation, USA) was used to analyze the size of the nanoparticles.

Adsorbate
Phenol was obtained from Merck, and its 500 ppm stock solution was prepared in double distilled water. Solution of 200 ppm phenol used in the experiments was prepared from stock solution and double distilled water was used for necessary dilutions. All reagents used in the investigation were of analytical grade.

Analytical measurements
Phenol was analyzed using the Amino-Antipyrine method [17]. For phenol analysis, 3 ml of distilled water was added with 1 ml of centrifuged sample from each reactor to make the sample to 4 ml. The sample vial was added with 0.60 ml of K_3 Fe CN_6, 0.20 ml of NH_4OH and 0.60 ml of 4-amino antipyrine. The absorbance of the sample was measured with UV–VIS spectrophotometry (Lab India). Theoretical density values were measured at 406 nm to yield the concentration of phenol. Chemical oxygen demand (COD) was measured using a HACH Colorimeter (DR 890). The COD solution (HR grade – 0–1500 ppm) was prepared by mixing 0.25 ml of COD solution A and 2.8 ml of COD solution B. To this solution, 2 ml of centrifuged sample (include dilution) was added. The digestion was done at 1500 C in a HACH COD digester for two hours using HACH COD vials. Final COD value after air cooling was measured in a HACH-DR/890 colorimeter. The toxicity of phenols before and after treatments was realized using Resazurin reduction method [18]. Pre-cultivated *Bacillus cereus* culture was used as the test organisms and was cultivated on nutrient broth medium. Resazurin changes color in the presence of dehydrogenase enzyme activity resulting from microorganisms actively growing in a culture medium. In the presence of an active bacterial culture, resazurin changes color from blue to pink. If bacterial growth is inhibited, no reduction of the resazurin occurs, and such a sample would remain blue. The reaction time for assay was 20 min and would be inhibited by the addition of 50 μl of $HgCl_2$. The colour intensity of centrifuged samples is determined using the UV–VIS spectrophotometer at 610 nm. Resazurin solution was prepared by dissolving 1 tablet of resazurin (5 mg in 100 ml) in 100 ml of distilled water. The resazurin solution was stored in a brown bottle and kept in a temperature of 4 °C.

NANF experiments
All experiments were conducted at 200 ppm phenol concentration and at pH 7 by the addition of NaOH or HCl. Experiments were conducted in a stirred membrane reactor. The schematic of the reactor is shown in Fig. 1. The details of the reactor is given elsewhere [19]. The effective permeation area of the NF membrane was 1.77 cm^2, and the volumes of both compartments were 35.0 mL. Mixing was obtained using magnetic stirrers, which were present in both compartments.

Experimental procedure
Commercial PCTE membranes (Poretics, Inc.) were used as NF membranes [20]. Membranes are numbered according to their diameter. Membrane with 10 nm pore size was referred as M1, whereas, membranes with 80 nm pore size

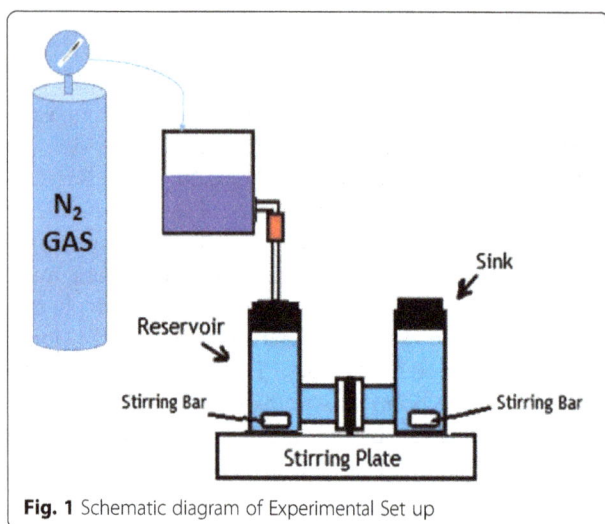

Fig. 1 Schematic diagram of Experimental Set up

was referred as M4. The intermediate membranes were referred as M2 and M3 respectively. The pressure is obtained through compressed N_2 gas. One hour before experiment, 0.5 mL of nanoparticles solution was mixed with 35 mL of neutral solution containing 200-ppm phenol. The solution was then moved to the membrane reactor. Additionally, each experiment was conducted without any nanoparticles to comprehensively evaluate the potential adsorption and filtration performance of NANF.

Theoretical surface area calculation

The surface area calculation was performed as given below.

a) Surface area of a single nanoparticles of 10 nm diameter (5 nm radius) is given by

$$= 4 \times 3.14 \times 5^2 \text{ nm}^2$$
$$= 314 \text{ nm}^2$$

b) Number of particles is obtained from
   i) Concentration of the nanoparticles solution, i.e. 60, 120 and 180 ppm
   ii) Nanoparticle average size
   iii) Total volume of nanoparticle solution used, i.e. 0.5 ml
   iv) Density of silver metal (10.49 g/cc)

From i), we calculate the total weight of the nanoparticles used for adsorption, for example60 ppm means 60 g in 10^6 ml of water, so for 0.5 ml of water

$$= 60 \times 0.5 / 10^6$$
$$= 30 \times 10^{-6} \text{ g} \text{-------x}$$

From ii) we can calculate the volume of individual particles, for example for 10 nm particle

$$= 1.33 \times 3.14 \times 5^3 \times (10^{-7})^2 \text{cm}^3$$
$$= 523 \times 10^{-21} \text{cm}^3$$

Weight of an individual particle can be calculated from the density value of silver metal (10.49 g/cc)

$$= 10.49 \times 523 \times 10^{-21} / 1$$
$$= 548 \times 10^{-20} \text{g} \text{------- y}$$

Hence, total number of particles can be obtained from

$$= x/y$$
$$= 30 \times 10^{-6} \text{g} / 548 \times 10^{-20} \text{g}$$
$$= 3000 \times 10^{12} / 548$$
$$= 5.47 \times 10^{12}$$
$$\approx 5 \times 10^{12}$$

c) Total surface area is obtained by

$$= \text{Surface area of individual particle}$$
$$\times \text{ Total number of particles}$$
$$= 314 \text{ nm}^2 \times 5 \times 10^{12}$$
$$= 1570 \times 10^{12} \text{nm}^2$$

Results and discussion

The current study combines two materials, i.e. the membrane filters and the nanomaterials, to exploit their inherent characteristics for the removal of toxic materials from water. For this purpose, membranes of various pore sizes are used along with silver nanoparticles with various average particle sizes.

Silver nanoparticles

Figure 2 shows the transmission electron micrograph of the silver nanoparticles, which were prepared using three distinct combinations of the bottom-up nanoparticles synthesis process for obtaining nanoparticles of three different average nanoparticles sizes i.e. 10 nm (Fig. 2a), 40 nm (Fig. 2b) and 70 nm (Fig. 2c). Silver nanoparticles with average size of 10 nm were obtained with 10 °C synthesis temperature, Polyvinyl Pyrolidone (PVP) as the dispersing and stabilizing agent and hydrazine as the reducing agent; 40 nm were obtained with 40 °C synthesis temperature, PVP and hydrazine; and 70 nm were obtained with 40 °C synthesis temperature, oligo condensate of naphthalene sulfonic acid (OCNS) and D-glucose combinations. It can be observed from the transmission electron micrographs (Fig. 2) that the particles have been formed with proper geometrical shape, almost spherical for most particles. It can be observed that the particles prepared with PVP have less variation in the particle size distribution (narrow) while those prepared with OCNS comparatively have a wide particle

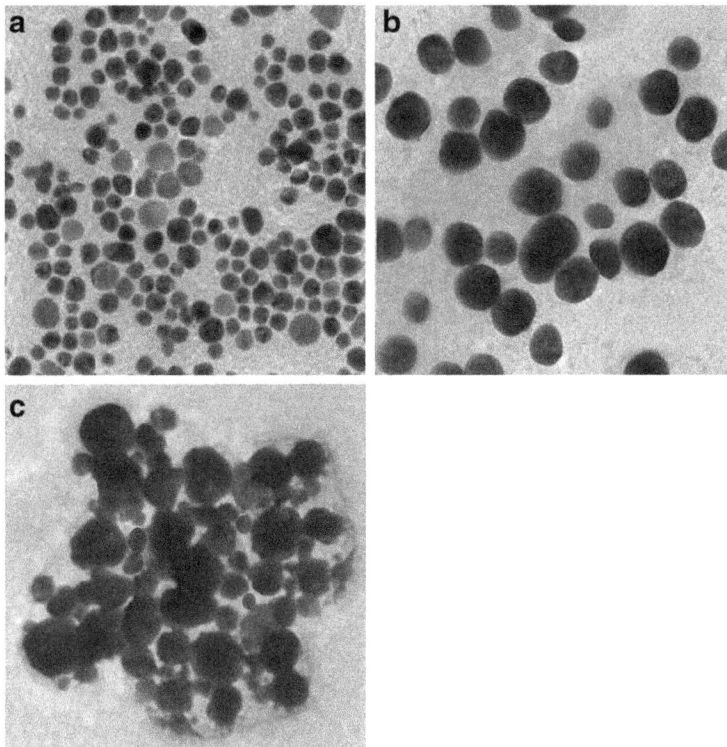

Fig. 2 Transmission electron micrograph of silver nanoparticles with average particle size of **a**) 10 nm, **b**) 40 nm and **c**) 70 nm

size distribution (broad). The mechanism of such particles formation with varying average particle size and distribution are discussed elsewhere [13].

Nanoparticle assisted nano filtration (NANF) effect on phenol removal from water

The work started with adsorption of phenols using silver nanoparticles. Since three distinct average size particle samples were prepared, the three particles batches are used in equal proportion, with the total particle concentrations at 60 ppm. Gulrajani et al. [16] have shown that at 60 ppm particle concentration good particle performance was obtained in their study. The effects of nanoparticle size and membrane pore size on filtration of the phenol are presented in Table 1. It can be observed from Table 1 that for all particle sizes the presence of the nanoparticles show a significant improvement in filtering the phenol with reference to the respective counterpart, i.e. in the absence of the particles for the respective membrane. The results clearly show the effectiveness of the NANF process for phenol removal. As high as 675 % improvement is observed for NANF process (membrane filtration with nanoparticles), compared to the process without nanoparticles. These results can be explained based on the mechanism of phenol adsorption followed by agglomeration of the nanoparticles, which results in a net increase in the particle size, which then cannot pass through the membrane pores and hence are filtered. Such effect is schematically shown in Fig. 3. In the absence of nanoparticles (Fig. 3a) the molecular phenolic compounds can easily pass through the pores of the membranes by diffusion. However, in the presence of nanoparticles (Fig. 3b) whose surface area is very large, there exists a spontaneous attraction of phenolic compounds towards the surface of the particles due to i) electrostatic charge attraction between the nanoparticles and the phenolic compounds (Fig. 3c) and ii) aggregate formation to obtain a lower energy state of the solute and the squeezing out of solvent in support of solute separation [21].

Table 1 Phenol removal percentage with and without nanoparticles for mixtures of nanoparticles

Membrane size (nm)	Phenol Removal (%)		% improvement
	Nanoparticles	No Nanoparticle	
10	68	27	152
30	57	18	216
50	42	11	281
80	31	4	675

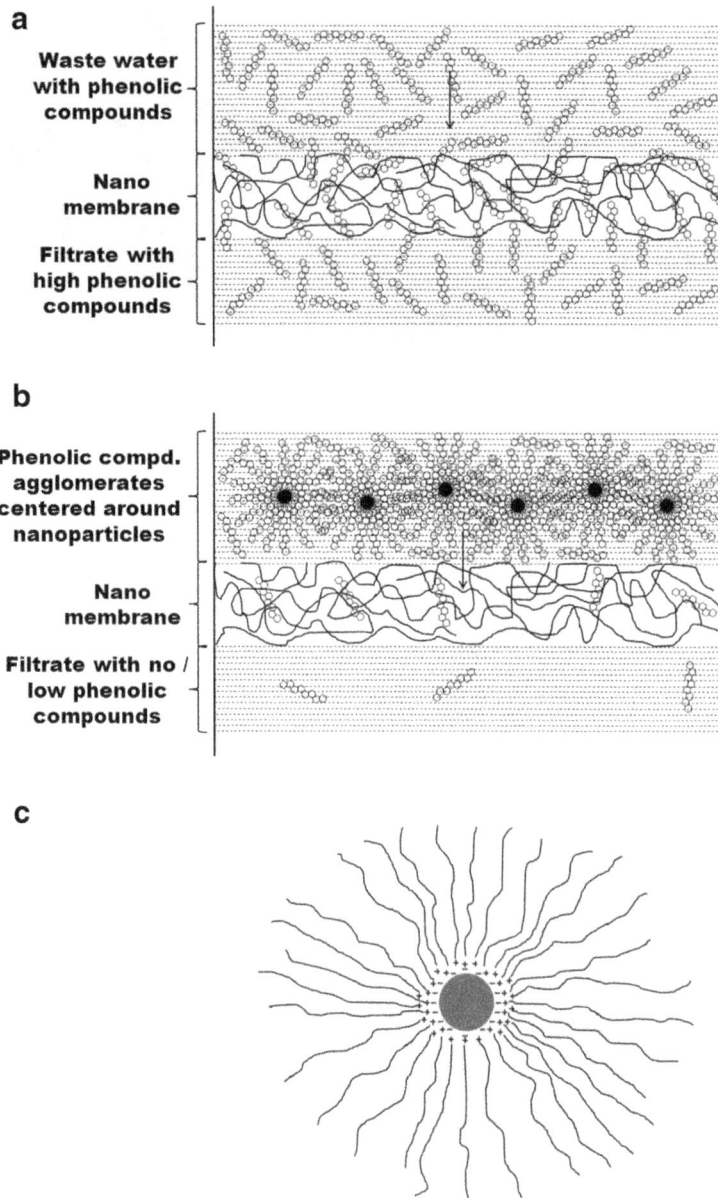

Fig. 3 Schematic of filtration of model phenolic compound through nano membrane **a**) without nano particles, **b**) with nanoparticles (NANF) and **c**) electrostatic adsorption

The effect of membrane pore size on NANF is observed to be significant. Filtration effects are large for the 80 nm pore size membranes (M4), i.e. up to a 675 % improvement (obtained from the phenol removal % with the presence and absence of nanoparticles [((31–4) × 100)/4]). It is surprising to note that such % improvement is low for 10 nm pore size membranes (M1), i.e. only a 151 % improvement. This may be explained to the fact that for large pore size almost all the phenolic compounds pass through the pores in the absence of the particles and the phenol removal % is very low; but with the low value in the denominator, for the case of the presence of nanoaparticles, the % improvement becomes large, as shown in the calculation above. However, in contrast with the small pore size membrane, i.e. 10 nm, there is a considerable amount of phenol removal even in the absence of particles, i.e. 27 %. Hence, even though the actual amount of phenol removal is high in the combined nanoparticles and membrane approach, i.e. 68 %, the % improvement in the presence of nanoparticles is only 151 %. It can be noticed that the best results are obtained for the small pore size membrane (10 nm) and the nanoparticles combinations. This can be explained to the fact that the nanoparticles have enormous surface area particularly below the 100 nm size. The specific surface area of the particles increases exponentially with decreasing size (Theodore, 2005). Hence, availability of large surface areas contributes to surface adsorption of the toxic compounds, which are then filtered by membrane filtration. The low phenol removal %, i.e. 31 % with the 80 nm membrane, even in the presence of the nanoparticles, is due to the presence of nanoparticles whose size is less than 80 nm. These particles pass through the pores resulting in lower nanoparticles concentration and a consequent decrease in phenol removal. This indicates that the smaller particles play an important role in adsorption and hence the phenol removal %. The effectiveness of the hybrid process is also reported by Geckeier et al. [22]. They studied polymer enhanced ultrafiltration (PEUF), which was helpful in removing metal from wastewater by complexing of polymers with the metals and subsequent removal from the wastewater. Similar study was reported by Li et al. [23].

Effect of concentration and particle size of silver nanoparticles

Silver nanoparticles with three different average particle sizes, i.e. 10, 40 and 70 nm, each with three different particle concentrations, 60, 120 and 180 ppm, are used for phenol removal study on membranes (M1 – M5). The results are depicted in Fig. 4 and in Fig. 5. Figure 4a shows the effects of the three particle concentrations, for 10 nm silver nanoparticles. Similarly, Figs. 4b and c show the effects of the three particle concentrations for 40 and 70 nm nanoparticles, respectively. Inversely, Fig. 5 shows the effect of particle size for a given concentration, i.e. Figure 5a shows the effect of the three nanoparticles sizes for the 60 ppm particle concentration. Similarly, Figs. 5b and c show the effect of the three nanoparticles size for the 120 and 180 ppm concentrations, respectively.

From Figs. 4-5, it can be inferred that for the 10 nm pore size membrane that about 100 % phenol removal is obtained for 120 and 180 ppm concentrations for all the three nanoparticles sizes. Further, it can be observed that for the 60 ppm particle concentration that, for almost for all particle sizes (Fig. 5a), the results are less than 100 % phenol removal on the 10 nm pore size membrane. Further, it can be observed that the phenol removal % decreases with increase in pore size of the membrane for almost all particle sizes, but with a varying slopes (Fig. 4a to c). The decrease is sharp after the 30 nm membrane and is well noticed for 60 ppm particle concentration (Fig. 4a to c and Fig. 5a). However, for higher particle concentrations, the decrease in phenol removal % with increasing membrane pore size is low. For the 10 nm size nanoparticles, a notable decrease is observed both for 120 (Fig. 5b) and 180 ppm (Fig. 5c)

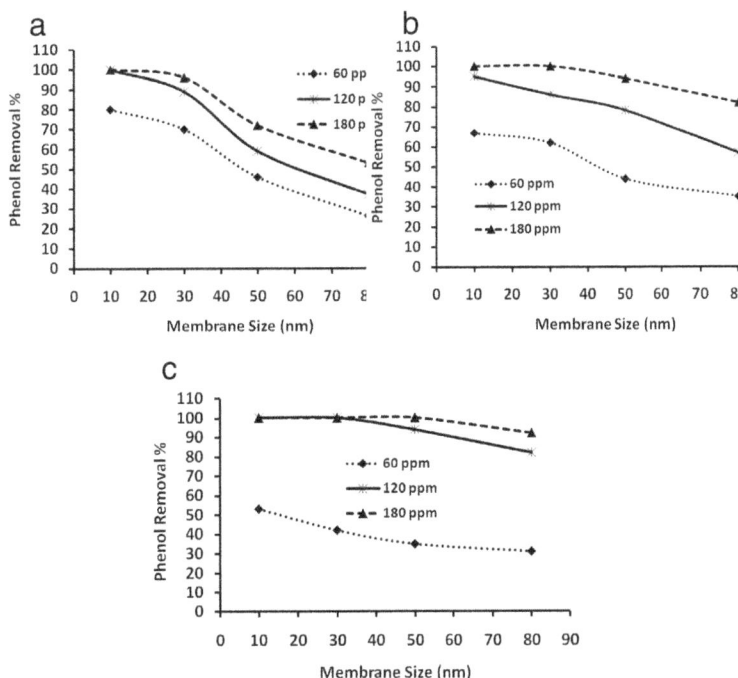

Fig. 4 Phenol removal %, through membranes having various average pore sizes, as an effect of concentration of silver nanoparticles of a) 10 nm, b) 40 nm and c) 70 nm

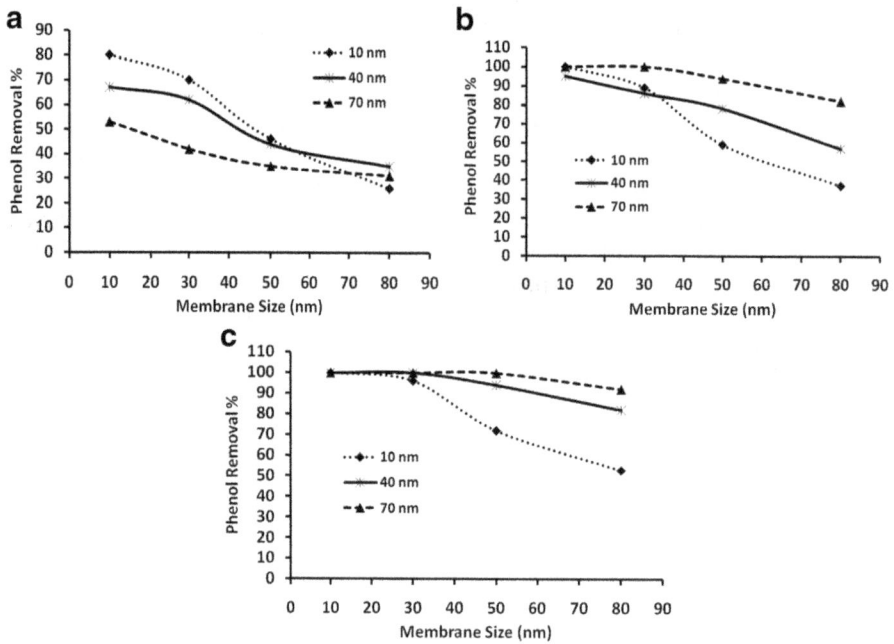

Fig. 5 Phenol removal %, through membranes having various average pore sizes, with silver nanoparticles as an effect of the particles sizes, **a**) 60 ppm, **b**) 120 ppm and **c**) 180 ppm

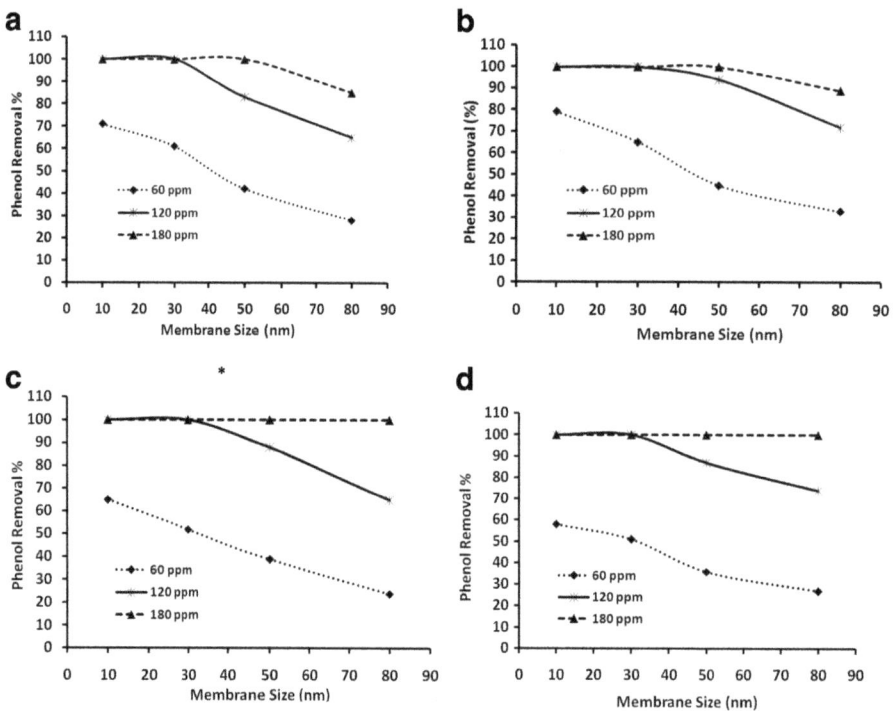

Fig. 6 Phenol removal % through membranes having various average pore sizes with mixture of nanoparticles, **a**) 1:1:1 (10 nm, 40 nm, 70 nm), **b**) 1:1 (10 nm, 40 nm), **c**) 1:1 (10 nm, 70 nm) and **d**) 1:1 (40 nm, 70 nm)

above the 30 nm membrane size. However, the decrease is less for the particle size of 40 nm both for 120 (Fig. 5b) and 180 ppm (Fig. 5c) after the 30 nm membrane size and is even less, almost reaching 100 % for the particle size of 70 nm both for 120 (Fig. 5b) and 180 ppm (Fig. 5c) above the 30 nm membrane size (Fig. 4c).

These observations can be explained based on nanoparticles size, membrane pore size and the surface area available of the nanoparticles for adsorption [24]. The surface area of the individual nanoparticles, total number of nanoparticles available for a given particle size and concentration and hence the total surface area (TSA) available for the adsorption studies have been discussed previously and the values are presented in Table 2. It can be observed from Table 2 that the highest TSA is obtained for the 10 nm particle size at 180 ppm concentration, i.e. 4710×10^{12} nm^2, and the lowest TSA of 2461×10^{11} nm^2 for the particle size of 70 nm at 60 ppm concentration with the total % difference of 1813 [$(4710 \times 10^{12} - 2461 \times 10^{11}) \times 100/2461 \times 10^{11}$]. For the other particle sizes and concentrations, the TSAs of the particles lie in between these two limits. Additionally, it can be observed from Table 2 that for the 60 ppm concentration the TSA for 10, 40 and 70 nm particles are 1570×10^{12}, 4521×10^{11} and 2461×10^{11}, respectively, and so on for the other concentrations. The percentage increases in TSA between the particles with various concentrations are shown separately in Table 3. The general formula for calculating % increase in TSA between various samples is shown as a foot note under Table 3. It can be observed from Table 3 that between samples 1 & 4 (between 10 and 40 nm both at 60 ppm conc.) the TSA increase is 247 % while between samples 4 & 7 (between 40 and 70 nm both at 60 ppm conc.) the TSA increase is only 84 %. However, the TSA increase between samples 1 & 7 (between 10 and 70 nm both at 60 ppm conc.) is a maximum of 537 %. Further it could be carefully noted that similar comparisons for the higher concentrations, i.e.

120 ppm and 180 ppm, also result in similar TSA % increases i.e. 247 %, 84 % and 537 %. However, in such trials when compared for the smallest particle (10 nm) with highest concentration (180 ppm) (sample 3) and largest particle (70 nm) at the lowest concentration (60 ppm) (sample 7) the TSA increase is the highest of all, about 1813 %.

It can be observed from Figs. 4-6 that the 10 nm particles at 180 ppm concentration with the 10 nm pore membrane result in 100 % phenol removal (Fig. 4a), while the 70 nm at 60 ppm with the 10 nm pore size membrane results in phenol removal of only about 50 % (Fig. 4c), which can be attributed to the large difference in the total surface areas available for adsorption, between these two samples, i.e. 1813 % difference. However, the phenol removal % for 10 nm particles on the larger pore size membranes, shows less than 100 % and is dramatic above the 30 nm pore size for all three concentrations (Fig. 4a), which could be because of the loss of nanoparticles through the large pore size membranes. Contrastingly, with 70 nm particle such decrease is observed to be low for 60 ppm and very low for the higher concentrations i.e. 120 and 180 ppm (Fig. 4c). In fact, at these concentrations almost 100 % phenol removal is observed for 70 nm particles except with the 80 nm pore size membrane, which may be due to some loss of the 70 nm particles with the 80 nm pore size membrane. The observations of phenol removal by the other particle sizes and concentrations lie in between these two limits of smallest particle and highest concentrations with the smallest membrane pore size (Fig. 4a & Fig. 5c) and the largest particle and lowest concentration with the largest membrane pore size (Fig. 4c & Fig. 5a). The results can be attributed to the corresponding TSA along with the loss of particles through the large pore size membrane, i.e., small particles of high concentrations might have large TSA but fail to show 100 % phenol removal with the large pore size membranes, which might be by virtue of their small size because they pass

Table 2 Theoretical calculation of total available surface area of the silver nanoparticles

Sample	Particle size (nm)	Surface Area (nm^2)	Con. in ppm	No. of Particles	Total Surface Area (TSA) (nm^2)	Comments
1	10	314	60	5×10^{12}	1570×10^{12}	Equal to 6
2	10	314	120	10×10^{12}	3140×10^{12}	In between
3	10	314	180	15×10^{12}	4710×10^{12}	Highest
4	40	5024	60	9×10^{10}	4521×10^{11}	Equal to 8
5	40	5024	120	18×10^{10}	9043×10^{11}	In between
6	40	5024	180	27×10^{10}	1356×10^{12}	Close to 1
7	70	15386	60	16×10^{9}	2461×10^{11}	Lowest
8	70	15386	120	32×10^{9}	4923×10^{11}	Close to 4
9	70	15386	180	48×10^{9}	7385×10^{11}	Close to 5

Table 3 Percentage increase in total surface area (TSA) between particles of various sizes and concentrations

Between Sample # and #	Increase in TSA (%)[a]
1 & 4	247
4 & 7	84
1&7	537
2&5	247
5&8	84
2&8	537
3&6	247
6&9	84
3&9	537
3 & 7	1813

[a](TSA of 1 – TSA of 2) × 100/TSA of 2

through the membrane pores and are lost (Fig. 4a & Fig. 5a to Fig. 5c).

The effect of the particle size upon phenol removal can be realized at 60 ppm concentration and with the 10 nm pore size membrane (Fig. 5a) where the loss of small particles should be very low. It can be observed from Fig. 4a that the 10, 40 and 70 nm size particles give about 82 %, 68 % and 52 % phenol removal which might be due to the difference in the TSA i.e. 1570×10^{12} nm^2 (sample 1), 4521×10^{11} nm^2 (sample 4) and 2461×10^{11} nm^2 (sample 7) for 10, 40 and 70 nm size particles, respectively. The corresponding % TSA increase between the samples are found to be 247 %, 84 % and 537 % for sample 1 & 4, sample 4 & 7 and sample 1 & 7, respectively. Though % increase between sample 1 & 7 is very large, the % phenol removal is only about 82 % for the 10 nm particles as against 52 % for the 70 nm particles. This is because the 10 nm particle is only the average size and so there may be particle loss for those whose size is smaller than 10 nm. This is also reported by Sarkar and Acarya [25]. They observed that phenol removal was maximum at lower particle size. Furthermore, Roostaei and Tezel [26], noted that adsorption capacity decreased by increasing the particle size. It can however, be concluded that the particle loss through the membrane impacts the phenol removal even though small size particle yields high TSA.

Effect of nanoparticles mixtures

In the individual particle study discussed in the previous section, either there was a particle loss effect for small particles or there was a low total surface area (TSA) effect for large particles which were particularly noticed at 60 ppm concentration. Hence we studied the effect of particle mixtures in various proportions for three concentrations, 60 ppm, 120 ppm and 180 ppm, for phenol

removal. Results of the studies are shown in Fig. 6. Figure 6a depicts the effect of the three different particles i.e. 10 nm, 40 nm and 70 nm mixtures in equal proportions (1:1:1) for the three different concentrations. Accordingly 20 ppm of 10 nm particles, 20 ppm of 40 nm particles and 20 ppm of 70 nm particles were taken for forming a silver nanoparticles mixture solution. Similarly this was done for the other two concentrations i.e. for 120 ppm 40 ppm of each particle solutions and for 180 ppm 60 ppm of each particle solutions were taken. The effect of equal proportions (1:1:1) of the three nanoparticles (triparticle mixture) for three concentrations on various membrane are shown in Fig. 6a. Similarly, the other possible proportions of particles i.e. diparticle mixtures such as 1:1 (10 nm, 40 nm), 1:1 (10 nm, 70 nm,) and 1:1 (40 nm, 70 nm) were carried out. The results of such diparticle mixtures are shown in Fig. 6b, c and d, respectively.

It can be observed from Fig. 6a, b, c, d that the triparticle mixture and all the diparticle mixtures at 60 ppm result in less than 100 % phenol removal ranging from 70 % for triparticle mixture (Fig. 6a) to 58 % for diparticle (40 nm, 70 nm) (Fig. 6d) for the 10 nm pore membrane. The phenol removal % decreases drastically on higher pore size membranes (M2-M4) for the triparticle and all the diparticle combinations for 60 ppm concentrations. This clearly brings out the effect of particle loss, particle size and the TSA effect. The effect of particle size can be observed from Fig. 6a in that the phenol removal of the triparticle mixture for the 10 nm pore membrane is 70 % however the same is about 80 % for pure 10 nm particles for 60 ppm (Fig. 4a). This indicates mixing the 40 nm & 70 nm particles results in a net decrease in TSA and so a decrease in phenol removal. The particle size and the mixture effects can be perceived from Figs. 6b to c, in that for the diparticle mixture of 10 nm and 40 nm the phenol removal is again 80 % (Fig. 6b). This result can be explained that without 40 nm particles, there would be some loss of 10 nm particles and in this case that would be minimal as there would be 40 nm particles which would prevent the loss of particles. Hence, the decrease in TSA of pure 60 ppm 10 nm particle is countered by the 40 nm particles to prevent their loss and, additionally, the TSA contribution from the 40 nm particles together might contribut for the equivalent phenol removal % to that of pure 10 nm at 60 ppm silver nanoparticles. The particle loss factor and particle size can also be perceived at high concentrations of 180 ppm from Fig. 6b and c. It can be observed from these Figures that for the diparticle mixture of 10 nm and 40 nm the phenol removal is only about 90 % whereas in the other diparticle mixture of 10 nm and 70 nm it is 100 % for the 80 nm pore membrane. In the former case there would be notable loss of particles, though there might be some agglomeration effect, resulting in less than 100 % phenol removal. However, in the latter case, though there might be loss of 10 nm particles, the contribution of TSA by 70 nm particles coupled with the TSA of 10 nm particles (which are retained by the 70 nm particles and by agglomeration effects) is sufficiently high to attain 100 % phenol retention. Likewise, the combined particle based adsorption effect would be techno-economically and ecologically better for efficient removal of the toxic contents rather than using the individual small and large particles for a given pore size membrane. For example, it can be seen in Figs. 4 and 5 that 100 % phenol removal was not observed even at 180 ppm concentration for 70 nm particles. However, in the case of diparticle mixtures of 10 nm and 70 nm and 40 nm and 70 nm for the same 180 ppm concentration, 100 % phenol removal has been obtained for all pore size membranes including the 80 nm pore size. Though the phenol removal % could be increased with individual particles just by increasing the concentration above 180 ppm, it would not be economical. Similarly, low particle size would be preferred because, generally, it is difficult to synthesize smaller particles and for smaller the particles there are chances that these might interfere with the ecological system. Further, although small particles lead to higher surface areas, it also creates operational issues as small particles would not be stable in a treatment system.

They could be easily swept away by the feed, thus posing difficulties with the maintenance of the system. Hence, for a given set of membranes, a combined particle mixture would yield a better toxic content removal due to the control of the loss of small particles by larger particles by a blocking effect and also by a particle agglomeration effect. This allows for exploiting the large total surface area (TSA) of small particles.

COD removal & toxicity study

As a further examination to the phenol removal % study, the filtrates were subjected to chemical oxygen demand (COD) studies as described in the experimental section. For this study, only four selected combinations were chosen from inference of the detailed phenol removal study done in the previous sections. The effect on COD removal by low and high particle concentrations i.e. 60 ppm and 180 ppm were studied for both individual particle (Fig. 7a & b) and for particle mixtures (Fig. 7c & d). It can be observed from these Figures that the COD removal % almost reflects the phenol removal %. It can be observed from Fig. 7a and b that the 10 nm particle is effective even at 60 ppm particle concentration, but only for the 10 nm & 30 nm membrane pore sizes. Beyond the 30 nm membrane pore size, the COD removal % decreases notable due to particle loss through the larger size pores. With high particle size, i.e. 70 nm, drastic

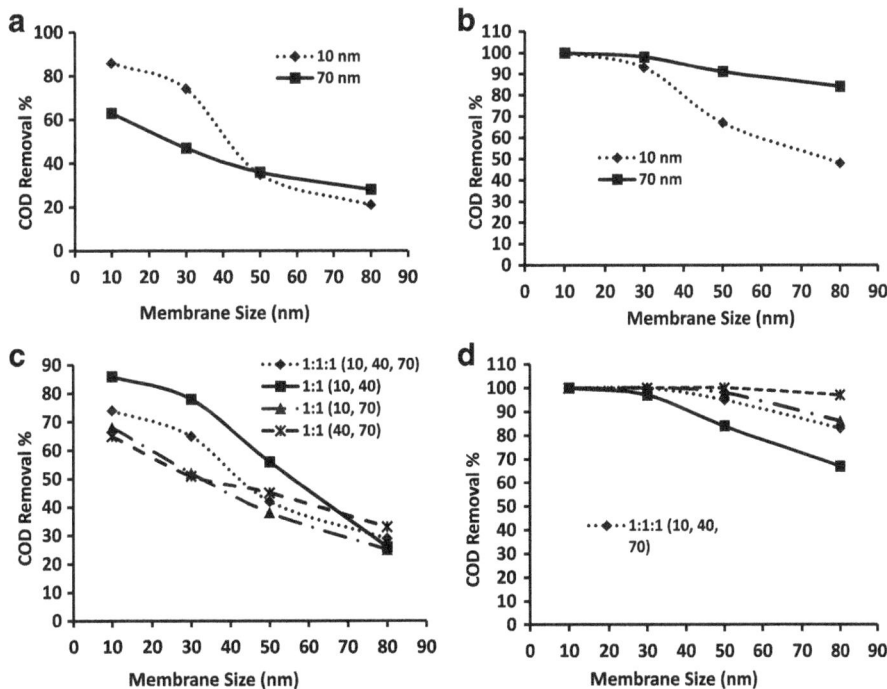

Fig. 7 COD removal % through membranes having various average pore sizes for individual particles **a**) 60 ppm, **b**) 180 ppm and for particle mixtures **c**) 60 ppm and **d**) 120 ppm

change is not noticed with membrane pore size, there is only a steady decrease. However, with the 180 ppm concentration, such effects are unnoticed due to the high concentration, which results in 100 % COD removal with only a slight decrease for the 50 nm and 80 nm membrane pore size due to phenol and particle loss effects. For the same study particle mixtures, it is interesting to note that the mixture yields a better effect than the results for individual particle sizes. It can be observed from Fig. 6c that the best result is achieved for a 10 nm and 40 nm diparticle mixture. The particle loss effect becomes dominant above the 30 nm membrane pore size. However, it can be observed from the same Figure that the 40 nm and 70 nm diparticle mixture performs better than the other particle mixtures for the 80 nm membrane pore size due to low particle loss and agglomeration effects. With high concentration, i.e. 180 ppm, almost all the diparticle mixtures give close to 100 % COD removal for most membrane pore sizes whereas the triparticle mixture has shows good results only up to the 30 nm membrane pore size and then shows a notable decrease due to the particle loss effects.

To further elucidate the mineralization potential of NANF, toxicity studies with resazurin were conducted for the four combinations for COD removal: the effect of low and high concentrations i.e. 60 ppm and 180 ppm were studied for both the individual particle study (Fig. 7a & b) and also for the particle mixture study (Fig. 7c & d). The results of this study are similar to those observed for the COD reductions and hence the results are not separately presented.

Thus it can be inferred that the small particles perform well even at low concentration for small membrane pore size while the large particles perform well only at high particle concentration for all membrane pore sizes. The diparticle mixture is most effective in obtaining good filtration results at low particle concentrations for all the membrane pore sizes.

Conclusion

This study explores the possibilities of improving toxic chemicals/particles removal through novel nanoparticles assisted nanofiltration (NANF). For this purpose, silver nanoparticles were prepared with three distinct average particle sizes of 10 nm, 40 nm and 70 nm and they were characterized in size and morphology by transmission electron microscopy (TEM). A model phenol compound was prepared and was used for filtration studies through nanoporous membranes with four distinct average pore sizes of 10 nm, 30 nm, 50 nm and 80 nm. This innovative approach yielded as high as 675 % improvement with respect to membrane filtration without nanoparticles. Further the effects of particle size and particle concentrations on filtration were studied. Triparticle

mixtures in equal proportion (1:1:1) showed 100 % phenol removal at 180 ppm up to the 50 nm pore size membranes and was reduced to 85 % phenol removal for the 80 nm pore membrane. At 120 ppm, the triparticle mixture showed 100 % phenol removal up to the 30 nm pore size membrane and then decreased notably. The 10 nm and 40 nm diparticle mixture showed > 95 % phenol removal for both 120 ppm and 180 ppm up to the 50 nm pore size membrane and then decreased, whereas 40 nm with 120 ppm particles showed only 80 % phenol removal for the 50 pore membrane. Both the 10 nm & 70 nm and 40 nm & 70 nm diparticle mixtures at 180 ppm showed 100 % phenol removal on all the membranes. The triparticle and diparticle studies reveal that the particle mixtures result in more phenol removal % than that of the individual particles particularly for small particles for large pore size membranes and larger particles at low particle concentrations. Similar trends were observed both with the COD and toxicity reduction which shows the effectiveness of using particle mixtures. Overall, from this study it can be inferred that NANF has very good potential for treating waste water and the removal of the toxic contents.

Abbreviations
NANF: nanoparticle assisted nano filtration; M1: membrane with 10 nm pore size; M2: membrane with 30 nm pore size; M3: membrane with 50 nm pore size; M4: membrane with 80 nm pore size.

Acknowledgement
The corresponding author thanks MHRD (Ministry of Human Resource Development), Government of India, for providing financial support through their institute, National Institute of Technology, Trichy (in the form of seed money for research scholars).

Authors' contributions
LDN designed the entire experimental set-up and conducted the experiments. SS coordinated along with MG to interpret the data. MG also assisted LDN in analysis and nanoparticle preparation. SPS unique contribution is in nanoparticle preparation and nanoparticle analysis. PS exceptional input is in nanomembrane preparation and thorough verification of the results. All authors read and approved the manuscript.

Competing interests
The authors declare that they have no competing interests.

Author details
[1]Department of Chemical Engineering, National Institute of Technology, Tiruchirapalli 15, India. [2]Center for Environmental Engineering, PRIST University, Thanjavur, India. [3]Department of Textile Technology, PSG College of Technology, Coimbatore, India. [4]Department of Chemical Engineering, University of California Davis, Davis, CA 95616, USA.

References
1. Goel M, Chovelon JM, Ferronato C, Bayard R, Sreekrishnan TR. The remediation of wastewater containing 4-chlorophenol using integrated photocatalytic and biological treatment. Journal of Photochemistry & Photobiology, B: Biology. 2010;98:1–6.
2. Jin XW, Li EC, Lu SG, Qiu ZF, Sui Q. Coking wastewater treatment for industrial reuse purpose: combining biological processes with ultrafiltration, nanofiltration and reverse osmosis. J Environ Sci. 2013;25:1565–74.

3. Kamani H, Nasseri S, Khoobi M, Nodehi RN, Mahvi AH. Sonocatalytic degradation of humic acid by N-doped TiO2 nano-particle in aqueous solution. J Env Health Sc Engg. 2016;14:3–10.

4. Khazaei M, Nasseri S, Ganjali MR, Khoobi M, Nabizadeh R, Mahvi AH, Nazmara S, Gholibegloo E. Response surface modeling of lead (II) removal by graphene oxide-Fe3O4nanocomposite using central composite design. J Env Health Sc Engg. 2016;14:2.

5. Naghizadeh A, Nasseri S, Mahvi AH, Nabizadeh R, Kalantary RR, Rashidi A. Continuous adsorption of natural organic matters in a column packed with carbon nanotubes. J Env Health Sc Engg. 2013;11:14–9.

6. Serrano D, Suárez S, Lema JM, Omil F. Removal of persistent pharmaceutical micropollutants from sewage by addition of PAC in a sequential membrane bioreactor. Water Res. 2011;45:5323–33.

7. Sahar E, David I, Gelman Y, Chikurel H, Aharoni A, Messalem R, Brenner A. The use of RO to remove emerging micropollutants following CAS/UF or MBR treatment of municipal wastewater. Desalination. 2011;273:142–7.

8. Chon K, Kyongshon H, Cho J. Membrane bioreactor and nanofiltration hybrid system for reclamation of municipal wastewater: removal of nutrients, organic matter and micropollutants. Bioresour Technol. 2012;122:181–8.

9. Yang X, Flowers RC, Weinberg HS, Singer PC. Occurrence and removal of pharmaceuticals and personal care products PPCPs in an advanced wastewater reclamation plant. Water Res. 2011;45:5218–28.

10. Kuncoro EP, Roussy J, Guibal E. Mercury recovery by polymer-enhanced ultrafiltration: comparison of chitosan and polyethyleneimine used as macroligand. Sep Sci Technol. 2005;40:659–84.

11. Korus L, Bodzek M, Loska K. Removal of zinc and nickel ions from aqueous solutions by means ofthe hybrid complexation-ultrafiltration process. Separation and Purification Technology. 1999;172:111–6.

12. Cañizares P. Effect of polymer nature and hydrodynamic conditions on a process of polymer enhanced ultrafiltration. Journal of Membrane Science. 2005;2531–2:149–63.

13. Llorens J, Pujóla M, Sabate J. Separation of cadmium from aqueous streams bypolymer enhanced ultrafiltration: a two-phase model for complexation binding. Journal of Membrane Science. 2004;2392:173–81.

14. Cordova RSM, Dams RI, Cordova REV, Radetski MR, Corrêa AXR, Radetski CM. Remediation of phenol-contaminated soil by a bacterial consortium and Acinetobacteria calcoaceticus isolated from an industrial wastewater Treatment plant. Journal of Hazardous Materials. 2009;164:61–6.

15. Arutchelvan V, Kanakasabai V, Elangovan R, Nagarajan S, Muralikrishnan V. Kinetics of high strength phenol degradation using Bacillus brevis. Journal of Hazardous Materials. 2005;B17:234–40.

16. Gulrajani ML Gupta D, Periyasamy S, Muthu SG. Preparation and application of silver nanoparticles on silk for imparting antimicrobial properties. Journal of Applied Polymer Science. 2008;108(1):614–23.

17. Dannis M. Determination of Phenols by the Amino-Antipyrine Method. Sewage and Industrial wastes. 1951;23:1516–22.

18. Strotmann UJ, Butz B, Bias WR. The dehydrogenase assay with resazurin – practical performance as a monitoring system and pH-dependent toxicity of phenolic compounds. Ecotoxicologyand Environmental Safety. 1993;25:79–89.

19. Stroeve P, Rahman M, Naidu LD, Gilbert B, Morteza M, Ramireze P, Salvador M. Protein diffusion through charged nanopores with different radii at low ionic strength. PhysChemChemPhys. 2014;16:21570.

20. Chun KY, Stroeve P. Protein Transport in Nanoporous Membranes Modified with Self-Assembled Monolayers of Functionalized Thiols. Langmuir. 2002; 18:4653–8.

21. Periaysamy S. Studies on Modification of Silk using UV Radiation, PhD thesis, Department of Textile Technology, Indian Institute of Technology Delhi, New Delhi, India. 2007.

22. Geckeier KE. Polymer-metal complexes for environmental protection. Chemoremediation in the aqueous homogeneous phase. Pure and Applied Chemistry. 2001;73(1):129–36.

23. Li CW. Polyelectrolyte enhanced ultrafiltration (PEUF)for the removal ofCd(II): Effects oforganic ligands and solution pH. Chemosphere. 2008;72(4):630–5.

24. Theodore L. Nanotechnology: Basic Calculations for Engineers and Scientists, Chapter 5: Particle Size, Surface Area and Volume. New York: John Wiley & Sons; 2006.

25. Sarkar M, Acharya KP. Use of fly ash for the removal of phenol and its analogues from contaminated water. Waste Manage. 2006;26:559–70.

26. Roostaei N, Tezel FH. Removal of phenol from aqueous solutions by adsorption. J Environ Manage. 2004;70(2):157–64.

17

Removal of micropollutants through a biological wastewater treatment plant in a subtropical climate, Queensland-Australia

Miguel Antonio Reyes Cardenas[1], Imtiaj Ali[2], Foon Yin Lai[3], Les Dawes[4], Ricarda Thier[5] and Jay Rajapakse[4*]

Abstract

Background: Municipal wastewaters contain a multitude of organic compounds derived from domestic and industrial sources including active components of pharmaceutical and personal care products and compounds used in agriculture, such as pesticides, or food processing such as artificial sweeteners often referred to as micropollutants. Some of these compounds or their degradation products may have detrimental effects on the environment, wildlife and humans. Acesuflame is one of the most popular artificial sweeteners to date used in foodstuffs. The main objectives of this descriptive study were to evaluate the presence of micropollutants in both the influent and effluent of a large-scale conventional biological wastewater treatment plant (WWTP) in South-East Queensland receiving wastewater from households, hospitals and various industries.

Methods: Based on USEPA Method 1694: Filtered samples were spiked with mass-labelled chemical standards and then analysed for the micropollutants using liquid chromatography coupled with tandem mass spectrometry.

Results: The presence of thirty-eight compounds were detected in the wastewater influent to the treatment plant while nine of the compounds in the categories of analgesic, anti-inflammatory, alkaloid and lipid/cholesterol lowering drugs were undetectable (100 % removed) in the effluent. They were: **Analgesic:** Paracetamol, Salicylic acid, Oxycodone; **Anti-inflammatory:** Naproxen + ve, Atorvastatin, Indomethacin, Naproxen; **Alkaloid:** Caffeine; **Lipid/cholesterol lowering:** Gemfibrozol.

Conclusions: The study results revealed that the micropollutants removal through this biological treatment process was similar to previous research reported from other countries including Europe the Americas and Asia, except for acesulfame, a highly persistent artificial sweetener. Surprisingly, acesulfame was diminished to a much greater extent (>90 %) than previously reported research for this type of WWTPs (45–65 %) that only include physical removal of objects and solids and a biodegradation step.

Keywords: Biological wastewater treatment, Micropollutants removal, Sub-tropical climate

Background

Health and environmental concerns about the effects of micropollutants in wastewater have become increasingly important in wastewater management. The term micro-constituents includes pharmaceuticals and personal care products (PPCPs), and other compounds that may be found in wastewater in small amounts including compounds used in food processing and agriculturally used pesticides. Public concern increases particularly in situations where wastewater effluent is released into the environment (e.g., streams and rivers) that are then used as a raw potable water source for communities located downstream [1]. This especially concerns countries in Europe, where recycled water is used not only for agricultural purposes but also for preparation of drinking water. Water usage has become highly critical in other countries as well. For example, several states in Australia are committed to recycle more water in the future due to Australia's general arid climate, frequent droughts and other pressures. Local governments are searching for strategies to

* Correspondence: jay.rajapakse@qut.edu.au
[4]Science and Engineering Faculty, School of Earth, Environment and Biological Sciences, Queensland University of Technology, QLD 4001 Brisbane, Australia
Full list of author information is available at the end of the article

minimise micropollutant release into surface waters and/or increase of removal from wastewater, in order to ensure the health of humans and their ecosystem [2].

As of 2006, there are about 50,000 chemicals used for industrial, agricultural and veterinary purposes in Australia [3]. Since the early 1990s, chemical assessments have taken place which resulted in the creation of different strategies and regulations for the utilisation and manipulation of pesticides, medicines, and so forth [4]. The National Industrial Chemicals Notification and Assessment Scheme (NICNAS) and the Therapeutic Goods Administration are examples of regulators created to control the use and disposal of industrial chemicals and pharmaceuticals, respectively. Significant questions remain about the types and levels of monitoring of treatment processes required in order to adequately protect human health and the environment.

The three largest sources of PPCPs include industry, hospitals, and private homes. PPCPs enter into the sanitary sewer primarily through excretion of un-metabolised pharmaceuticals [5]. Table 1 shows urinary excretion percentage as parent compound of some of the most common pharmaceuticals found in sewerage systems. In addition, residual products are frequently discarded via the sewerage system. For example, the results from a survey of the American public found that only 1.4 % of the surveyed people returned unused medication to the pharmacy, whereas 54 % threw them away and 35.4 % disposed them in the sink or the toilet [6]. Another source of these compounds are uncontrolled landfill sites where they and their chemical or biological degradation products reach nearby rivers or groundwater as surface run off or leachate [7]. Surface run-off may also contain chemicals from agricultural activity such as pesticides and animal medicines [8].

Recently, artificial sweeteners (ASs) have been identified as emerging environmental contaminants [9–13]. ASs are widely used in foods, particularly beverages, as sugar substitutes and are excreted mainly unmetabolised. They are excreted via the kidney and reach surface waters of the environment mainly in this unchanged form [11–13].

Table 1 Percentage of drug found in urinary excretion as parent compound for common medicines

Drug	Class	Compound excreted (%)	References
Ibuprofen	Non-steroidal anti-inflammatory (NSAID)	10	[53]
Paracetamol	Painkiller	4	[54]
Erythromycin	Antibacterial	25	[54]
Sulfamethoxazole	Antibacterial	15	[55]
Atenolol	β - blocker	90	[53]
Metoprolol	β - blocker	10	[54]
Carbamazepine	Antiepileptic	3	[54]

Acesulfame (ACE) is one of these sugar substitutes found in the aquatic environment as a result of effluents containing ACE from wastewater treatment plants being discharged into water courses. Acesulfame concentrations in the wastewater treatment plant effluents were reported from as low as 20 µg/L [14] up to very high values as 2.5 mg/L [11–13, 15]. It is so persistent that on the one hand, it has raised concern as an environmental pollutant but on the other hand, it is appreciated as a marker for contamination of e.g., groundwater with domestic wastewater [12, 16]. However, as Lange et al. [12] identified the actual knowledge about environmental persistence and potential hazard of ACE is not well understood.

The impact of most micropollutants on human health and environment is not well understood. However, effects of some PPCPs on human health and environment are known. For example triclosan, an antibacterial and antifungal used in personal care products such as soap and tooth paste, is called an endocrine disrupting compound (EDC) because it interferes with natural hormonal functions, potentially altering metabolism, development, reproduction and growth [17] including decline in reproductive function in men [18]. Effects of EDCs on the environment comprise birth abnormalities and feminisation of organisms including fish, frogs, birds and mammals [4]. Effects on aquatic organisms have also been documented for certain herbicides, such as 2,4-Dichlorophenoxyacetic acid (2,4 D), which alters the shell formation of the bivalve *Anodonta cygnea* [16]. This compound has also been investigated intensely for chronic toxicity in humans but results remained inconclusive [19].

Some water micro-constituents, such as the antibiotic ciprofloxacin and the artificial sweetener acesulfame, have been shown to be degraded during certain treatment steps, which can lead to conversion into more toxic compounds [9, 20]. Antibacterials are of concern not only because of their toxicity but also as harbinger of bacterial resistance. Bacteria in wastewater comprise very high levels and varieties of resistance genes, which may be disseminated to human and animal pathogens [21–23]. Multi-resistance in bacteria is a major global health issue that has further restricted treatment options for already limited options of infectious diseases [24].

There is only limited knowledge around the accumulative effects of individual compounds and combination effects of mixtures of micropollutants in wastewater. Some advances in this field have been made with the consideration of toxic equivalent concentrations and the use of mode of action based test batteries where concentrated water samples are tested and the risk is assessed by comparing the results of e.g., environmental water samples to specific reference compounds for each test [25–27]. Advanced methods of water treatment have been designed for reclamation of wastewater for drinking purposes including reverse osmosis, ozonation, UV-irradiation, nano- and ultrafiltration, activated carbon filter and biofiltration. When applied after traditional wastewater treatment these techniques reduce a wide variety of biological effects including estrogenic, genotoxic, neurotoxic and phyto- toxic effects. These reduction of these effects varied depending on the treatment combined with the compound composition of the water [28–30].

In summary, it becomes clear that additional steps for wastewater treatment are essential to decrease the dis- charge of micropollutants [31] into Australian rivers and estuaries as current traditional wastewater treatment is insufficient to avoid their release into the environment. In South-East Queensland, several studies have investi- gated the removal of micropollutants in modern state of the art water reclamation plants [27–30]. These studies have focussed on the removal of micropollutants with sophisticated methods. Furthermore, in the South-East Queensland region, Shareef et.al [32] investigated the re- moval of EDCs and PPCPs at Oxley Creek and Luggage Point which received wastewater from domestic and domestic/industrial sources respectively. Ying et al. [33] studied the estrogens and xenoestrogens removal in the final effluents of five wastewater treatment plants from South-East Queensland. Also, Tan et al. [34] did a com- prehensive study of the removal of 15 EDCs and estrogen equivalent (EEqs) of five wastewater treatment plants from South-East Queensland. Two other studies have investi- gated the fate of antibacterials and these were used for comparison where appropriate [35, 36].

This paper assesses the efficiency of conventional waste- water management practices on the removal of PCPPs and other micropollutants in a traditional 3-step WWTP in South-East Queensland to inform future individual wastewater management plans for this WWTP. The con- centrations of 95 different micropollutants in wastewater influent and effluent were determined. Removal rates for the biodegradation step were calculated for compounds that were found in the influent.

Methods
Description of the wastewater treatment plant
The municipal wastewater treatment plant (WWTP) receives wastewater mainly from households, hospitals and various manufacturing and service industries includ- ing meat manufacturing, automobile repairing and maintenance, fuel retailing, laundry and dry cleaning and spirit manufacturing.

The WWTP is equipped with a conventional 3-step treatment process including physical removal of objects and solids, biological oxidation and a chlorine based disinfection process (Fig. 1). Preliminary treatment in- volves three band screens and two grit chambers, which removes grit and screenings. Sodium hypochlorite (NaOCl) and sodium hydroxide (NaOH) are used in the preliminary treatment for odour control and pH adjustments. The sys- tem has a fully automated wet weather by pass system to send excess flow into environment (a receiving stream). The biological process integrates four oxidation ditches (OD) with an aerobic and anaerobic zone followed by clari- fiers. The sludge dewatering facility consists of a gravity

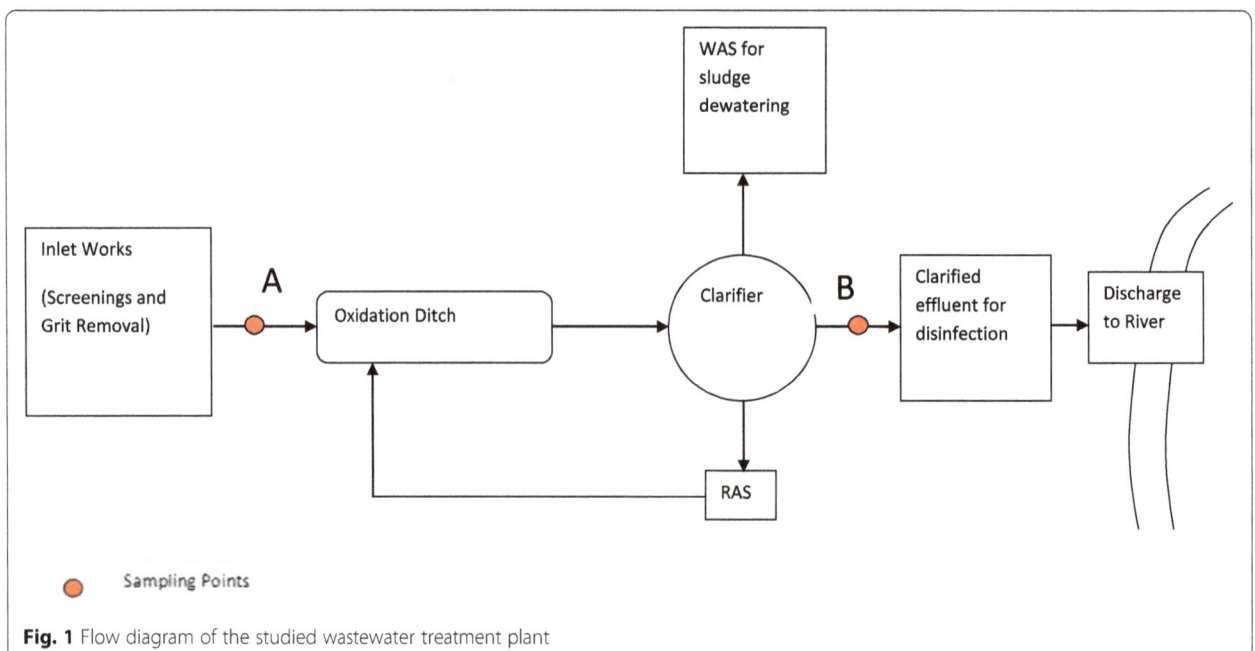

Fig. 1 Flow diagram of the studied wastewater treatment plant

drainage deck (GDD) and a belt filter (BFP) process. The disinfection process uses chlorination. Current average flow rate of influent is 45 ML/day.

Sampling points are shown in Fig. 1 as A and B. Sampling Point A: The wastewater influent, the raw wastewater, was collected after screenings and grit removal. Sampling Point B: The effluent was collected after the biological oxidation ditch and clarifier, but before disinfection of the final effluent. WAS: Waste Activated Sludge, RAS: Return Activated Sludge.

Sampling and analysis

During September 2012, grab samples were collected in alignment with the Water Monitoring Data Collection Standards of Queensland Government using the methodology described in AS/NZS 5667.1 Water Quality – Sampling – Guidance on the design of sampling programs, sampling techniques and the preservation and handling of samples [37–39]. Two sets of samples were collected at two different locations (influent A and effluent B) of the WWTP on 2 different days. The sampling points were located at two different stages of the treatment plant process: Point A was located after primary treatment and before entering the biological treatment unit (oxidation ditch), while Point B after the biological treatment and clarification process (Fig. 1). In total eight grab samples were collected in 500 ml plastic bottles following the hydraulic retention time (average HRT = 24 h; i.e., 20 h in oxidation ditch and 4 h in clarifier) of the WWTP and together with blank samples were stored at 4 °C until analysis.

The samples in this study were analysed for a total of 95 compounds, including PPCPs, agricultural, food processing and other micropollutants, which are commonly found in wastewater, using an in-house validated analytical method from Queensland Health Forensic and Scientific Services. This method has been optimised according to the USEPA Method 1694 [40]. A 1 mL filtered-sample was spiked with mass-labelled chemical standards (compensating for any instrumental variations during analysis) and analysed using liquid chromatography (Shimadzu Prominence, Shimadzu Corp., Kyoto, Japan) coupled with Tandem mass spectrometry (LC-MS/MS; Applied Biosystem/Sciex API 4000Q system). Separation of the compounds was performed on a C18 analytical column (Luna C18, 150X2.1 mm, 3 μm, Phenomenex) at 45 °C with a gradient mobile phase (A: 1 % acetonitrile, 99 % Milli-Q water and 0.1 % formic acid; B: 95 % acetonitrile, 5 % Milli-Q water and 0.1 % formic acid) programmed as: 8 % B at start; ramped up to 35 % B at 3.5 min; increased to 100 % B at 11 min; held 100 % B for 4 mins; equilibration of the column for 3 mins. The mass spectrometry was operated in a multiple reaction monitoring mode to identify and quantify the micropollutants in the samples.

Results

Wastewater samples were taken at two sampling points before and after the biodegradation step, which has an aerobic and an anaerobic zone and analysed 95 compounds including PPCPs, compounds used in food processing and agriculture by LC-MS/MS. From the two sets of samples collected on two different days (with 5-days apart), during the month of September 2012, the arithmetic average pollutant levels of the two samples were calculated. Out of the 95 compounds tested, we found 38 compounds in the influent, mainly drugs or drug metabolites as shown in Table 2. There were three pesticides and two food components, the artificial sweetener acesulfame and caffeine. Nine of the identified compounds were undetectable in the effluent, but 29 were still present to various degrees in the effluent samples after the biodegradation step in the oxidation ditch.

The majority were removed to at least 80 % (Fig. 2). The most persistent compounds included MCPA (2-methyl-4-chlorophenoxyacetic acid), 2,4-D (2,4-Dichloro-phenoxyacetic acid), desmethyl citalopram, phenytoin and carbamazepine (Fig. 2). Surprisingly, acesulfame was removed in the conventional treatment process by 92 % (Fig. 3). Three compounds paracetamol, salicylic acid and caffeine found at concentrations of 289.40 μg/L, 32.74 μg/L, and 78.09 μg/L respectively in the influent were undetectable in the effluent (Fig. 3).

Tables 3 and 4 present the characteristics of raw wastewater (influent) received by the treatment plant and the quality of treated effluent before disinfection and discharge respectively. The data are typical for water quality of this type of traditional treatment plants. The treatment plant total nitrogen (TN) discharged in the adjacent licensed discharge point is 53 t per year, keeping acceptable levels according to the maximum permitted by regulator (Environment and Heritage Protection) for the river (98 t per year).

Discussion

The concentrations of observed micropollutants in the wastewater of the investigated WWTP in South-East Queensland were generally within the lower range of previously reported values from around the world [36, 41–45]. This is in agreement with the comparatively low population density in the catchment of the observed WWTP. Where applicable, all values were at least one order of magnitude below the Australian Drinking Water Guidelines 6 [46] and the Australian Guidelines for Water Recycling [47]. In addition, the differences between concentrations of micropollutant parent compounds in influent and effluent water observed in our study fit well within the ranges of current reports from other studies [12, 43].

In the present study, caffeine is at about 80 μg/L in the influent samples. This is about 10 times higher than

Table 2 List of 95 compounds tested in the influent (The 38 compounds detected in the influent are in bold with 9 compounds absent in effluent shaded)

Atenolol	**Phenytoin**	Paraxanthine
Ranitidine	**Oxazepam**	Asulam
Codeine	Desmethyldiazepam	**Diatriozate**
Gabapentin	Sulfsalazine	Iopromide
Lincomycin	**Temazepam**	**Hydrochlorthiazide**
Oxycodone	**Naproxen**	**Acesulfame**
Iopromide	Praziquantel	Acetylsalicylic acid
Trimethoprim	Diazepam	Chloramphenicol
Doxylamine	Atorvastatin	**Salicylic acid**
Paracetamol	**Indomethacin**	**Frusemide**
Ciprofloxacin	**Diclofenac**	**Naproxen +Ve**
Metoprolol	Desisopropyl Atrazine	Warfarin
Tramadol	Desethyl Atrazine	**Atorvastatin**
Caffeine	Tebuthiuron	Fluvastatin
Sulphadiazine	Hexazinone	Diclofenac
Sulphathiazole	Ametryn	Ibuprofen
Cephalexin	Bromacil	**Gemfibrozol**
Venlafaxine	Simazine	Dalapon
Propranolol	Prometryn	Picloram
Primidone	Terbutryn	Fluroxypyr
Desmethyl Citalopram	Flumeturon	Dicamba
Dapsone	Atrazine	**MCPA**
Erythromycin	Diuron	**24 D**
Erythromycin hydrate	3,4 DiCl Aniline	Triclopyr
Sulphamethoxazole	Metolachlor	24 DP
Tylosin	Haloxyfop methyl	Mecoprop
Ifosfamide	Haloxyfop-ethoxyethyl	24 DB
Cyclophosphamide	Propoxur	MCPB
Fluoxetine	Carbaryl	Haloxyfop
Sertraline	Diazinon	Asulam
Roxithromycin	Chlorpyriphos	Bromoxynil
Carbamazepine	Flamprop-M-methyl	

the study by Shareef et al., (8 μg/L = 8000 ng/L) [32]. The difference is reasonable due to the difference of business and human activities, lifestyle, and probably season of the year. Our studied WWTP serves the catchment with about 180,000 people (2011 census). For Oxley and Luggage point, the WWTPs serves about 85,000 and 300,000 people respectively.

With an influent concentration of about 43 μg/L, the surprisingly higher reduction (about 92 %) of acesulfame at the Queensland WWTP is the first report of reduction of this artificial sweetener in a traditional 3-step WWTP at this high level. The QA/QC details are presented below:

Internal standard recovery: 100 % ± 3.16 % (mean ± S.D.) for caffeine and 103 % ± 6.84 % for acesulfame.
Variation of duplicate analysis: <10 % variation (CV%) for both caffeine and acesulfame.

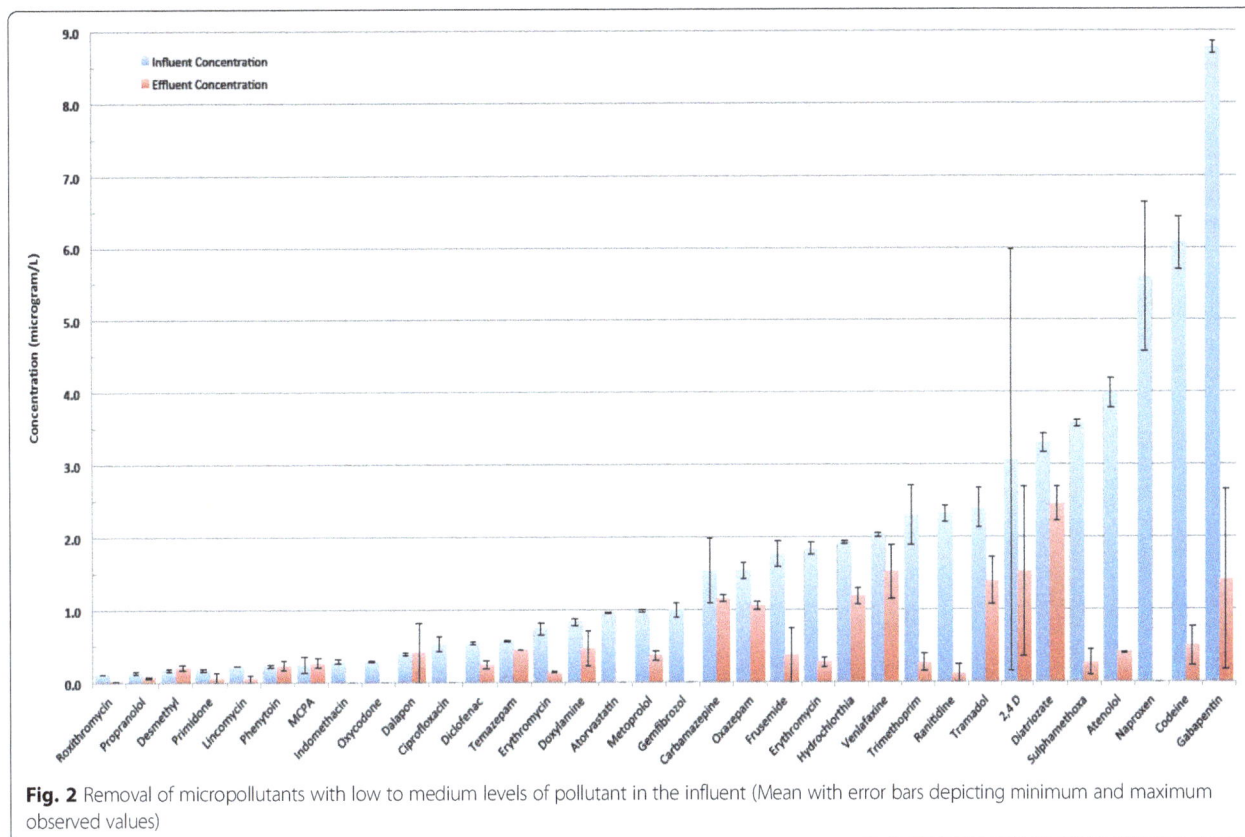

Fig. 2 Removal of micropollutants with low to medium levels of pollutant in the influent (Mean with error bars depicting minimum and maximum observed values)

Inter-day variation: <20 % (CV%) for caffeine and <5 % for acesulfame.

Milli-Q blank samples were included in the batch of analysis. No contamination of these chemicals was found in these blank samples.

Removal rates vary from study to study, which is anticipated as they depend on various conditions including the physico-chemical properties of the compounds and the treatment process itself [12, 43]. As noted earlier, acesulfame concentrations in the wastewater treatment plant effluents were reported from as low as 20 μg/L (Scheurer et al., 2009) [14] up to higher values as 2.5 mg/L (Loos et al., 2013) [15]. In the German study [14] acesulfame concentration in the influent to the wastewater treatment plant ranged from 34 to 50 μg/L and up to 41 % removal was observed.

Some compounds are biodegraded by bacteria and in these cases the removal rates can depend on their initial concentration as the removal depends partially on enzyme kinetics [42]. These effects are, however, unlikely to explain sufficiently the extremely high removal rate of ACE found in this study. In previous studies, ACE was removed consistently at 45–65 % by conventional WWT and only additional treatment with ozone or UV light increased the removal rate up to 90 % [10, 16, 40, 48]. But ozonation

and/or UV light treatment of wastewater as currently practiced in water treatment plants, which prepare drinking water in Switzerland and Germany, reduced the amount of ACE by only 30 % [10, 41, 48]. The detection of a significant amount of unexpected compounds in the wastewater system, in particular the anomolous level of acesulfame removal suggests the need for a detailed assessment. Further investigation of these should use improved analytical protocols.

Our findings also raise questions with regard to the impact of the remaining ACE and its water-soluble degradation products on the environment as well as downstream users of the surface waters. However, degradation of ACE depends on the decomposition process and conditions and many transformation products that have been identified from different processes [9, 10, 48]. It is important to investigate whether the degradation products of ACE under conventional WWT in SE-Qld conditions are identical to the intermediates and products found in these studies and to assess their toxicological impact on humans and the environment.

While ACE has been substantially tested for its lack of adverse effects to humans before its registration as food additive, only limited data are available about its ecotoxic potential. Ecological toxicity tests using duckweed, *Lemna minor*, green algae, *Scenedesmus vacuolatus*, and

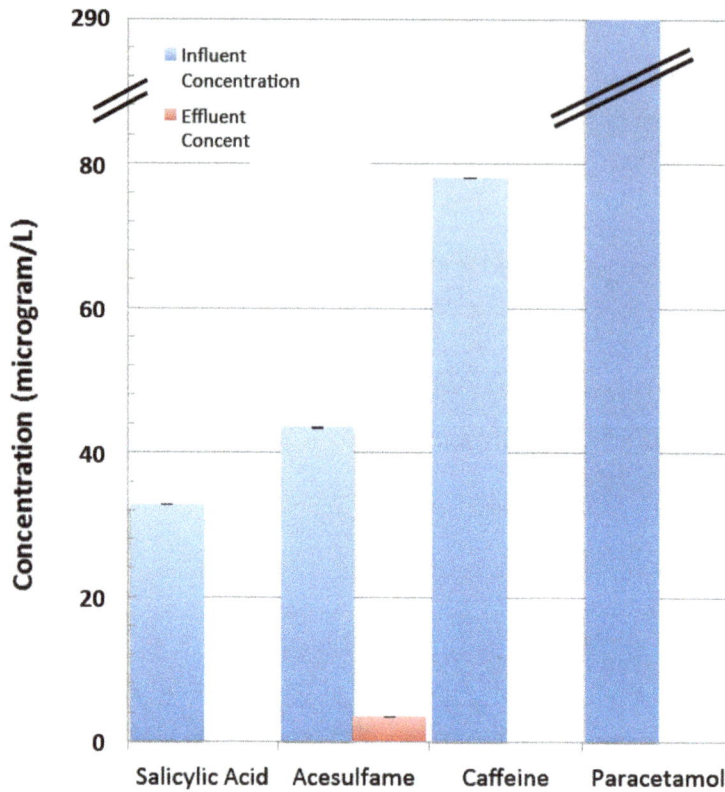

Fig. 3 Removal of micropollutants with high levels of pollutant in the influent (Mean with error bars depicting minimum and maximum observed values)

water fleas, *Daphina magna,* revealed very high Lowest Observed Effect Concentrations (LOEC) for ACE [49].

In recent years, detection of micropollutants in wastewater and surface waters has inspired the search for practicalities beyond monitoring of their discharge into ecosystems. Monitoring the use of specific drugs, including illicit drugs, through wastewater analysis in regions and countries has become common, for example to estimate illicit drug use per capita or to monitor increases over holiday periods [50, 51]. Individual micro-constituents, such as the drugs valsartan acid, carbamapezine and the artificial sweetener acesulfame, were identified to have physico-chemical properties, which makes them resistant to degradation in wastewater treatment plants. This persistence has led to the concept of using them as tracers and markers in water bodies e.g., for identification of groundwater contamination with urban wastewater [52]. In fact, removal rate of carbamazepine in our study was also poor

Table 3 Inlet wastewater quality

Date	pH	NH$_3$ mg/L	NOx-N mg/L	ortho-PO$_4$-P mg/L	TP mg/L	TN mg/L	TSS mg/L	TKN mg/L	COD mg/L	BOD mg/L	Alkalinity mg/L (CaCO$_3$)
3/09/2012	7.5	66	0.2	7.7	8.7	71	330	71	760	460	340
5/09/2012	7.1	74	0.1	7.7	8.2	76	370	76	800	440	350
10/09/2012	7.9	49	0.4	7.0	7.0	73	560	73	750	620	330
12/09/2012	7.5	54	0.2	8.0	11.0	87	580	87	1180	700	370
17/09/2012	7.6	46	0.1	6.8	10.0	70	390	70	830	480	340
19/09/2012	7.6	58	0.1	7.4	7.9	83	950	83	950	618	370
24/09/2012	7.7	50	0.1	6.9	10.0	69	420	69	810	540	370
26/09/2012	7.5	63	0.1	7.7	10.0	80	510	80	960	540	370

TN total nitrogen, *TP* total phosphorus, *TSS* total suspended solids, *TKN* total Kjedahl nitrogen, *COD* chemical oxygen demand, *BOD* biochemical oxygen demand

Table 4 Quality of treated effluent after clarification

Date	NH$_3$ mg/L	NOx mg/L	ortho PO$_4$-P mg/L	pH	TSS mg/L	TP mg/L
3/09/2012	2.1	0.7	3.2	7.6	4	3.4
5/09/2012	1.7	0.8	3.7	7.6	5	3.9
10/09/2012	0.8	0.7	2.9	7.8	6	3.1
12/09/2012	0.7	0.7	4.3	7.6	6	4.6
17/09/2012	0.9	0.7	4.1	7.7	7	4.2
19/09/2012	0.1	1.8	5.6	7.5	3	5.8
24/09/2012	0.4	0.4	2.6	7.7	7	2.8
26/09/2012	0.3	0.7	3.4	7.7	7	3.6

TN total nitrogen, *TP* total phosphorus, *TSS* total suspended solids

(~20 %). In light of our results with acesulfame it appears relevant to consider environmental conditions on local removal efficiencies of such tracers for appropriate interpretation of quantitative results in particular.

Conclusions

The presence of 95 common micropollutants of domestic, industrial and agricultural, including pharmaceuticals, personal care products (PCPPs) and food components was determined in the wastewater influent and effluent of a particular WWTP in SE-Qld. Thirty eight compounds were found in the influent. Although this is a relatively simple conventional wastewater treatment, the levels of most of these chemicals were reduced similarly to the extent in more sophisticated WWTP, particularly regarding anti-inflammatory drugs, analgesics and antibacterials. Surprisingly, more than 90 % of the artificial sweetener acesulfame was removed in this WWTP. This anomalous level of acesulfame removal suggests the need for a detailed assessment.

Nine of the compounds in the categories of analgesic, anti-inflammatory, alkaloid and lipid/cholesterol lowering drugs were undetectable (100 % removed) in the effluent. They were: **Analgesic**: Paracetamol, Salicylic acid, Oxycodone; **Anti-inflammatory**: Naproxen + ve, Atorvastatin, Indomethacin, Naproxen; **Alkaloid**: Caffeine; **Lipid/cholesterol lowering**: Gemfibrozol.

Abbreviations

ACE: Acesulfame; AS: Artificial sweeteners; EDC: Endocrine disrupting compounds; HRT: Hydraulic retention time; LC-MS/MS: Liquid chromatography coupled with Tandem mass spectrometry; PPCPs: Pharmaceuticals and personal care products; WWTPs: Wastewater treatment plants

Acknowledgements

We greatly acknowledge the support provided by the management and operational staff of the WWTP in the investigations and the City Council for granting permission to publish the findings.
Steve Carter (Queensland Health Forensic Scientific Services, Special Services, Organic Chemistry) provided access to LC-MS/MS for the sample analysis.
The support of Professor Jochen Mueller (EnTox) and Dr Phong Thai (EnTox) with assistance in sample collection and preparation is greatly acknowledged. Thanks are due to Dr Paul Burrell of QUT for proof reading the manuscript.

Funding

Financial support provided by the School of Earth, Environmental and Biological Sciences (EEBS) of the Queensland University of Technology (QUT), under special projects, BEN910/2012.

Authors' contributions

MA, IA, JR, RT and LD contributed to drafting and editing the manuscript. YL contributed to sample analysis, interpretation of results and editing the manuscript. All authors read and approved the final manuscript.

Competing interests

The authors declare that they have no competing interests.

Author details

[1]MI Murrumbidgee Irrigation, Research Station Road, Hanwood, NSW 2680, Australia. [2]Treatment Program, Logan City Council, Logan City DC, QLD 4114, Australia. [3]National Research Centre for Environmental Toxicology (EnTox), The University of Queensland, Brisbane, QLD 4108, Australia. [4]Science and Engineering Faculty, School of Earth, Environment and Biological Sciences, Queensland University of Technology, QLD 4001 Brisbane, Australia. [5]Faculty of Health, Queensland University of Technology, QLD 4001 Brisbane, Australia.

References

1. Jones OAH, Lester JN, Voulvoulis N. Pharmaceuticals: a threat to drinking water? Trends Biotechnol. 2005;24:163–7.
2. Australian Government: Land & Water Australia.Targeting Endocrine Disruptors in Australia's Water Ways. 2008; www.lwa.gov.au: Product code PN22054: ISBN 9781 92 15444262 Available at: http://pandora.nla.gov.au/pan/103522/20091110-1348/lwa.gov.au/files/products/innovation/pn22054/pn22054.pdf. Accessed 08 May 2015.
3. Environment Protection and Heritage Council. National Framework for Chemicals Management in Australia – Discussion Paper [Online]. Australian Government, 2006. Available at: http://www.nepc.gov.au/system/files/resources/74b7657d-04ce-b214-d5d7-51dcbce2a231/files/cmgt-nchem-discppr-national-framework-chemicals-management-australia-200607.pdf. Accessed 20 October 2012.
4. EPA South Australia. Risks from endocrine disrupting substances in the South Australian aquatic environment. Adelaide: Environmental Protection Authority, 2008 Available at: http://www.epa.sa.gov.au/xstd_files/Water/Report/risks_endocrine.pdf. Accessed 20 October 2012.
5. Helland J. Endocrine Disrupters as Emerging Contaminants in Wastewater. Minnesota Minnessota House of Representatives: Research Department, 2006. Available at: http://www.house.leg.state.mn.us/hrd/pubs/endodis.pdf.
6. Kuspis D, Krenzelok E. What happens to expired medications? a survey of community medication disposal. US national library of medicine. Vet Hum Toxicol. 1996;38(1):48–9.
7. Ahel M, Mikac N, Cosovic B, Prohic E, Soukup V. The impact of contamination from a municipal solid waste landfill (zagreb, croatia) on underlying soil. Water Sci Technol. 1998;37:203–10.
8. WHO. International Program of Chemical Safety: Global Assessment of the State of the Science of Endocrine Disruptors World Health Organization. 2002. Available at: http://www.who.int/ipcs/publications/new_issues/endocrine_disruptors/en/.
9. Sang Z, Jiang Y, Tsoi YK, Leung KS. Evaluating the environmental impact of artificial sweeteners: a study of their distributions, photodegradation and toxicities. Water Res. 2014;52:260–74.
10. Scheurer M, Schmutz B, Happel O, Brauch HJ, Wülser R, Storck FR. Transformation of the artificial sweetener acesulfame by UV light. Sci Total Environ. 2014;15:425–32.
11. Kokotou MG, Asimakopoulos AG, Thomaidis NS. Artificial sweeteners as emerging pollutants in the environment: analytical methodologies and environmental impact. Anal Methods. 2012;4(10):3057–70.
12. Lange FT, Scheurer M, Brauch HJ. Artificial sweeteners–a recently recognized class of emerging environmental contaminants: a review. Anal Bioanal Chem. 2012;403(9):2503–18.
13. Gan Z, Sun H, Feng B, Wang R, Zhang Y. Occurrence of seven artificial sweeteners in the aquatic environment and precipitation of Tianjin, China.

Water Res. 2013;47(14):4928–37.

14. Scheurer M, Brauch HJ, Lange FT. Analysis and occurrence of seven artificial sweeteners in German waste water and surface water and in soil aquifer treatment (SAT). Anal Bioanal Chem. 2009;394(6):1585–94. doi:10.1007/s00216-009-2881-y. Epub 2009 Jun 16.

15. Loos R, Carvalho R, António DC, Comero S, Locoro G, Tavazzi S, Paracchini B, Ghiani M, Lettieri T, Blaha L, Jarosova B, Voorspoels S, Servaes K, Haglund P, Fick J, Lindberg RH, Schwesig D, Gawlik BM. EU-wide monitoring survey on emerging polar organic contaminants in Wastewater treatment plant effluents. Water Res. 2013;47(17):6475–87. doi:10.1016/j.watres.2013.08.024. Epub 2013 Sep 15.

16. Buerge IJ, Buser H-R, Kahle M, Müller MD, Poiger T. Ubiquitous occurrence of the artificial sweetener acesulfame in the aquatic environment: an ideal chemical marker of domestic wastewater in groundwater. Environ Sci Technol. 2009;43(12):4381–5.

17. Jackson J, Sutton R. Sources of endocrine-disrupting chemicals in urban wastewater, Oakland, CA. Sci Total Environ. 2008;405:153–60.

18. Evans R, Kortenkamp A, Martin O, McKinlay R, Orton F & Rosivatz E. State of the art assessment of Endocrine Disrupters, 2nd Interim Report. 2011; Project Contract Number 070307/2009/550687/SER/D3.

19. Burns CJ, Swaen G. Review of 2,4-dichlorophenoxyacetic acid (2,4-D) biomonitoring and epidemiology. Crit Rev Toxicol. 2012;42(9):768–86.

20. Garcia-Käufer M, Haddad T, Bergheim M, Gminski R, Gupta P, Mathur N, Kümmerer K, Mersch-Sundermann V. Genotoxic effect of ciprofloxacin during photolytic decomposition monitored by the in vitro micronucleus test (MNvit) in HepG2 cells. Environ Sci Pollut Res. 2012;19:1719–27.

21. Ahmad M, Khan AU, Wahid A, Ali AS, Ahmad F. Role of untreated waste water in spread of antibiotics and antibiotic resistant bacteria in river. Pak J Sci. 2013;65(1):10–4.

22. Thuy HTT, Nguyen TD. The potential environmental risks of pharmaceuticals in Vietnamese aquatic systems: case study of antibiotics and synthetic hormones. Environ Sci Pollut Res. 2013;20:8132–40.

23. Vaz-Moreira I, Nunes OC, Manaia CM. Bacterial diversity and antibiotic resistance in water habitats: searching the links with the human microbiome. FEMS Microbiol Rev. 2014;38:761–78.

24. WHO. Antimicrobial resistance: global report on surveillance 2014, World Health Organization. 2014. ISBN 978 92 4 156474 8. Available at: http://www.who.int/drugresistance/documents/surveillancereport/en/. Accessed 6 Oct 2016.

25. Escher BI, Bramaz N, Mueller JF, Quayle P, Rutishauser S, Vermeirssen EL. Toxic equivalent concentrations (TEQs) for baseline toxicity and specific modes of action as a tool to improve interpretation of ecotoxicity testing of environmental samples. J Environ Monit. 2008;10(5):612–21.

26. Escher BI, Bramaz N, Quayle P, Rutishauser S, Vermeirssen EL. Monitoring of the ecotoxicological hazard potential by polar organic micropollutants in sewage treatment plants and surface waters using a mode-of-action based test battery. J Environ Monit. 2008;10(5):622–31.

27. Macova M, Escher BI, Reungoat J, Carswell S, Chue KL, Keller J, Mueller JF. Monitoring the biological activity of micropollutants during advanced wastewater treatment with ozonation and activated carbon filtration. Water Res. 2010;44:477–92.

28. Reungoat J, Escher BI, Macova M, Mueller JF, Carswell S, Keller J. Removal of micropollutants and reduction of biological activity in a full scale reclamation plant using ozonation and activated carbon filtration. Water Res. 2010;44:625–37.

29. Reungoat J, Escher BI, Macova M, Keller J. Biofiltration of wastewater treatment plant effluent: Effective removal of pharmaceuticals and personal care products and reduction of toxicity. Water Res. 2011;45:2751–62.

30. Reungoat J, Escher BI, Macova M, Argaud FX, Gernjak W, Keller J. Ozonation and biological activated carbon filtration of wastewater treatment plant effluents. Water Res. 2012;46:863–72.

31. Jekel M, Dott W, Bergmann A, Dünnbier U, Gnirß R, Haist-Gulde B, Hamscher G, Letzel M, Licha T, Lyko S, Miehe U, Sacher F, Scheurer M, Schmidt CK, Reemtsma T, Ruhl AS. Selection of organic process and source indicator substances for the anthropogenically influenced water cycle. Chemosphere. 2015;125:155–67.

32. Shareef A, Williams M, Kookana R. Concentration of selected endocrine disrupting chemicals and pharmaceutical and personal care products entering wastewater treatment plants in South East Queensland. Urban Water Security Research Alliance, 2010; Technical Report No. 20.

33. Ying GG, Kookana RS, Kumar A, Mortimer M. Occurrence and implications of estrogens and xenoestrogens in sewage effluents and receiving waters from South East Queensland. Science of The Total Environment. 2009;407(18):5147–55.

34. Tan BLL, Hawker DW, Muller JF, Leusch FDL, Tremblay LA, Chapman HF. Comprehensive study of endocrine disrupting compounds using grab and passive sampling at selected wastewater treatment plants in South East Queensland, Australia. Environ Int. 2007;33:654–69.

35. Watkinson AJ, Murby EJ, Costanzo SD. Removal of antibiotics in conventional and advanced wastewater treatment: Implications for environmental discharge and wastewater recycling. Water Res. 2007;41:4164–76.

36. Watkinson AJ, Murby EJ, Kolpine DW, Costanzo SD. The occurrence of antibiotics inan urban watershed: From wastewater to drinking water. Sci Total Environ. 2009;407:2711–23.

37. Queensland Government. In: Department of Natural Resources and Water, editor. Water monitoring data collection standards. Brisbane: Queensland Government; 2007.

38. Standards Australia. Water quality - Sampling - Guidance on the design of sampling programs, sampling techniques and the preservation and handling of samples. Canberra: Standards Australia; 1997.

39. Standards Australia. Water quality—Sampling—Guidance on sampling of waste waters. Canberra: Standards Australia; 1998.

40. U.S. Environmental Protection Agency. Method 1694: Pharmaceuticals and Personal Care Products in Water, Soil, Sediment, and Biosolids by HPLC/MS/MS; Office of Water, Office of Science and Technology, Engineering and Analysis Division (4303 T): Washington DC, 2007; p77.

41. Ndler K, Hillebrand O, Idzik K, Strathmann M, Schiperski F, Zirlewagen J, Licha T. Occurrence and fate of the angiotensin II receptor antagonist transformation product valsartan acid in the water cycle – A comparative study with selected β-blockers and the persistent anthropogenic wastewater indicators carbamazepine and acesulfame. Water Res. 2013;47:6650–9.

42. Onesios-Barry KM, Berry D, Proescher JB, Sivakumar IK, Bouwer EJ. Removal of pharmaceuticals and personal care products during water recycling: microbial community structure and effects of substrate concentration. Appl Environ Microbiol. 2014;80(8):2440–50.

43. Luo Y, Guoa W, Ngo HH, Nghiemb LD, Hai FI, Zhang J, Liang S, Wang XC. A review on the occurrence of micropollutants in the aquatic environment and their fate and removal during wastewater treatment. Sci Total Environ. 2014;473–474:619–41.

44. Mohapatra DP, Brar SK, Tyagi RD, Picard P, Surampalli RY. Analysis and advanced oxidation treatment of a persistent pharmaceutical compound in wastewater and wastewater sludge-carbamazepine. Sci Total Environ. 2014;470–471:58–75.

45. Yang X, Riley C, Flowers B, Howard S, Weinberg B, Philip C, Singer B. Occurrence and removal of pharmaceuticals and personal care products (PPCPs) in an advanced wastewater reclamation plant. Water Res. 2011;45:5218–28.

46. NHMRC, NRMMC. Australian Drinking Water Guidelines Paper 6 National Water Quality Management Strategy. National Health and Medical Research Council, National Resource Management Ministerial Council, Commonwealth of Australia, Canberra. 2011. Version 2 (Updated and published 2013) Available at: https://www.clearwater.asn.au/user-data/resource-files/Aust_drinking_water_guidelines.pdf. Accessed 05 May 2015.

47. NRMMC–EPHC–NHMRC. Australian Guidelines for Water Recycling: Augmentation of Drinking Water Supplies. Natural Resource Management Ministerial Council-Environment, Protection and Heritage Council-National Health and Australian Medical Research Council, Canberra. 2008. Available at: http://www.environment.gov.au/system/files/resources/9e4c2a10-fcee-48ab-a655-c4c045a615d0/files/water-recycling-guidelines-augmentation-drinking-22.pdf. Accessed 05 May 2015.

48. Scheurer M, Storck FR, Brauch HJ, Lange FT. Performance of conventional multi-barrier drinking water treatment plants for the removal of four artificial sweeteners. Water Res. 2010;44(12):3573–84.

49. Stolte S, Steudte S, Schebb NH, Willenberg I, Stepnowski P. Ecotoxicity of artificial sweeteners and stevioside. Environ Int. 2013;60:123–7.

50. Lai FY, Ort C, Gartner C, Carter S, Prichard J, Kirkbride P, Bruno R, Hall W, Eaglesham G, Mueller JF. Refining the estimation of illicit drug consumptions from wastewater analysis: co-analysis of prescription pharmaceuticals and uncertainty assessment. Water Res. 2011;45(15):4437–48.

51. Ort C, Eppler JM, Scheidegger A, Rieckermann J, Kinzig M, Sörgel F. Challenges of surveying wastewater drug loads of small populations and generalizable

aspects on optimizing monitoring design. Addiction. 2014;109(3):472–81.

52. Stuart ME, Lapworth DJ, Thomas J, Edwards L. Fingerprinting groundwater pollution in catchments with contrasting contaminant sources using microorganic compounds. Sci Total Environ. 2014;15:468–9. 564–77.

53. Bound J, Voulvoulis N. Household disposal of pharmaceuticals as a pathway for aquatic contamination in the United Kingdom. Environ Health Perspect. 2005;113:1705–11.

54. Huschek G, Hansen P, Maurer H, Krengler D, Kayser A. Environmental risk assessment of medicinal products for human use according to European commission recommendations. Environ Toxicol. 2004;19:226–40.

55. Hirsch R, Ternes T, Haberer K, Kratz KL. Occurrence of antibiotics in the aquatic environment. Sci Total Environ. 1999;225:109–18.

18

One-Pot synthesis, characterization and adsorption studies of amine-functionalized magnetite nanoparticles for removal of Cr (VI) and Ni (II) ions from aqueous solution: kinetic, isotherm and thermodynamic studies

Abbas Norouzian Baghani[1,2], Amir Hossein Mahvi[2,3], Mitra Gholami[4,5*], Noushin Rastkari[6] and Mahdieh Delikhoon[7]

Abstract

Background: Discharge of heavy metals such as hexavalent chromium (Cr (VI)) and nickel (Ni (II)) into aquatic ecosystems is a matter of concern in wastewater treatment due to their harmful effects on humans. In this paper, removal of Cr (VI) and Ni (II) ions from aqueous solution was investigated using an amino-functionalized magnetic Nano-adsorbent (Fe_3O_4-NH_2).

Methods: An amino-functionalized magnetic Nano-adsorbent (Fe_3O_4-NH_2) was synthesized by compositing Fe_3O_4 with 1, 6-hexanediamine for removal of Cr (VI) and Ni (II) ions from aqueous solution. The adsorbent was characterized by Scanning Electron Microscope (SEM), Transmission Electron Microscopy (TEM), powder X-Ray Diffraction (XRD), and Vibrating Sample Magnetometry (VSM). Also, the effects of various operational parameters were studied.

Results: According to our finding, Fe_3O_4-NH_2 could be simply separated from aqueous solution with an external magnetic field at 30 s. The experimental data for the adsorption of Cr (VI) and Ni (II) ions revealed that the process followed the Langmuir isotherm and the maximum adsorption capacity was 232.51 mg g^{-1} for Cr (VI) at pH = 3 and 222.12 mg g^{-1} and for Ni(II) at pH = 6 at 298 °K. Besides, the kinetic data indicated that the results fitted with the pseudo-second-order model (R^2: 0.9871 and 0.9947 % for Cr (VI) and Ni (II), respectively. The results of thermodynamic study indicated that: standard free energy changes ($\Delta G^{\ominus}$), standard enthalpy change ($\Delta H^{\ominus}$), and standard entropy change ($\Delta S^{\ominus}$) were respectively −3.28, 137.1, and 26.91 kJ mol^{-1} for Cr (VI) and −6.8433, 116.7, and 31.02 kJ mol^{-1} for Ni (II). The adsorption/desorption cycles of Fe_3O_4-NH_2 indicated that it could be used for five times.

Conclusions: The selected metals' sorption was achieved mainly via electrostatic attraction and coordination interactions. In fact, Fe_3O_4-NH_2 could be removed more than 96 % for both Cr (VI) and Ni (II) ions from aqueous solution and actual wastewater.

Keywords: Fe_3O_4-NH_2, Heavy metal ions, Amino-functionalized, 1, 6 hexanediamine, Thermodynamics, Adsorption/desorption

* Correspondence: gholamim@iums.ac.ir; gholamimitra32@gmail.com
[4]Research Center for Environmental Health Technology, Iran University of Medical Sciences, Tehran, Iran
[5]Environmental Health Department, School of Public Health, Iran University of Medical Sciences, Tehran, Iran
Full list of author information is available at the end of the article

Background

Discharge of heavy metals into aquatic ecosystems is a matter of concern in wastewater treatment due to their harmful effects on humans even at low concentrations [1, 2]. Among heavy metals, Cr (VI) is among the toxic elements that may enter the environment due to effluent discharge by some industries, such as tanning, textile, wood preservations, paint, metal and mineral processing, pulp, and paper industries [3, 4]. Evidence has shown that these elements can be carcinogenic and mutagenic to living organisms [5]. Nickel is also another heavy metal used in different industries, such as porcelain enameling, electroplating, storage batteries, dying, steel manufacturing, and pigment industries. The acceptance tolerance of nickel has been reported to be 0.01 mgL^{-1} and 2.0 mgL^{-1} in drinking water and industrial wastewater, respectively [6]. Due to the problems remarked above, some effective wastewater treatment approaches have to be employed for Cr (VI) and Ni (II) removal. Up to now, many methods have been used in this regard, including chemical precipitation, ion exchange, membrane technologies, coagulation, electrocoagulation, reduction, bio sorption, filtration, adsorption, reverse osmosis, foam flotation, granular ferric hydroxide, electrolysis, and surface adsorption [7–11]. Most of these methods have economic and technical disadvantages and could not achieve the discharge standards. Yet, adsorption is an effective and flexible method, generating high-quality treated effluent [12]. Until now, many adsorbents have been grown, including maple sawdust, walnut, hazelnut, almond shell [2], carbon nanotubes [13], amino-functionalized polyacrylic acid (PAA) [14], and Lewatit FO36 Nano [15]. However, in many cases, these materials do not have the sufficient adsorption efficiency because of not having enough active surface sites. Furthermore, these materials have a lot of problems, including high cost, difficulty in separation, desorption, and regeneration of adsorbents, and secondary wastes. Therefore, new materials, such as various functional groups, including amide, amino groups, and carboxyl, are to develop new adsorbents that have high selectivity toward toxic metals [16–18]. In this respect, amino-groups have attracted more attention as chelation sites due to their large specific surface areas. Thus, amino-groups are capable of adsorbing a number of metal anions and cations from aqueous solution [19].

As described above, due to the high specific surface area created through grafting of appropriate organic amino-groups on inorganic magnetic Fe_3O_4 particles, with strong magnetic properties, low toxicity, and easy separation, it could be used as a sorbent for removing heavy metals [18, 20]. Another advantage is that it is useful for recovery or reuse of the magnetite nanoparticles modified with amino-groups [17, 21].

In this study, we prepared a novel amino-functionalized magnetic Nano-adsorbent (Fe_3O_4-NH_2) developed by grafting amino-groups onto the surfaces of Fe_3O_4 nanoparticles and used nanocompostie as the adsorbent for removal of Cr (VI) and Ni (II) from aqueous solution. The adsorbent was characterized by Transmission Electron Microscopy (TEM), powder X-Ray Diffraction (XRD), Vibrating Sample Magnetometry (VSM), Scanning Electron Microscope (SEM), and zeta-potential measurement. The effects of pH, initial concentrations of Cr (VI) and Ni (II), adsorption kinetics, thermodynamics, and adsorption isotherm were studied, as well.

Methods

Chemicals

Anhydrous sodium acetate, iron (III) chloridehexahydrat ($FeCl_3 \cdot 6H_2O$), potassium dichromate, ethanol, 1,6-hexanediamine, ethylene glycol, nickel (II) chloride hexahydrat ($NiCl_2.6H_2O$), sodium hydroxide, hydrogen chloride, which were of analytical grade, were purchased from Merck, Germany and were used without further purification. Potassium dichromate (99 %) and nickel (II) chloride hexahydrat (99 %) were used for preparation of Cr (VI) and Ni (II) solution. Additionally, doubly distilled deionized water was used throughout the work.

Synthesis of amino-functionalized magnetic Nano-adsorbent (Fe₃O₄-NH₂) by one-pot synthesis

Amino-functionalized magnetic Nano-adsorbent (Fe_3O_4-NH_2) was prepared according to hydrothermal reduction method. In doing so, a solution of 1, 6-hexanediamine (13 g), anhydrous sodium acetate (4.0 g), and $FeCl_3 \cdot 6H_2O$ as a single Fe ion source (2.0 g) was added to ethylene glycol (80 mL). The above mixture was stirred at 50 °C under vigorous stirring for 30 min. Then, this solution was heated at 198 °C in a Teflonlined autoclave for 6 h. Thereafter, the mixture was cooled down to room temperature. The magnetite nanoparticles were collected with a magnet and were then washed with water and ethanol (3 times) to effectively remove the solvent and unbound 1, 6-hexanediamine. Finally, the amino-functionalized magnetic Nano-adsorbent (Fe_3O_4-NH_2) was dried in a vacuum oven at 50 °C before characterization and application [22]. The size and morphology of the Fe_3O_4-NH_2 were showed by SEM (Holland, company: Philips). Besides, the magnetic property (M–H loop) of the typical magnetic nanoparticles bound with 1, 6-hexadiamine at 25 °C was characterized by VSM (MDKFD, Iran). The crystal structure and phase purity of Fe_3O_4-NH_2 were also examined by XRD (Philips, Holland) using Cu Kα radiation ($\lambda = 0.1541$ nm) at 2^θ, 30 kV, and 30 mA. Finally, the TEM image of Fe_3O_4-NH_2 was examined using TEM, Model EM10C-100KV (Zeiss, Germany).

Adsorption experiments

The absorption experiments were conducted in 1000 ml Erlenmeyer flasks containing 50 ml Ni (II) and Cr (VI) solutions at 5 to 100 mg L^{-1} concentrations and 0.05 g of Fe_3O_4-NH_2. The mixtures were stirred (200 rpm) at room temperature from 10 to 90 min. After adsorption, Fe_3O_4-NH_2 with adsorbed Cr (VI) and Ni (II) was separated from the solution under the external magnetic field. The concentrations of Cr (VI) and Ni (II) ions in the solutions were measured by an Inductive Coupled Plasma (ICP-OES, Spectro arcos, Germany (Company: SPECTRO)).

In order to determine the effects of various factors, the experiments were performed at different Fe_3O_4-NH_2 doses (0.1 to 0.3 g/L), initial concentrations of Cr (VI) and Ni (II) (5 to 100 mg/L), and temperatures (298.15 to 338.15 °K). Besides, each experiment was carried out in duplicate. The removal of Cr (VI) and Ni (II) by Fe_3O_4-NH_2 and removal efficiency have been figured by equations in Table 1 [18].

Results and Discussion
Characterization of Fe_3O_4-NH_2
The SEM, VSM, XRD, and TEM of Fe_3O_4-NH_2 were recorded. The SEM of Fe_3O_4-NH_2 image has been shown in Fig. 1. Based on the results, the SEM image indicated that the size of Fe_3O_4-NH_2 was much smaller than that of naked particles, confirming the coating of 1, 6 hexanediamine [18].

The magnetic hysteresis loops measured at room temperature has been illustrated in Fig. 2. The M–H curves showed that Fe_3O_4 and Fe_3O_4-NH_2 were essentially super-paramagnetic. Fe_3O_4-NH_2 and Fe_3O_4 have a magnetization saturation value of 73.25 and 91.57 emu g^{-1}, respectively. According to Fig. 3, the magnetic Fe_3O_4-NH_2 was dispersed in water. In addition, it could be collected by external magnetic field and be re-dispersed through slight shaking, making the solid and liquid phases separate easily.

The XRD patterns of Fe_3O_4-NH_2 have been shown in Fig. 4. In this study, the crystal structure and phase purity of Fe_3O_4-NH_2 were examined by XRD using a Cu Kd radiation ($\lambda = 0.1541$ nm) at 2^θ of 30.1°, 35.5°, 43.1°, 53.4°, 57.0°, and 62.6° corresponding to their indices; i.e., 220, 311, 400, 422, 511, and 440, at 30 kV and 30 mA.

The particle size was obtained via XRD analysis through Debye-Sherrer's formula [23]: $D = K \lambda/\beta \cos \theta$.

Where λ is the wavelength of the X-rays, θ is the diffraction angle, and β is the corrected full width. The result of size distribution demonstrated that the size of prepared Fe_3O_4-NH_2 was under 90 nm. Additionally, the sharp and strong peaks of the products revealed its appropriate crystallinity. Moreover, the six characteristic peaks of Fe_3O_4 showed that amino-groups did not cause any measureable alter in the phase property of Fe_3O_4 cores. Therefore, the amino-groups were fixed on the surface of Fe_3O_4 cores, making a core-shell structure. In other words, binding and amino-functionalization (NH_2) occurred only on the surface of Fe_3O_4 cores to form a core–shell structure [22].

The TEM image of Fe_3O_4-NH_2 has been shown in Fig. 5. Accordingly, Fe_3O_4-NH_2 particles were multidispersed with an average diameter of around 25 nm. It has been reported that magnetic particles of less than 30 nm would show paramagnetism [24].

The effect of initial concentration and pH on the adsorption properties and zeta potential analyses
The effect of initial concentration on the adsorption properties was intensively studied for Fe_3O_4-NH_2 by varying C_0 of Cr (VI) and Ni (II) ions at 5, 25, 50, and 100 mg L^{-1}. The results have been presented in Figs. 6 and 7. Under corresponding pH values from 2.0 to 9.0, the adsorption efficiency of Cr (VI) and Ni (II) respectively decreased and increased with increase in the initial Cr (VI) and Ni (II) concentrations. Accordingly, the percentage of uptake of Cr (VI) and Ni (II) ions at the Fe_3O_4-NH_2 concentration of 5 mg L^{-1} decreased from 98.02 to 36.85 % for Cr (VI) and increased from 46.21 to 93.03 % for Ni (II) with increasing the pH from 2.0 to 9.0. This can be justified by the fact that for a fixed adsorbent dosage, the total available adsorption sites would be relatively settled. Thus, increasing the initial Cr (VI) and Ni (II) concentrations led to a decrease in the adsorption percentage of the adsorbate [25].

Table 1 The kinetic, isotherm, and thermodynamic equations used for adsorption of Cr (VI) and Ni (II) onto Fe_3O_4-NH_2

Kinetic models	Isotherm equations	Thermodynamic equations	Removal efficiency and equilibrium adsorption capacity	Ref.
Pseudo-first-order $\ln(q_e - q_t) = \ln q_e - k_1 t$	Freundlich isotherm $\ln q_e = \ln K_F + \frac{1}{n} \ln Ce$	Van' t Hoff $\Delta G^\theta = -RT \ln K$	equilibrium adsorption capacity $q_e = \frac{(C_0 - Ce)V}{m}$	[31–38]
Pseudo-second-order $\frac{t}{qt} = \frac{1}{K_2 q_{e,c}^2} + \frac{t}{q_{e,c}}$	Langmuir adsorption model $\frac{Ce}{q_e} = \frac{1}{K_l q_m} + \frac{Ce}{q_m}$	Free energy of adsorption $\ln K = -\frac{\Delta H\theta}{RT} + \frac{\Delta S\theta}{R}$	Removal efficiency (%) $= \frac{(C_i - C_o)}{C_i} \times 100$	[27, 31, 33–37, 39]
	Separation factor (R_L) $R_L = 1/(1 + K_L C_0)$			[18, 25, 26]

qe (mg/g), K_1 (1/min), K_2 (g/mgmin), qm (mg/g), K_F [(mg g^{-1}) (mgL^{-1}) n], Kp (mg/g min$^{-0.5}$), C_0 (mg/L), ΔH^θ (kJ/mol), ΔS^θ (J/mol.K), T (K), R (8.314 J/mol.K), V (mL), m (mg)

Fig. 1 SEM image of Fe_3O_4 -NH_2

To assess the effect of pH, the study was conducted from pH 2 to 9 for both Cr (VI) and Ni (II) ions. The maximum sorption was perceived at pH = 6 for Ni (II), but at pH = 3 for Cr (VI). The adsorption of Cr (VI) at lower pH levels was also observed in other magnetic materials, such as the mesoporous magnetic ɤ-Fe_2O_3 [26]. pH value affected the adsorption efficiency due to its influence on the amino-groups modified on the surface of Fe_3O_4-NH_2. The plot of pH initial vs. pH final depicted that the pHzpc was 5.8 for Fe_3O_4–NH_2. Hence, at pH >5.8, the surface charge of Fe_3O_4-NH_2 was negative and the electrostatic interactions between the metal ions and the adsorbent enhanced. Considering Ni (II), the interaction between the adsorbents and the Ni (II) ions might be defined by Equations 1–5 [14, 27].

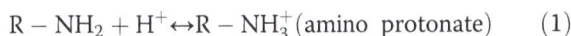

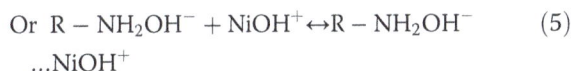

$$R - NH_2 + H^+ \leftrightarrow R - NH_3^+ \text{(amino protonate)} \quad (1)$$

$$R - NH_2 + Ni^{2+} \leftrightarrow R - NH_2Ni^{2+} \quad (2)$$

$$R - NH_2 + OH^- \leftrightarrow R - NH_2OH^- \quad (3)$$

$$R - NH_2OH^- + Ni^{2+} \leftrightarrow R - NH_2OH^- ...Ni^{2+} \quad (4)$$

$$\text{Or } R - NH_2OH^- + NiOH^+ \leftrightarrow R - NH_2OH^- \quad (5)$$
$$...NiOH^+$$

The protonation/deprotonation reactions of the Fe_3O_4-NH_2 amino-groups in the solution have been presented in Equation 1. Based on Equation 2, the ability of NH_2 to be protonated was weakened at higher pH levels, resulting in more –NH_2 on the surface of the adsorbent

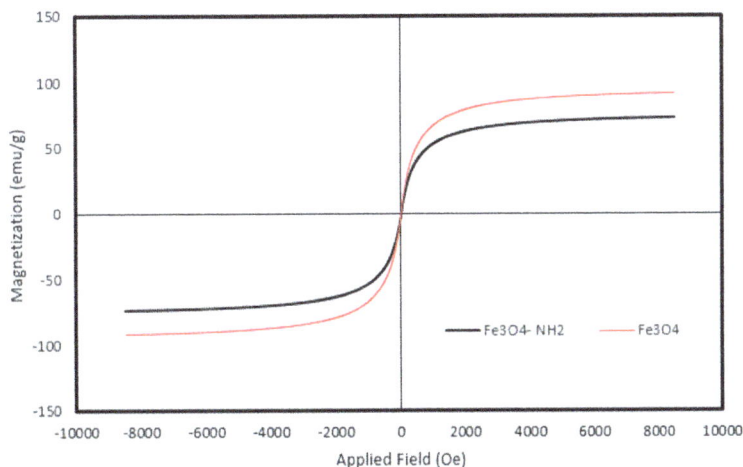

Fig. 2 VSM magnetization curves of Fe_3O_4-NH_2 and Fe_3O_4

Fig. 3 Demonstration of magnetic separation at 30 s

to coordinate with Ni (II). At higher pH levels, OH^- in the solution is competitively adsorbed by amino-groups ($-NH_2$), and the electrostatic adsorption is prevailed gradually compared to coordination. Considering Cr (VI), a large number of H^+ exists under acidic conditions (pH levels: 2–3.5), causing amino-groups ($-NH_2$) to be protonated to NH^{3+} more easily and electrostatic attraction to occur between these two oppositely charged ions (Equation (6)) [14, 27].

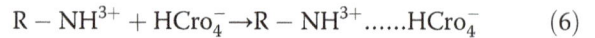

$$R - NH^{3+} + HCro_4^- \rightarrow R - NH^{3+} ......HCro_4^- \qquad (6)$$

Kinetic, equilibrium, and thermodynamic studies

Adsorption isotherms of $Fe_3O_4-NH_2$ were gained at pH = 3 for Cr (VI) and pH = 6 for Ni (II) with the initial concentrations of 5 to 100 mg L^{-1}. The relevant equations for kinetic, equilibrium, and thermodynamic studies have been shown in Table 1 [18]. Besides, the Langmuir and

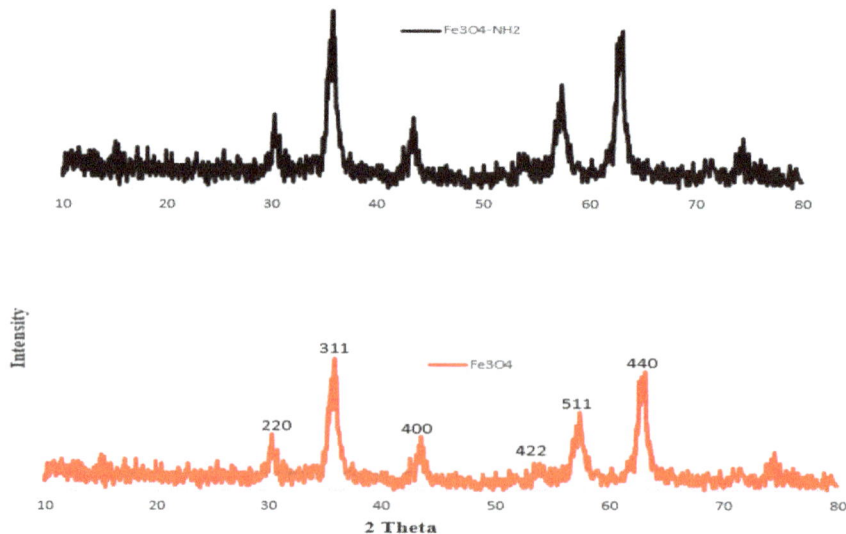

Fig. 4 XRD for $Fe_3O_4-NH_2$ and Fe_3O_4

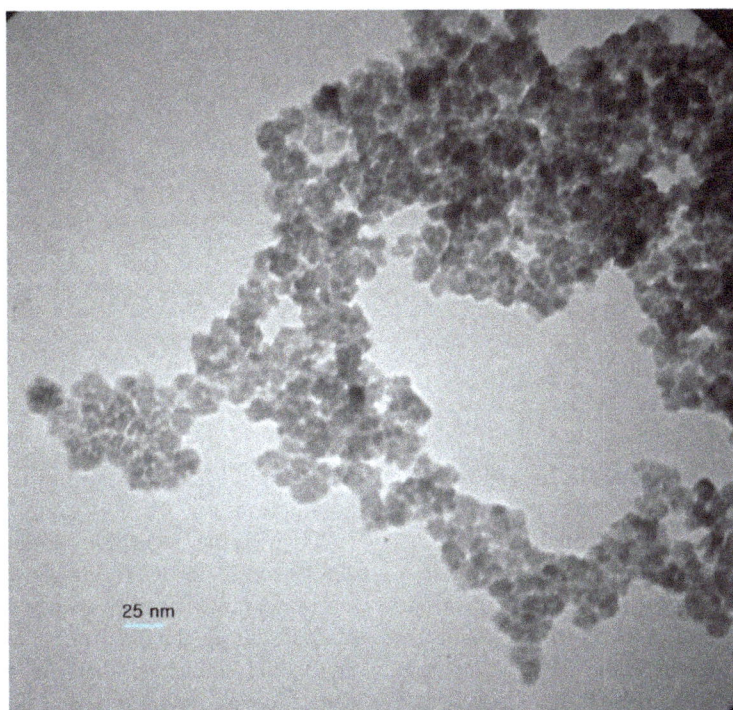

Fig. 5 TEM image of Fe_3O_4-NH_2

Freundlich parameters, correlation coefficients (R^2), and separation factor (R_L) for the adsorption of Cr (VI) and Ni (II) on Fe_3O_4-NH_2 have been summarized in Table 2. The essential characteristics of the Langmuir isotherm can be expressed in terms of a dimensionless separation factor (R_L). The plot of C_e vs. (C_e/q_e) for Fe_3O_4-NH_2 gave a straight line with correlation coefficients of 0.998 and 0.994 for Cr (VI) and Ni (II), respectively. The q_{max} and K_L were derived from the slope and intercept of the line, respectively. The adsorption capacities (q_m, mg/g) using the Langmuir isotherm equation were as follows: q_m Ni (II) (232.15) > q_m Cr (VI) (222.12). Considering the larger adsorption capacity of Fe_3O_4-NH_2 attributed to the amino-groups modified on the surface of Fe_3O_4-NH_2, the amino-groups played a very important role in the adsorption process of Cr (VI) and Ni (II) in aqueous solution. This implies that increasing the percentage of nitrogen in the Fe_3O_4-NH_2 could increase the value of q_m. The calculated R_L values for adsorption of Cr (VI) and Ni (II) were also 0.03–0.39 and 0.02–0.34, respectively, which fall between 0 and 1. Thus, the adsorption of Cr (VI) and Ni (II) onto NH_2- Fe_3O_4 was favorable. However, the correlation coefficients ($R^2 > 0.99$) and R_L ($0 < R_L < 1$) proved that the Langmuir isotherm fitted better for adsorption of Cr (VI) and Ni (II) on NH_2- Fe_3O_4. On the other hand, the value of $1/n$ shows whether the adsorption is suitable for the

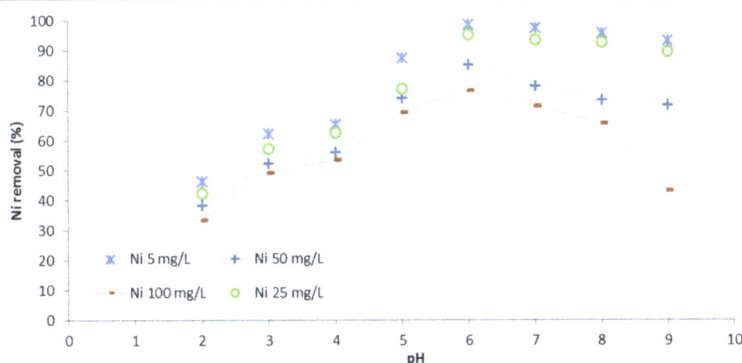

Fig. 6 The effect of pH on the adsorption of Ni (II) onto Fe_3O_4-NH_2 at different initial concentrations

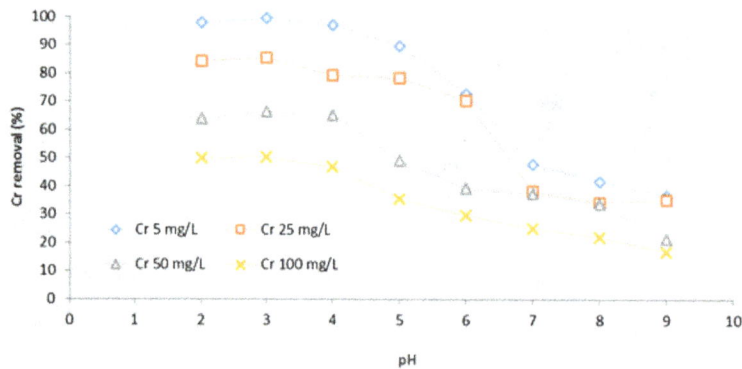

Fig. 7 The effect of pH on the adsorption of Cr (VI) onto Fe_3O_4-NH_2 at different initial concentrations

Freundlich isotherm [18]. The value of $1/n$ reported in Table 2 was less than 1; hence, adsorption by the Freundlich model was unfavorable. Kinetics of the adsorption process is essential for aqueous solution since it gives essential information on the rate of adsorbate uptake on the adsorbent and controls the equilibrium time. The results presented in Table 2 indicated that the adsorption capacity of Fe_3O_4-NH_2 for Cr (VI) and Ni (II) was high (q_m for Ni (II) = 232.51 mg/g^{-1} at pH = 6 and q_m for Cr (VI) = 222.12 mg/g^{-1} at pH = 3) compared to other adsorbents. Afkhami et al. also reported that the adsorption capacity of DNPH-γ-Al_2O_3 for Ni (II) was 18.18 (mg g^{-1}) at pH = 5 and that the process followed the Langmuir isotherm [26]. In another study, the experimental data for the adsorption of Ni (II) on Fe_3O_4-GS revealed that the process followed the Langmuir isotherm and that the maximum adsorption capacity was 158.5 mg g^{-1} at pH = 6 [28]. The parameters of the pseudo-first-order and pseudo-second-order sorption kinetic models have been presented in Table 3. In order to evaluate the applicability of these kinetic models to fit the experimental data, K_1 and K_2 constants were determined experimentally from the slope and intercept of straight-line plots. The value of qe.c earned from the pseudo-second-order model was 28.25 mg g^{-1} for Cr (VI) and 25.97 mg g^{-1} for Ni (II) ions, which perfectly corresponded to the experimental values of qe (24.25 and 25.12 mg g^{-1}) for Cr (VI) and Ni (II) ions. Overall, the pseudo-second-order model (R^2: 0.9871 and 0.9947 % for Cr (VI) and Ni (II), respectively) was more efficient compared to the pseudo-first-order model (R^2: 0.8422 and 0.8862 % for Cr (VI) and Ni (II), respectively). Because all the correlation coefficients were higher than 0.98 %, Cr (VI) and Ni (II) adsorption onto Fe_3O_4-NH_2 might take place through a chemical process involving valence forces through sharing or exchange of electrons [25]. In another kinetic study using Fe_3O_4-adsorbent, R^2 value of Ni (II) was 0.998 at the optimum pH of 5.5. Therefore, the results showed that Ni (II) adsorption on Fe_3O_4 could be followed by the Freundlich

model [29]. In order to measure the thermodynamic parameters for Cr (VI) and Ni (II) adsorption on Fe_3O_4-NH_2, the adsorption studies were accomplished at 298.15 to 313.15 °K. The negative values of ΔG^θ at different temperatures, positive value of ΔS^θ, and positive value of ΔH^θ during the adsorption of Cr (VI) and Ni (II) on Fe_3O_4-NH_2 indicated that the adsorption was spontaneous, increased randomness at the solid-solution interface, and was endothermic in nature [18]. In addition, the slope and intercept of the plot of lnK vs. $1/T$ indicated the ΔH^θ and ΔS^θ values [18]. The values of standard enthalpy change (ΔH^θ) and standard entropy change (ΔS^θ), which were related to distribution coefficient (K_D), were calculated and presented in Table 2. Using the ΔH^θ and ΔS^θ values, standard free energy changes (ΔG^θ) for Fe_3O_4-NH_2 were estimated. The results indicated that adsorption of Cr (VI) and Ni (II) on Fe_3O_4-NH_2 could be followed spontaneously, was endothermic, and was entropy favored in nature. The positive value of ΔS^θ proved increase in the randomness at the solid-solution interface during the adsorption of Cr (VI) and Ni (II) on Fe_3O_4-NH_2. This indicated that the amino-functionalized magnetic Nano-adsorbent (Fe_3O_4-NH_2) could be regarded as an efficient and low cost adsorbent. The results of thermodynamic study in our research indicated that ΔG, ΔH, and ΔS were respectively –3.28, 137.1, and 26.91 kJ mol^{-1} for Cr (VI) and –6.8433, 116.7, and 31.02 kJ mol^{-1} for Ni (II). Shen et al. conducted a similar study using adsorbent DETA-NMPs and disclosed that ΔG, ΔH, and ΔS were –13.7, 8.41, and 72.83 kJ mol^{-1}, respectively for Cr (VI) [25]. Hence, the results indicated that adsorption of Ni (II) on DETA-NMPs could be followed spontaneously, was endothermic, and was entropy favored in nature [25]. One other study also reported that ΔG, ΔH, and ΔS were –1.599, 8.438, and 83.1, respectively for Ni (II) adsorption on Nano-HAP [30].

Overall, simple preparation, fast separation, and high adsorption capacity of Fe_3O_4-NH_2 make it a potential

Table 2 The kinetic and thermodynamic and isotherm constants for the adsorption of Cr (VI) and Ni (II) by Fe_3O_4 NH_2 and other adsorbents

Tem. (K)	Adsorbent	pH Cr (VI)	pH Ni (II)	Pseudo-second-order K_2 (g/mg^{-1})(min^{-1})	$q_{e,cal}$ (mg/g^{-1})	R^2	Thermodynamic parameters $\Delta G^\ominus$ (kJ mol^{-1})	$\Delta H^\ominus$ (kJmol^{-1})	$\Delta S^\ominus$ (J (mol K)$^{-1}$)	Ref.
308	EDA-NMPs	2.5	-	0.7862	-	1	-1.61	10.59	37.50	[25]
308	DETA-NMPs	2.5	-	0.3668	-	1	-1.76	9.67	34.64	[25]
308	TETA-NMPs	2.5	-	0.3219	-	1	-2.15	23.15	76.82	[25]
308	TEPA-NMPs	2.5	-	0.1042	-	1	-2.36	10.46	38.04	[25]
303	Activated Alumina	3	-	-	0.0757	0.999	9.78	-	-	[40]
298	LewaitMP 610	5	-	-	-	-	-10.40	-2.51	35.49	[5]
298	Fe_3O_4	-	6	0.004	-	0.998	-	-	-	[41]
298	ZnO	-	6	0.002	-	0.998	-	-	-	[41]
298	CuO	-	6	0.019	-	0.995	-	-	-	[41]
303	Bagass Fly ash	5	-	-	-	-	-1.46	14.24	49	[42]
293	Nano-HAP	-	6.6	-	-	-	-1.599	8.438	83.1	[30]
298	Superparamagnetic Iron Oxide	-	5.5	-	-	-	27.9	7.8	110	[29]
293	Fe_3O_4-GS	2	-	0.055	17.29	0.999	-4.182	76.63	18.28	[28]
293	Fe_3O_4-GS	-	6	0.0203	22.07	0.998	-3.456	31.86	5.965	[28]
323	Fe_3O_4-TW	-	6	-	-	-	10.02	33.41	0.5799	[6]
298	Waste tea	-	4	-	-	-	-3.82	17.07	20.92	[43]
313	DETA-NMPs	-	6	1.03	9	0.999	-13.7	8.41	72.83	[25]
298	Fe_3O_4-NH_2 (Cr VI)	3	-	0.002	28.25	0.987	-3.2891	137.1	$\Delta S^\ominus$ 26.91 (R^2 0.975)	This study
303	Fe_3O_4-NH_2 (Ni II)	-	6	0.008	25.97	0.994	-6.8433	116.7	31.02 (R^2 0.960)	This study

This study — $\Delta G^\ominus$ temperature series:

Tem. (K)	$\Delta G^\ominus Cr$ (VI)	$\Delta G^\ominus Ni$ (II)	Ref.
298	-3.2891	-6.8433	[17, 44, 45]
303	-5.5038	-8.424	
308	-9.7477	-10.045	
313	-13.234	-12.934	

If: RL > 1, the adsorption is unfavorable. RL = 1, the adsorption is linear.
0 < RL < 1, the adsorption is favorable. RL = 0, the adsorption is irreversible.
If: 1/n < 1, the adsorption is unfavorable. If: 0.1 < 1/n < 1, the adsorption is favorable.

Tem. (K)	Adsorbent	pH Cr (VI)	pH Ni (II)	Freundlich constants K_F (Lg^{-1})	n	R^2	Langmuir constants q_{max} (mg g^{-1})	K_L (Lmg^{-1})	R^2	Ref.
308	EDA-NMPs	2.5	-	-	-	-	136.98	0.1648	0.999	[25]
308	DETA-NMPs	2.5	-	-	-	-	149.25	0.4467	0.999	[25]
308	TETA-NMPs	2.5	-	-	-	-	204.08	0.075	0.998	[25]

Table 2 The kinetic and thermodynamic and isotherm constants for the adsorption of Cr (VI) and Ni (II) by Fe_3O_4 NH_2 and other adsorbents *(Continued)*

T (K)	Adsorbent	pH		n		R^2	q_{max}	K_L	R_L	R^2	Reference
308	TEPA- NMPs	2.5	–	–	–	–	370.37	0.1233		0.999	[25]
298	Walnut	3.5	0.244	–	3.36	0.989	8.01	2.98		0.964	[2]
298	Hazelnut	3.5	0.386	–	2.38	0.992	8.28	4.42		0.976	[2]
298	Almand Shells	3	0.153	–	2.68	0.984	3.40	0.580		0.972	[2]
303	Activated Alumina	3	2.84	–	1.80	09880	25.57	0.467		0.991	[40]
298	Mag	2.5	4.90	–	2.94	0.729	20.16	0.262		0.998	[47]
298	MagDt-H	2.5	1.62	–	2.63	0.984	13.88	0.030		0.965	[47]
298	Lewait MP 610	5	–	–	–	–	.41	–		0.99	[5]
298	PAC	–	0.02	8	2.85	0.708	31.08	0.27		0.98	[48]
298	Bagass	–	1.4E-03	8	0.868	0.868	0.03	0.95		0.97	[48]
298	Fly ash	–	2.03E-04	8	0.696	0.811	0.001	0.95		0.98	[48]
298	Fe_3O_4	–	1.550	6	0.996	–	–	–		–	[41]
298	ZnO	–	0.319	6	0.991	–	–	–		–	[41]
298	CuO	–	0.162	6	0.992	–	–	–		–	[41]
303	Bagass Fly ash	5	1.86	–	12.05	–	4.35	0.014		0.987	[42]
298	NH_2-MCM-41	5	2.759	–	2.2	0.900	12.36	0.2245		0.956	[49]
293	Nano-HAP	–	8.87	6.6	2.74	0.934	46.17	0.07		0.995	[30]
298	DNPH-γ-Al_2O_3	–	1.95	5	2.037	0.926	18.18	1.426		0.985	[26]
293	Fe_3O_4-GS	2	30.58	–	3.01	0.997	39.92	4.08		0.959	[28]
293	Fe_3O_4-GS	–	3.801	6	2.56	0.993	158.5	0.2830		0.966	[28]
323	Fe_3O_4-TW	–	4.85	6	1.78	0.975	38.30	0.085		0.996	[6]
298	Superparamagnetic Iron Oxide	–	0.113	5.5	0.213	0.986	0.189 mmol/g	1.39 L/mmol		0.999	[29]
298	Waste tea	–	0.258	4	0.93	0.922	15.26	0.088		0.996	[43]
333	Fe_3O_4 -CNTs	–	7.23	2	3.05	0.981	65.96	0.42		0.997	[44]
313	DETA-NMPs	–	1.304	6	2.54	09026	43.24	0.288		0.999	[25]
298	Fe_3O_4-NH_2	3	1.13	–	2.05	0.940	222.12	KL 0.314	RL 0.03–0.39	0.995	This study
298	Fe_3O_4-NH_2	–	1.44	6	1.83	0.979	232.51	0.383	0.02–0.34	0.988	This study [17, 44, 45]

If: RL > 1, the adsorption is unfavorable. RL = 1, the adsorption is linear.
0 < RL < 1, the adsorption is favorable. RL = 0, the adsorption is irreversible.

If: 1/n < 1, the adsorption is unfavorable. If: 0.1 < 1/n < 1, the adsorption is favorable. [46]

Table 3 Kinetic adsorption parameters obtained using Pseudo-first-order and Pseudo-second-order models

Fe_3O_4-NH_2	Pseudo-first-order				Pseudo-second-order		
Metals	K_1 (min^{-1})	$q_{e,exp}$ (mg/g^{-1})	$q_{e,cal}$ (mg/g^{-1})	R^2	K_2 (g/mg^{-1}) $(min)^{-1}$	$q_{e,cal}$ (mg/g^{-1})	R^2
Cr (VI)	0.06	24.25	06.41	0.8422	0.002	28.25	0.9871
Ni (II)	0.03	25.12	14.64	0.8862	0.008	25.97	0.9947

applicant for Cr (VI) and Ni (II) removal. Considering the larger adsorption capacity of Fe_3O_4-NH_2 attributed to the amino-groups modified on the surface of Fe_3O_4-NH_2, the amino-groups played a very important role in the adsorption process of Cr (VI) and Ni (II) in aqueous solution. This indicated that the increase of nitrogen percentage in Fe_3O_4-NH_2 could result in an increase in the value of q_m. Similar results were also obtained by Shen et al. [27] and Zhao et al. [25].

Desorption and reusability of Fe_3O_4-NH_2

For practical application of a cost-effective adsorbent for Cr (VI) and Ni (II) removal, desorption of metal ions from adsorbent and regeneration of Fe_3O_4-NH_2 is of particular importance. Since the adsorption of Cr (VI) and Ni (II) onto Fe_3O_4-NH_2 highly depends on the solution pH, desorption of the two heavy metals can be achieved by adjusting the pH. In the present study, the adsorption reversibility of Cr (VI)-laden Fe_3O_4-NH_2 and Ni (II)-laden Fe_3O_4-NH_2 was examined using NaOH (0.01, 0.05, 0.1, 0.2, and 0.3 mol L^{-1}) and HNO_3 (0.001, 0.005, 0.01, 0.05, and 0.1 mol L^{-1}). For desorption studies, the metal-adsorbed modified Fe_3O_4 nanoparticles were first washed by ultrapure water for three times to remove the unadsorbed metals loosely appended to the adsorbent. When the concentration of NaOH and HNO_3 was increased, the removal efficiency of desorption increased, as well. The best result was achieved with 2 min sonication in the presence of 0.2 mol L^{-1} NaOH and 0.05 mol L^{-1} HNO_3. Finally, the Fe_3O_4-NH_2 was dried in an oven (at 50 °C) during regeneration. In our study, each

sorption/desorption process experienced a base and a heat treatment. The adsorption/desorption cycle results showed that the Fe_3O_4-NH_2 could be reused for 5 times. Besides, the results presented in Fig. 8 indicated that at the end of the fifth cycle, the Fe_3O_4-NH_2 maintained more than 76.19 % of its original Cr (VI) adsorption capacity and 77.13 % of its original Ni (II) adsorption capacity. Therefore, the great reusability Fe_3O_4-NH_2 demonstrated its good potential for practical application.

The effect of real water matrix

In this study, 1.0 and 5 mg L^{-1} Cr (VI) and Ni (II) were spiked with tap water and industrial wastewater for evaluating the practical application of Fe_3O_4-NH_2. The initial concentration, pH, and removal efficiencies of Cr (VI) and Ni (II) after treatment with Fe_3O_4-NH_2 have been presented in Table 4. According to the results, the removal efficiency of Cr (VI) at the concentration of 1.0 mg L^{-1} was 97.94 and 98.56 % for tap water and industrial wastewater, respectively. These measures were respectively obtained as 96.12 and 97.24 % for Ni (II) at the concentration of 1.0 mg L^{-1}. This implies the excellent potential of Fe_3O_4-NH_2 in water and wastewater treatment.

Conclusion

In this study, Fe_3O_4-NH_2 was prepared using a simple, cost-effective, and environmentally friendly method for the removal of Cr (VI) and Ni (II) ions from aqueous solution and was characterized by SEM, TEM, XRD, and VSM. The effects of controlling parameters, such as

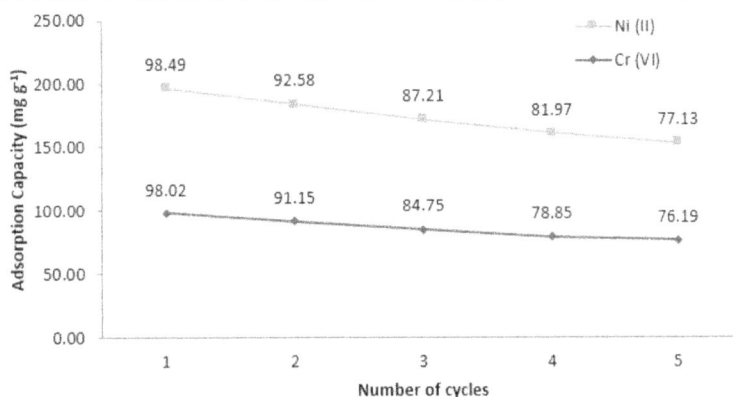

Fig. 8 Adsorption and desorption efficiency of Cr (VI) and Ni (II) by Fe_3O_4-NH_2 in adsorption—desorption cycle

Table 4 The adsorption efficiencies of Cr (VI) and Ni (II) by Fe$_3$O$_4$-NH$_2$ from tap water and industrial wastewater

Matrix	pH $_{Cr(VI)}$ and pH $_{Ni(II)}$	Cr (VI) $_{initial}$ and Ni (II) $_{initial}$ (mg L^{-1})	Cr (VI) $_{removal}$ (%)	Ni (II) $_{removal}$ (%)
Tap water	6.6	1	97.94	98.56
Tap water	6.6	5	96.85	97.61
Industrial wastewater	6.2	1	96.12	97.74
Industrial wastewater	6.2	5	95.25	96.42

contact time, temperature, pH, Fe$_3$O$_4$-NH$_2$ dose, and initial concentration of both heavy metals, were studied, as well. Based on the results, the Langmuir model fitted the isotherm data for both heavy metals and the maximum sorption capacity was 232.51 mg g^{-1} at pH = 3 for Cr (VI) and 222.12 mg g^{-1} at pH = 6 for Ni (II). Moreover, the adsorption kinetic data for Cr (VI) and Ni (II) were based on the assumption of a pseudo-second-order model and thermodynamic parameters showed that the adsorption process was endothermic, spontaneous, and entropy favored in nature. In addition, this nano-adsorbent was able to remove over 96 % of both heavy metals from tap water and industrial wastewater. The Fe$_3$O$_4$-NH$_2$ could be regenerated with acid after adsorption and the adsorption capabilities only decreased with 6-7 % for both metal ions after five cycles. Overall, this study indicated that an amino-functionalized magnetic nano-adsorbent was promising for removal of Cr (VI) and Ni (II) ions in field application.

Highlights

A sensitive method was developed for removal of Cr (VI) and Ni (II) from aqueous solution.

In-lab synthesized magnetic nanoadsorbent was developed by grafting amino-groups onto the surfaces of Fe$_3$O$_4$ nanoparticles.

The adsorbent was characterized by Transmission Electron Microscopy (TEM), powder X-Ray Diffraction (XRD), Vibrating Sample Magnetometry (VSM), and Scanning Electron Microscope (SEM).

The effects of pH, initial concentrations of Ni (II) and Cr (VI), adsorption kinetics, thermodynamics, and adsorption isotherm were studied.

Abbreviations

SEM, scanning electron microscope; TEM, transmission electron microscopy; VSM, vibrating sample magnetometry; XRD, powder X-Ray diffraction

Acknowledgements

This research was financially supported by Tehran University of Medical Sciences (TUMS), the Institute for Environmental Research (IER) (project No. 92-01-46-20998) which is highly appreciated. Thanks also go to Ms. A. Keivanshekouh at the Research Improvement Center of Shiraz University of Medical Sciences for improving the use of English in the manuscript.

Funding

This research was financially supported by Tehran University of Medical Sciences (TUMS), the Institute for Environmental Research (IER) (project No. 92-01-46-20998).

Authors' contributions

MG supervised of this study, and participated in its design and coordination and helped to draft the manuscript. ANB participated in the design of the study, synthesized magnetic nanoadsorbent and prepared to draft the manuscript. AHM participated in the design of the study and in the sequence alignment. NR carried out to synthesize magnetic nanoadsorbent. MD performed the statistical analysis. All authors read and approved the final manuscript.

Competing interests

The authors declare that they have no competing interests.

Author details

[1]Center for Water Quality Research, Institute for Environmental Research (IER), Tehran University of Medical Sciences, Tehran, Iran. [2]Department of Environmental Health Engineering, School of Public Health Science, Tehran University of Medical Sciences, Tehran, Iran. [3]Center for Solid Waste Research, Institute for Environmental Research (IER), Tehran University of Medical Sciences, Tehran, Iran. [4]Research Center for Environmental Health Technology, Iran University of Medical Sciences, Tehran, Iran. [5]Environmental Health Department, School of Public Health, Iran University of Medical Sciences, Tehran, Iran. [6]Center for Air Pollution Research (CAPR), Institute for Environmental Research (IER), Tehran University of Medical Sciences, Tehran 1417613151, Iran. [7]Environmental Health Department, School of Public Health, Shiraz University of Medical Sciences, Shiraz, Iran.

References

1. Aguado J, Arsuaga JM, Arencibia A, Lindo M, Gascón V. Aqueous heavy metals removal by adsorption on amine-functionalized mesoporous silica. J Hazard Mater. 2009;163(1):213–21.
2. Pehlivan E, Altun T. Biosorption of chromium (VI) ion from aqueous solutions using walnut, hazelnut and almond shell. J Hazard Mater. 2008;155(1):378–84.
3. Kumar PA, Ray M, Chakraborty S. Hexavalent chromium removal from wastewater using aniline formaldehyde condensate coated silica gel. J Hazard Mater. 2007;143(1):24–32.
4. López-Téllez G, Barrera-Díaz CE, Balderas-Hernández P, Roa-Morales G, Bilyeu B. Removal of hexavalent chromium in aquatic solutions by iron nanoparticles embedded in orange peel pith. Chem Eng J. 2011;173(2):480–5.
5. Gode F, Pehlivan E. Removal of Cr (VI) from aqueous solution by two Lewatit-anion exchange resins. J Hazard Mater. 2005;119(1):175–82.
6. Panneerselvam P, Morad N, Tan KA. Magnetic nanoparticle (Fe 3 O 4) impregnated onto tea waste for the removal of nickel (II) from aqueous solution. J Hazard Mater. 2011;186(1):160–8.
7. Hua M, Zhang S, Pan B, Zhang W, Lv L, Zhang Q. Heavy metal removal from water/wastewater by nanosized metal oxides: a review. J Hazard Mater. 2012;211:317–31.
8. Miretzky P, Cirelli AF. Cr (VI) and Cr (III) removal from aqueous solution by raw and modified lignocellulosic materials: a review. J Hazard Mater. 2010;180(1):1–19.
9. Owlad M, Aroua MK, Daud WAW, Baroutian S. Removal of hexavalent chromium-contaminated water and wastewater: a review. Water Air Soil Pollut. 2009;200(1–4):59–77.
10. Asgari A, Vaezi F, Nasseri S, Dördelmann O, Mahvi A, Fard ED. Removal of hexavalent chromium from drinking water by granular ferric hydroxide. Iranian J Environ Health Sci Eng. 2008;5(4):277–82.
11. Bazrafshan E, Mahvi AH, Naseri S, Mesdaghinia AR. Performance evaluation of electrocoagulation process for removal of chromium (VI) from synthetic chromium solutions using iron and aluminum electrodes. Turk J Eng Environ Sci. 2008;32(2):59–66.
12. Dakiky M, Khamis M, Manassra A, Mer'eb M. Selective adsorption of

chromium (VI) in industrial wastewater using low-cost abundantly available adsorbents. Adv Environ Res. 2002;6(4):533–40.

13. Li Y-H, Ding J, Luan Z, Di Z, Zhu Y, Xu C, et al. Competitive adsorption of Pb 2+, Cu 2+ and Cd 2+ ions from aqueous solutions by multiwalled carbon nanotubes. Carbon. 2003;41(14):2787–92.

14. Huang S-H, Chen D-H. Rapid removal of heavy metal cations and anions from aqueous solutions by an amino-functionalized magnetic nano-adsorbent. J Hazard Mater. 2009;163(1):174–9.

15. Rafati L, Mahvi A, Asgari A, Hosseini S. Removal of chromium (VI) from aqueous solutions using Lewatit FO36 nano ion exchange resin. Int J Environ Sci Technol. 2010;7(1):147–56.

16. Chang Y-C, Chang S-W, Chen D-H. Magnetic chitosan nanoparticles: Studies on chitosan binding and adsorption of Co (II) ions. React Funct Polym. 2006;66(3):335–41.

17. Hao Y-M, Man C, Hu Z-B. Effective removal of Cu (II) ions from aqueous solution by amino-functionalized magnetic nanoparticles. J Hazard Mater. 2010;184(1):392–9.

18. Tan Y, Chen M, Hao Y. High efficient removal of Pb (II) by amino-functionalized Fe 3 O 4 magnetic nano-particles. Chem Eng J. 2012;191:104–11.

19. Sayar O, Amini MM, Moghadamzadeh H, Sadeghi O, Khan SJ. Removal of heavy metals from industrial wastewaters using amine-functionalized nanoporous carbon as a novel sorbent. Microchim Acta. 2013;180(3–4):227–33.

20. Cui Y, Liu S, Hu Z-J, Liu X-H, Gao H-W. Solid-phase extraction of lead (II) ions using multiwalled carbon nanotubes grafted with tris (2-aminoethyl) amine. Microchim Acta. 2011;174(1–2):107–13.

21. Yuan P, Fan M, Yang D, He H, Liu D, Yuan A, et al. Montmorillonite-supported magnetite nanoparticles for the removal of hexavalent chromium [Cr (VI)] from aqueous solutions. J Hazard Mater. 2009;166(2):821–9.

22. Wang L, Bao J, Wang L, Zhang F, Li Y. One-Pot Synthesis and Bioapplication of Amine-Functionalized Magnetite Nanoparticles and Hollow Nanospheres. Chemistry-A Eur J. 2006;12(24):6341–7.

23. Zhou Y, Rahaman M. Hydrothermal synthesis and sintering of ultrafine CeO 2 powders. J Mater Res. 1993;8(07):1680–6.

24. Watson J, Cressey B, Roberts A, Ellwood D, Charnock J, Soper A. Structural and magnetic studies on heavy-metal-adsorbing iron sulphide nanoparticles produced by sulphate-reducing bacteria. J Magn Magn Mater. 2000;214(1):13–30.

25. Zhao Y-G, Shen H-Y, Pan S-D, Hu M-Q, Xia Q-H. Preparation and characterization of amino-functionalized nano-Fe3O4 magnetic polymer adsorbents for removal of chromium (VI) ions. J Mater Sci. 2010;45(19):5291–301.

26. Afkhami A, Saber-Tehrani M, Bagheri H. Simultaneous removal of heavy-metal ions in wastewater samples using nano-alumina modified with 2, 4-dinitrophenylhydrazine. J Hazard Mater. 2010;181(1):836–44.

27. Shen H, Pan S, Zhang Y, Huang X, Gong H. A new insight on the adsorption mechanism of amino-functionalized nano-Fe 3 O 4 magnetic polymers in Cu (II), Cr (VI) co-existing water system. Chem Eng J. 2012;183:180–91.

28. Guo X, Du B, Wei Q, Yang J, Hu L, Yan L, et al. Synthesis of amino functionalized magnetic graphenes composite material and its application to remove Cr (VI), Pb (II), Hg (II), Cd (II) and Ni (II) from contaminated water. J Hazard Mater. 2014;278:211–20.

29. Nassar NN. Kinetics, equilibrium and thermodynamic studies on the adsorptive removal of nickel, cadmium and cobalt from wastewater by superparamagnetic iron oxide nanoadsorbents. Can J Chem Eng. 2012;90(5):1231–8.

30. Mobasherpour I, Salahi E, Pazouki M. Removal of nickel (II) from aqueous solutions by using nano-crystalline calcium hydroxyapatite. J Saudi Chem Soc. 2011;15(2):105–12.

31. Abbott M, Smith J, Van Ness H. Introduction to chemical engineering thermodynamics: McGraw-Hill. 2001.

32. Freundlich H, Heller W. The adsorption of cis-and trans-azobenzene. J Am Chem Soc. 1939;61(8):2228–30.

33. Ho Y, Ng J, McKay G. Removal of lead (II) from effluents by sorption on peat using second-order kinetics. Sep Sci Technol. 2001;36(2):241–61.

34. Ho Y-S, McKay G. Pseudo-second order model for sorption processes. Process Biochem. 1999;34(5):451–65.

35. Javadian H, Taghavi M. Application of novel Polypyrrole/thiol-functionalized zeolite Beta/MCM-41 type mesoporous silica nanocomposite for adsorption of Hg 2+ from aqueous solution and industrial wastewater: Kinetic, isotherm and thermodynamic studies. Appl Surf Sci. 2014;289:487–94.

36. Qu R, Zhang Y, Sun C, Wang C, Ji C, Chen H, et al. Adsorption of Hg (II) from an aqueous solution by silica-gel supported diethylenetriamine prepared via different routes: kinetics, thermodynamics, and isotherms. J Chem Eng Data. 2009;55(4):1496–504.

37. Rostamian R, Najafi M, Rafati AA. Synthesis and characterization of thiol-functionalized silica nano hollow sphere as a novel adsorbent for removal of poisonous heavy metal ions from water: Kinetics, isotherms and error analysis. Chem Eng J. 2011;171(3):1004–11.

38. Yuh-Shan H. Citation review of Lagergren kinetic rate equation on adsorption reactions. Scientometrics. 2004;59(1):171–7.

39. Najafi M, Yousefi Y, Rafati A. Synthesis, characterization and adsorption studies of several heavy metal ions on amino-functionalized silica nano hollow sphere and silica gel. Sep Purif Technol. 2012;85:193–205.

40. Bhattacharya A, Naiya T, Mandal S, Das S. Adsorption, kinetics and equilibrium studies on removal of Cr (VI) from aqueous solutions using different low-cost adsorbents. Chem Eng J. 2008;137(3):529–41.

41. Mahdavi S, Jalali M, Afkhami A. Removal of heavy metals from aqueous solutions using Fe3O4, ZnO, and CuO nanoparticles. J Nanopart Res. 2012;14(8):1–18.

42. Gupta VK, Ali I. Removal of lead and chromium from wastewater using bagasse fly ash—a sugar industry waste. J Colloid Interface Sci. 2004;271(2):321–8.

43. Malkoc E, Nuhoglu Y. Investigations of nickel (II) removal from aqueous solutions using tea factory waste. J Hazard Mater. 2005;127(1):120–8.

44. Chen J, Hao Y, Chen M. Rapid and efficient removal of Ni2+ from aqueous solution by the one-pot synthesized EDTA-modified magnetic nanoparticles. Environ Sci Pollut Res. 2014;21(3):1671–9.

45. Chen R, Chai L, Li Q, Shi Y, Wang Y, Mohammad A. Preparation and characterization of magnetic Fe3O4/CNT nanoparticles by RPO method to enhance the efficient removal of Cr (VI). Environ Sci Pollut Res. 2013;20(10):7175–85.

46. Malay DK, Salim AJ. Comparative study of batch adsorption of fluoride using commercial and natural adsorbent. Res J Chem Sci ISSN. 2011;2231:606X.

47. Yuan P, Liu D, Fan M, Yang D, Zhu R, Ge F, et al. Removal of hexavalent chromium [Cr (VI)] from aqueous solutions by the diatomite-supported/ unsupported magnetite nanoparticles. J Hazard Mater. 2010;173(1):614–21.

48. Rao M, Parwate A, Bhole A. Removal of Cr 6+ and Ni 2+ from aqueous solution using bagasse and fly ash. Waste Manag. 2002;22(7):821–30.

49. Heidari A, Younesi H, Mehraban Z. Removal of Ni (II), Cd (II), and Pb (II) from a ternary aqueous solution by amino functionalized mesoporous and nano mesoporous silica. Chem Eng J. 2009;153(1):70–9.

Permissions

List of Contributors

Samaneh Saber-Samandari and Jalil Heydaripour
Department of Chemistry, Eastern Mediterranean University, TRNC via Mersin 10, Gazimagusa, Turkey

Seyed Ahmad Mirbagheri, Majid Bagheri, Siamak Boudaghpour and Majid Ehteshami
Department of Civil Engineering, K.N. Toosi University of Technology, Vanak square, Tehran, Iran

Zahra Bagheri
Department and Faculty of Basic Sciences, PUK University, Kermanshah, Iran

Alireza Mesdaghinia, Simin Nasseri and Mahdi Hadi
Center for Water Quality Research (CWQR), Institute for Environmental Research (IER), Tehran University of Medical Sciences, Tehran, Iran

Amir Hossein Mahvi
Department of Environmental Health Engineering, Faculty of Health, Tehran
University of Medical Sciences, Tehran, Iran
Center for Solid Waste Research (CSWR), Institute for Environmental Research (IER), Tehran University of Medical Sciences, Tehran, Iran

Hamid Reza Tashauoei
Department of Environmental Health Engineering, School of Public Health, Islamic Azad University-Tehran Medical Branch, Tehran, Iran

Neda Molaee
Department of Microbiology and Immunology, Arak University of Medical sciences, Arak, Iran

Hamid Abtahi
Molecular and Medicine Research Center, Arak University of Medical Sciences, Arak, Iran

Mohammad Javad Ghannadzadeh
Department of Environmental Health, Faculty of Health, Arak University of Medical Sciences, Arak, Iran

Masoude Karimi
Department of Medical Microbiology and Immunology, Arak University of Medical sciences, Arak, Iran

Ehsanollah Ghaznavi-Rad
Molecular and Medicine Research Center, Arak University of Medical Sciences, Arak, Iran
Department of Medical Microbiology and Immunology, Faculty of Medicine, Arak University of Medical Sciences, Arak, Iran

Seyyed Mohammad Emadian and Behnam Khoshandam
Department of Chemical Engineering, Semnan University, Semnan, Iran

Mostafa Rahimnejad and Morteza Hosseini
Department of Chemical Engineering, Babol University of Technology, Babol, Iran

Ahed Zyoud, Amani Zu'bi and Hikmat S. Hilal
SSERL, Department of Chemistry, An-Najah National University, Nablus, Palestine

Muath H. S. Helal
College of Pharmacy and Nutrition, University of Saskatchewan, 116 Thorvaldson Building, Saskatoon S7N 5C9, Canada

DaeHoon Park
Dansuk Industrial Co, LTD. #1239-5, Jeongwang-Dong, Shiheung-Si, Kyonggi-Do 429-913, South Korea

Guy Campet
Institut de Chimie de la Matie're Condense'ıe de Bordeaux (ICMCB), 87 Avenue du Dr. A Schweitzer, Pessac 33608, France

Ali Jafari and Ramin Nabizadeh
Department of Environmental Health Engineering, School of Public Health, Tehran University of Medical Sciences, Tehran, Iran

Amir Hossein Mahvi
Center for Water Quality Research (CWQR), Institute for Environmental Research (IER), Tehran
Department of Environmental Health Engineering, School of Public Health, Tehran Tehran University of Medical Sciences, Tehran, Iran
Center for Solid Waste Research (CSWR), Institute for Environmental Research (IER), Tehran University of Medical Sciences, Tehran, Iran

Simin Nasseri
Center for Water Quality Research (CWQR), Institute for Environmental Research (IER), Tehran University of Medical Sciences, Tehran, Iran
Department of Environmental Health Engineering, School of Public Health, Tehran University of Medical Sciences, Tehran, Iran

Alimorad Rashidi
Nanotechnology Research Institute of Petroleum Industry (RIPI), West Blvd. Azadi Sport Complex, Tehran, Iran

Reza Rezaee
Kurdistan Environmental Health Research Center, Kurdistan University of Medical Sciences, Sanandaj, Iran

Mohammad Delnavaz
Civil and Environmental Engineering Faculty, Tarbiat Modares University, Tehran, Iran
Civil Engineering Department, Faculty of Engineering, Kharazmi University, Tehran, Iran

Bita Ayati and Hossein Ganjidoust
Civil and Environmental Engineering Faculty, Tarbiat Modares University, Tehran, Iran

Sohrab Sanjabi
Material Engineering Department, Nano Materials Division, Tarbiat Modares University, Tehran, Iran

Somayeh Piri, Farideh Piri and Mohamadreza Yaftian
Department of Chemistry, Faculty of Science, University of Zanjan, 45371-38791 Zanjan, Iran

Zahra Alikhani Zanjani and Abbasali Zamani
Department of Environmental Science, Faculty of Science, University of Zanjan, 45371-38791 Zanjan, Iran

Mehdi Davari
Iranian Research Organization for Science and Technology, Tehran, Iran

Ali Asghar Najafpoor
Health Sciences Research Center, Department of Environmental Health Engineering, School of Health, Mashhad University of Medical Sciences, Mashhad, Iran

Mojtaba Davoudi
Department of Environmental Health Engineering, School of Health, Torbat Heydariyeh University of Medical Sciences, Torbat Heydariyeh, Iran

Elham Rahmanpour Salmani
Student Research Committee, School of Health, Mashhad University of Medical Sciences, Mashhad, Iran

Masuma Moghaddam Arjmand and Abbas Rezaee
Department of Environmental Health Engineering, Faculty of Medical Sciences, Tarbiat Modares University, Tehran, Iran

Simin Nasseri
Department of Environmental Health Engineering, School of Public Health, and Center for Water Quality Research, Institute for Environmental Research, Tehran
University of Medical Sciences, Tehran, Iran

Said Eshraghi
Department of Pathobiology, School of Public Health, Tehran University of Medical Sciences, Tehran, Iran

Gholam Hossein Safari and Mahmood Alimohammadi
Department of Environmental Health Engineering, School of Public Health, Tehran University of Medical Sciences, Tehran, Iran

Simin Nasseri
Department of Environmental Health Engineering, School of Public Health, Tehran University of Medical Sciences, Tehran, Iran
Center for Water Quality Research, Institute for Environmental Research, Tehran University of Medical Sciences, Tehran, Iran

Amir Hossein Mahvi and Kamyar Yaghmaeian
Department of Environmental Health Engineering, School of Public Health, Tehran University of Medical Sciences, Tehran, Iran
Center for Solid Waste Research, Institute for Environmental Research, Tehran University of Medical Sciences, Tehran, Iran

Center for Air Pollution Research, Institute for Environmental Research,Tehran University of Medical Sciences, Tehran, Iran

Mohammad Hassan Ehrampoush, Mohammad Miria and Mohammad Hossien Salmani
Department of Environmental Health, School of Public Health, Shahid
Sadoughi University of Medical Science, Yazd, Iran

Amir Hossein Mahvi
Center for solid Waste Research, Institute for Environmental Research, Tehran University of Medical Science, Tehran, Iran

K. Thirugnanasambandham and V. Sivakumar
Department of Chemical Engineering, AC Tech Campus, Anna University, Chennai 600 025TN, India

Hossein Kamani and Ramin Nabizadeh Nodehi
Department of Environmental Health Engineering, School of Public Health, Tehran University of Medical Sciences, Tehran, Iran

Simin Nasseri
Department of Environmental Health Engineering, School of Public Health, Tehran University of Medical Sciences, Tehran, Iran
Center for Water Quality Research, Institute for Environmental Research, Tehran University of Medical Sciences, Tehran, Iran

Mehdi Khoobi
Department of Medicinal Chemistry, Faculty of Pharmacy and Pharmaceutical Sciences Research Center, Tehran University of Medical Sciences, Tehran, Iran

Amir Hossein Mahvi
Department of Environmental Health Engineering, School of Public Health, Tehran University of Medical Sciences, Tehran, Iran
Center for Solid Waste Research, Institute for Environmental Research, Tehran University of Medical Sciences, Tehran, Iran
National Institute of Health Research, Tehran University of Medical Sciences, Tehran, Iran

L. D. Naidu and S. Saravanan
Department of Chemical Engineering, National Institute of Technology, Tiruchirapalli 15, India

Mukesh Goel
Center for Environmental Engineering, PRIST University, Thanjavur, India

S. Periasamy
Department of Textile Technology, PSG College of Technology, Coimbatore, India

Pieter Stroeve
Department of Chemical Engineering, University of California Davis, Davis, CA 95616, USA

Miguel Antonio Reyes Cardenas
MI Murrumbidgee Irrigation, Research Station Road, Hanwood, NSW 2680, Australia

Imtiaj Ali
Treatment Program, Logan City Council, Logan City DC, QLD 4114, Australia

Foon Yin Lai
National Research Centre for Environmental Toxicilogy (EnTox), The University of Queensland, Brisbane, QLD 4108, Australia

Les Dawes and Jay Rajapakse
Science and Engineering Faculty, School of Earth, Environment and Biological Sciences, Queensland University of Technology, QLD 4001 Brisbane, Australia

Ricarda Thier
Faculty of Health, Queensland University of Technology, QLD 4001 Brisbane, Australia

Abbas Norouzian Baghani
Center for Water Quality Research, Institute for Environmental Research (IER), Tehran University of Medical Sciences, Tehran, Iran
Department of Environmental Health Engineering, School of Public Health Science, Tehran University of Medical Sciences, Tehran, Iran

Amir Hossein Mahvi
Department of Environmental Health Engineering, School of Public Health Science, Tehran University of Medical Sciences, Tehran, Iran
Center for Solid Waste Research, Institute for Environmental Research (IER), Tehran University of Medical Sciences, Tehran, Iran

Mitra Gholami
Research Center for Environmental Health Technology, Iran University of Medical Sciences, Tehran, Iran
Environmental Health Department, School of Public Health, Iran University of Medical Sciences, Tehran, Iran

Noushin Rastkari
Center for Air Pollution Research (CAPR), Institute for Environmental Research (IER), Tehran University of Medical Sciences, Tehran 1417613151, Iran

Mahdieh Delikhoon
Environmental Health Department, School of Public Health, Shiraz University of Medical Sciences, Shiraz, Iran

Index